Recent Advances in Functional Polymer Materials for Water Treatment

Recent Advances in Functional Polymer Materials for Water Treatment

Guest Editors

Zhiwei Wang
Tongtong Wang
Yanzhu Guo
Jian Zhang

MDPI

Basel • Beijing • Wuhan • Barcelona • Belgrade • Novi Sad • Cluj • Manchester

Guest Editors

Zhiwei Wang
Guangxi Key Laboratory of Clean Pulp & Papermaking and Pollution Control
Guangxi University
Nanning
China

Tongtong Wang
Institute for Interdisciplinary and Innovation Research
Xi'an University of Architecture and Technology
Xi'an
China

Yanzhu Guo
Liaoning Key Laboratory of Pulp and Paper Engineering
Dalian Polytechnic University
Dalian
China

Jian Zhang
College of Light Industry and Food Engineering
Guangxi University
Nanning
China

Editorial Office
MDPI AG
Grosspeteranlage 5
4052 Basel, Switzerland

This is a reprint of the Special Issue, published open access by the journal *Polymers* (ISSN 2073-4360), freely accessible at: https://www.mdpi.com/journal/polymers/special_issues/BS25C5T65G.

For citation purposes, cite each article independently as indicated on the article page online and as indicated below:

Lastname, A.A.; Lastname, B.B. Article Title. *Journal Name* **Year**, *Volume Number*, Page Range.

ISBN 978-3-7258-5915-3 (Hbk)
ISBN 978-3-7258-5916-0 (PDF)
https://doi.org/10.3390/books978-3-7258-5916-0

Contents

About the Editors

Zhiwei Wang

Zhiwei Wang, a Researcher at Guangxi University, specializes in papermaking wastewater treatment and fiber valorization. His work focuses on developing functional materials and technologies for sustainable wastewater management. He pioneered the application of microbial fuel cells for simultaneous wastewater treatment and energy recovery, and proposed a way to convert papermaking waste into high-value products. His research integrates science with industry, improving treatment efficiency and reducing costs. His practical solutions have been implemented in engineering projects, contributing to a more sustainable papermaking process.

Tongtong Wang

Tongtong Wang, an Assistant Researcher at Xi'an University of Architecture and Technology, specializes in photocatalysis, advanced oxidation water treatment, and functional materials for use in ecological environments. His work focuses on developing functional materials and technologies for sustainable wastewater treatment and pioneering their application in the removal of emerging pollutants or resource utilization. He has published 72 academic papers in journals such as *Coordination Chemistry Reviews*, *Environmental Chemistry Letters*, and *npj Clean Water*, among which he is the first or corresponding author for more than half of the SCI papers. There are two highly cited ESI papers among these papers, with a total of more than 1200 citations. Currently, his H-index is 21, and he has been awarded the title of Excellent Talent of Xi'an Area. He also serves as Director of the Ecological Society of Shaanxi Province and as a member of several societies, including the Chinese Chemical Society. He has led and participated in more than ten projects, such as the National Natural Foundation of China. He has been granted five national invention and utility model patents, and serves as an Editorial Board Member of journals such as *Recent Innovations in Chemical Engineering*. He has served as a reviewer more than 200 times for more than 70 well-known SCI journals and Chinese journals, including *Water Research*, *Chemical Engineering Journal*, and *Journal of Hazardous Materials*.

Yanzhu Guo

Yanzhu Guo, a Professor at Dalian Polytechnic University, majors in pulping and papermaking, green production, and biomass-based materials. She focuses on fabricating flocculants and adsorbents from lignin or cellulose to treat dye, oil, eutrophic ammonium, or phosphate-containing wastewater. She has also worked to develop fluorescent carbon dots from lignocellulose as sensors for determining metal ions and tetracycline antibiotics in wastewater. Her research provides insights into novel adsorbents for wastewater treatment and the creation of a feasible approach to analyzing trace pollutants in wastewater.

Jian Zhang

Jian Zhang, an Associate Professor at Guangxi University, is committed to researching pollution control and resource utilization in waste paper pulping wastewater treatment. His work focuses on elucidating the formation mechanisms of sludge calcification and developing advanced blocking and regulation technologies, with the objectives of enhancing treatment efficiency, ensuring system stability, and facilitating resource recovery. He has pioneered a variety of innovative solutions, including composite decalcifying agents with synergistic regeneration processes, a phosphate-based composite calcium inhibition system, and regulation strategies employing mildly acidic conditions to reduce extracellular polymeric substances' affinity for Ca^{2+}. These advancements substantially improve anaerobic systems' tolerance and treatment capacity for high-calcium wastewater, resolving critical engineering challenges such as activity loss and calcification-induced clogging in anaerobic granular sludge during waste paper pulping wastewater treatment. His research contributes significantly to achieving the stable operation of pulping wastewater treatment processes, enabling resource recycling, and promoting the industrial implementation of related technologies.

Preface

Water pollution has become a significant threat to the ecological environment and human health. Polymer materials show great potential in wastewater treatment because of their unique physical and chemical properties, such as their large molecular weight, controllable structure, large surface area, and easy modification. With the rapid development of materials science, new polymer materials with excellent performance have emerged, such as new flocculants, scale inhibitors, functional filtration membranes, ion-exchange resins, selective adsorption materials, and modified functional fibers, as well as nanocomposite materials and hybrid nanomaterials used in microbial fuel cell electrodes. Functional polymer materials have been widely used in the field of wastewater pollution control. Based on the above information, we published a Special Issue (SI) pertaining to recent advances in functional polymer materials for water treatment. This SI aims to provide a platform for researchers to disseminate recent advances in the fundamentals, science, and technology of polymeric materials for use in water treatment.

Zhiwei Wang, Tongtong Wang, Yanzhu Guo, and Jian Zhang
Guest Editors

polymers

Editorial

Recent Advances in Functional Polymer Materials for Water Treatment

Zhiwei Wang [1,*,†], Tongtong Wang [2,*,†], Yanzhu Guo [3] and Jian Zhang [1]

1 Key Laboratory of Clean Pulp & Papermaking and Pollution Control of Guangxi, College of Light Industrial and Food Engineering, Guangxi University, Nanning 530004, China; zhangjian@gxu.edu.cn
2 Institute for Interdisciplinary and Innovation Research, Xi'an University of Architecture and Technology, Xi'an 710055, China
3 The Liaoning Province Key Laboratory of Paper and Pulp Engineering, College of Light Industry and Chemical Engineering, Dalian Polytechnic University, Dalian 116034, China; guoyz@dlpu.edu.cn
* Correspondence: wangzhiwei@gxu.edu.cn (Z.W.); wangtongtong@xauat.edu.cn (T.W.)
† These authors contributed equally to this work.

1. Introduction

With the increasing severity of global water pollution, research on the application of functional polymer materials in the field of water treatment is receiving increasing attention [1]. Due to their unique physicochemical properties, such as their large molecular weight, controllable structure, large specific surface area, and easy modification, functional polymer materials have shown great potential in water treatment. These materials can not only effectively remove pollutants in water, such as heavy metal ions, organic dyes, and microorganisms, but also enhance their separation efficiency and stability through modification [2–4]. Recently, with the rapid development of materials science, new flocculants, scale inhibitors, functional filtration membranes, ion-exchange resins, selective adsorption materials, modified functional fibers, and other polymer materials have been emerging [2]. These materials allow for the efficient removal of pollutants and the recycling of resources. Moreover, the application of nanocomposites and hybrid nanomaterials in microbial fuel cell electrodes has received increasing attention, opening up new research directions in the field of water treatment [4–6]. Functional polymer materials have been widely used in the field of wastewater pollution control, and their diversity and tunability provide new ideas and methods for solving the increasingly serious water environment problems [3]. The research on such materials not only promotes the innovation of pollutant treatment technology, but also promotes the practice of green chemistry and the concept of sustainable development in the field of water treatment.

Moreover, research on the application of functional polymer materials in water treatment has made significant progress. Regarding membrane materials, researchers are committed to developing new polymer membranes with higher selectivity, flux, and stability, and exploring their applications in seawater desalination, industrial wastewater treatment, and other fields [3,4]. For example, porous core–shell nanofiber membranes exhibit highly efficient photocatalytic performance, while new functional membranes based on graphene oxide achieve excellent dye retention through layer spacing modulation [5]. As for adsorbent materials, the synthesis and modification of novel polymeric adsorbents are hot research topics, such as the efficient adsorption of heavy metal ions by ionic liquid cross-linked hydrogels and the green synthesis method of magnetic chitosan composites, which provide innovative solutions for the treatment of heavy metals and organic pollutants [4]. Furthermore, research on polymer-based photocatalytic materials and biodegradable polymer materials has also gained extensive attention in recent years, which has culminated in the efficient removal of pollutants in water through photocatalytic degradation, selective adsorption, and bio-enhanced action [6,7]. These research advances promote the innovation of water treatment technology and lay a solid foundation for developing more efficient and environmentally friendly solutions.

This Special Issue aims to collect and publish high quality research results on the latest advances in functional polymer materials for water treatment. We encouraged the submission of original articles, review articles, case studies, and newsletters on polymeric flocculants and adsorbents, polymeric separation membranes and ion exchange resins, polymeric materials for enhanced biological water treatment, polymeric materials for water treatment equipment, and the preparation and modification of polymeric materials. Through this Special Issue, we expect to provide the latest research results on functional polymer materials in water treatment and further promote the development of this field.

2. Overview of the Published Articles

A total of 19 submissions were received, and 10 articles were accepted for this Special Issue. These articles cover various aspects of membrane materials, adsorbent materials, photocatalytic materials, and biological treatment materials. From the optimized design of membrane materials to the preparation and performance study of adsorbents, and then to the exploration of the application of photocatalytic composites and biofilm carriers, these articles provide an in-depth discussion of the synthesis, modification, performance characterization, and practical application of the materials. This Special Issue focuses on the application of functional polymeric materials in the field of wastewater pollution control, discusses the characterization of the physicochemical properties of functional polymeric materials based on various advanced techniques, and provides insights into their removal effects and mechanisms. The articles in this Special Issue not only demonstrate the potential application of functional polymer materials in the treatment of polluted wastewater, but also provide researchers in the related fields with a theoretical basis and practical experience with important reference value. It is expected that this Special Issue will promote the further development of functional polymer materials in the field of water treatment and contribute to solutions to the global water shortage and water pollution problems.

This Special Issue systematically demonstrates the frontier breakthroughs of functional polymeric materials in the field of water treatment. In heavy metal treatment, Sun et al. (contribution 2) developed an ionic liquid cross-linked hydrogel (PAM/AA/[Vim]Br_2) that removed 98.1% of Cr^{3+} at 100 ppm via physico-chemical dual mechanism adsorption, and its XPS analysis revealed that the coordination of carboxylate groups with amino groups dominated the adsorption at high concentrations. Fakhry et al. (contribution 3) further extended the application of bio-based materials and found that the maximum adsorption capacity of Cr(VI) by fungal melanin (*Aureobasidium pullulans*) reached 595.974 mg/g, which was a 22.7% enhancement compared to the conventional biomass, and the prediction error of the adsorption process was reduced by 37% through decision tree modeling. Yadav et al. (contribution 4) focused on green synthesis techniques and prepared magnetic chitosan composites (MCSs) using *Myrica esculenta* leaf extracts with Cd^{2+} adsorption up to 426 mg/g, retaining 80.44% efficiency after seven cycles and a 69.5% increase in the BET specific surface area (178 m^2/g) compared to unmodified magnetic nanoparticles. In the field of membrane separation technology, Yang et al. (contribution 6) constructed AgCl/ZnO heterogeneous nodule shell nanofiber membranes via coaxial electro spinning, which achieved a 98% photodegradation of methylene blue in 70 min, whilst the efficiency still exceeded 95% after five cycles. Moreover, Meng et al. (contribution 9) modulated graphene oxide (GO) membrane layer spacing via ethylenediamine (EDA) cross-linking, which enhanced the dye retention to 96% and increased the permeate flux by 5-fold, whilst XPS confirmed that C-N covalent bonding inhibited membrane swelling. For biological treatment enhancement, Liu et al. (contribution 10) used a surface positively charged modified polyurethane carrier (+1.79 mV) to increase the abundance of anaerobic ammonia-oxidizing bacteria (AnAOB) by 2.3-fold, which resulted in a 78% removal of the total nitrogen in an SBR reactor, and reduced the nitrate concentration to 6.03 mg-N/L. These results confirm the theory of the "multi-mechanism synergistic removal of pollutants" and provide a new paradigm for the management of complex pollutants.

In the fields of engineering applications and smart material design, the articles in this Special Issue show remarkable innovations. Wei et al. (contribution 5) developed a cationic polyacrylamide (CPAM) suspension for oilfield wastewater treatment problems, which reduced the dissolution time by 81.4% (16 min) and achieved a 35.6% turbidity removal rate via a continuous grafting process with a dual initiator, which was 25.4% higher than the traditional solid CPAM. Xie et al. (contribution 7) focused on the decolorization of indigo wastewater. The PP-g-AA-MAH fiber prepared with a dual initiator (DBPH/BPO) had an adsorption capacity of 110.43 mg/g for indigo, which maintained 81.7% efficiency after eight cycles, and FTIR confirmed the successful grafting of acrylic acid with maleic anhydride (contribution 7). For scale inhibitor materials, Al-Hamzah et al. (contribution 8) systematically evaluated sulfonated polycarboxylic acid scale inhibitors (molar mass 2000–9500 g/mol). They found that 2000–2500 g/mol of low-molecular-weight materials inhibited calcium sulfate deposition with 85% efficiency in seawater at 125 °C, which provides key data for the optimization of the thermal desalination process. Smart responsive material design is a highlight: a review by Kumar et al. (contribution 1) shows that MOFs/COFs hybrid systems increase the membrane selective permeability to 3–5 times that of conventional polymers. Innovations in characterization techniques are also prominent, with Fakhry et al. (contribution 3) introducing decision tree modeling for the first time to optimize the Cr(VI) adsorption process. Additionally, Meng et al. (contribution 9) analyzed the amidation mechanism of EDA with GO via in situ FTIR-XPS coupling.

These results not only break through the theoretical framework of dynamic adaptive materials, but also promote the intelligent water treatment system on the ground. Future research can focus on material database construction and machine learning linkage, such as the neural network optimization of adsorption strategies (contribution 3) and full life cycle assessment (LCA) system, which together promote the development of water treatment technology towards precision and sustainability (contribution 2).

3. Conclusions and Outlook

This Special Issue focuses on the latest progress in functional polymer materials in the field of water treatment, and through the inclusion of 10 high-quality articles, it comprehensively demonstrates the remarkable effectiveness and innovative applications of functional polymer materials in solving water pollution problems. These papers not only cover research on polymer membranes, adsorbent materials, anti-scaling agents, and other aspects, but they also discuss the preparation technology, performance optimization, and practical application effects of the materials. By synthesizing the conclusions of the papers, it is clear that functional polymer materials, by virtue of their unique physicochemical properties, have demonstrated excellent performance in many aspects, such as heavy metal ion adsorption, organic dye degradation, and oilfield wastewater treatment. By optimizing the structure and function of the materials, the researchers have successfully improved water treatment efficiency and reduced treatment costs, providing strong support for the protection and sustainable use of water resources.

Future research should focus on three major directions. In terms of material performance optimization, it is necessary to break through the bottleneck of anti-pollution and the stability of membrane materials, develop smart response materials with high selectivity and low energy consumption, and explore the scale-up processing technology of MOFs/COFs hybrid systems. At the level of technology integration and innovation, we can deepen the integration of machine learning and experimental data, construct adsorption kinetic prediction models, and expand multi-material synergistic systems (such as magnetic separation and photocatalytic coupling) combined with green synthesis processes to reduce energy consumption and carbon footprint. Engineering applications need to strengthen the validation of complex scenarios: assessing the risk of particle shedding in the long-term operation of photocatalytic membranes, establishing molecular weight efficacy correlation models of scale inhibitors to guide industrial selection, and optimizing the replacement cycle of carriers to enhance system stability.

Furthermore, there is an urgent need to improve sustainability assessment systems, including full life cycle analysis of bio-based materials, environmental impact assessment of low-toxicity modification strategies, and research on the economics of recycling materials. Through interdisciplinary collaboration and data-driven design, functional polymer materials are expected to drive water treatment technologies toward the goals of high efficiency, intelligence, and carbon neutrality, and provide innovative solutions to the global challenge of clean drinking water outlined in the United Nations 2030 Sustainable Development Goals. In short, the application of functional polymer materials in the field of water treatment will continue to progress, contributing more wisdom and strength to address global water challenges.

Author Contributions: Z.W. conceptualization and writing—review and editing, T.W. conceptualization and writing—original draft preparation, Y.G. writing—review, J.Z. writing—review. All authors have read and agreed to the published version of the manuscript.

Funding: This research was funded by the following grants, including the Shaanxi Provincial Natural Science Basic Research Program for Youth Project (2025JC-YBQN-457), the Key Research and Development Program of Shaanxi Province (2023-LL-QY-42), the Guangxi Province Talent Project (GXR-1BGQ2424002), and the National Key Research and Development Plan Project (2022YFC2105500).

Institutional Review Board Statement: Not applicable.

Data Availability Statement: Not applicable.

Acknowledgments: The Special Issue entitled "Recent Advances in Functional Polymer Materials for Water Treatment" is a collection of ten articles, including nine original research articles and one review. We are confident that this Special Issue will inspire the scientific community on the topics of the design, development, and water treatment applications of functional polymer materials, covering a wide range of real-world applications. Finally, we would like to express our sincere gratitude to all (co-)authors and reviewers for their valuable contributions to this Special Issue. Finally, we would also like to thank the editorial team of *Polymers* for their continued support.

Conflicts of Interest: The authors declare no conflicts of interest.

List of Contributions:

1. Kumar, A.; Chang, D.W. Optimized Polymeric Membranes for Water Treatment: Fabrication, Morphology, and Performance. *Polymers* **2024**, *16*, 271. https://doi.org/10.3390/polym16020271.
2. Sun, J.; Jin, Z.; Wang, J.; Wang, H.; Zhang, Q.; Gao, H.; Jin, Z.; Zhang, J.; Wang, Z. Application of Ionic Liquid Crosslinked Hydrogel for Removing Heavy Metal Ions from Water: Different Concentration Ranges with Different Adsorption Mechanisms. *Polymers* **2023**, *15*, 2784. https://doi.org/10.3390/polym15132784.

3. Fakhry, H.; Ghoniem, A.A.; Al-Otibi, F.O.; Helmy, Y.A.; El Hersh, M.S.; Elattar, K.M.; Saber, W.I.A.; Elsayed, A. A Comparative Study of Cr(VI) Sorption by *Aureobasidium pullulans* AKW Biomass and Its Extracellular Melanin: Complementary Modeling with Equilibrium Isotherms, Kinetic Studies, and Decision Tree Modeling. *Polymers* **2023**, *15*, 3754. https://doi.org/10.3390/polym15183754.
4. Yadav, A.; Raghav, S.; Jangid, N.K.; Srivastava, A.; Jadoun, S.; Srivastava, M.; Dwivedi, J. Myrica esculenta Leaf Extract—Assisted Green Synthesis of Porous Magnetic Chitosan Composites for Fast Removal of Cd (II) from Water: Kinetics and Thermodynamics of Adsorption. *Polymers* **2023**, *15*, 4339. https://doi.org/10.3390/polym15214339.
5. Wei, Z.; Long, W.; Li, S.; Zhao, Y.; Yu, S.; Zhou, F. Preparation of Cationic Polyacrylamide Suspension and Its Application in Oilfield Wastewater Treatment. *Polymers* **2024**, *16*, 151. https://doi.org/10.3390/polym16010151.
6. Yang, Y.; Zhou, S.; Cao, X.; Lv, H.; Liang, Z.; Zhang, R.; Ye, F.; Yu, D. Coaxial Electrospun Porous Core–Shell Nanofibrous Membranes for Photodegradation of Organic Dyes. *Polymers* **2024**, *16*, 754. https://doi.org/10.3390/polym16060754.
7. Xie, S.; Fang, Z.; Lian, Z.; Luo, Z.; Zhang, X.; Ma, S. A Novel, Dual-Initiator, Continuous-Suspension Grafting Strategy for the Preparation of PP-g-AA-MAH Fibers to Remove of Indigo from Wastewater. *Polymers* **2024**, *16*, 2144. https://doi.org/10.3390/polym16152144.
8. Al-Hamzah, A.A.; Fellows, C.M.; Hamed, O.A. Methallylsulfonate Polymeric Antiscalants for Application in Thermal Desalination Processes. *Polymers* **2024**, *16*, 2838. https://doi.org/10.3390/polym16192838.
9. Meng, N.; Sun, X.; Liu, J.; Mi, J.; Rong, R. Effect of Addition Amount of Ethylenediamine on Interlayer Nanochannels and the Separation Performance of Graphene Oxide Membranes. *Polymers* **2024**, *16*, 3123. https://doi.org/10.3390/polym16223123.
10. Liu, Z.; Xu, Z.; Li, K.; Xie, L.; Han, B.; Wang, Q.; Song, H.; Zhang, J. Enhancement of Partial Nitrification–Anaerobic Ammonia Oxidation in SBR Reactors via Surface-Modified Polyurethane Sponge Biofilm Carrier. *Polymers* **2025**, *17*, 1145. https://doi.org/10.3390/polym17091145.

References

1. Wang, T.; Stadler, F.J.; Husein, D.Z.; Zhang, D.; Cai, J.; Wang, Y.; Li, M.; Qiang, Y.; Zheng, J. Comparative study of enhanced adsorption-photodegradation activity using activated biochar composited with Ag_3PO_4 or $Ag_6Si_2O_7$ in wastewater treatment and disinfection: Effects and mechanisms. *Colloids Surf. A* **2022**, *655*, 130235. [CrossRef]
2. Sharma, G.; Verma, A.; Wang, T.; Naushad, M.; Kumar, A.; Dhiman, P. Remediation of antibiotics using coordination polymers. *Coord. Chem. Rev.* **2024**, *519*, 216120. [CrossRef]
3. Zheng, D.; Wang, K.; Bai, B. A critical review of sodium alginate-based composites in water treatment. *Carbohydr. Polym.* **2024**, *331*, 121850. [CrossRef] [PubMed]
4. Zhi, J.; Bai, X.; Wang, Q.; Wang, T.; Verma, Y.; Sharma, G.; Kumar, A.; Dhiman, P. Natural gums-derived hydrogels for adsorptive removal of heavy metals: A review. *Int. J. Biol. Macromol.* **2025**, *310*, 143350. [CrossRef] [PubMed]
5. Cho, Y.; Beak, J.W.; Sagong, M.; Ahn, S.; Nam, J.S.; Kim, I.D. Electrospinning and Nanofiber Technology: Fundamentals, Innovations, and Applications. *Adv. Mater.* **2025**, 2500162. [CrossRef] [PubMed]
6. Shi, S.; Bai, W.; Chen, X.; Si, Y.; Zhi, C.; Wu, H.; Su, Y.; Cai, W.; Fei, B.; Kan, C.W. Advances in nanofiber filtration membranes: From principles to intelligent applications. *Adv. Funct. Mater.* **2025**, 2423284. [CrossRef]
7. Wang, T.; Kumar, A.; Sharma, G.; Wang, S.; Jia, J.; Zheng, J.; Shi, H. Synergistic mechanisms of novel Z-Scheme N,S co-doped biochar-based Ag_3PO_4 composites for efficient removal of norfloxacin. *npj Clean Water* **2024**, *7*, 97. [CrossRef]

polymers

MDPI

Article

Enhancement of Partial Nitrification–Anaerobic Ammonia Oxidation in SBR Reactors via Surface-Modified Polyurethane Sponge Biofilm Carrier

Zexiang Liu [1,†], Zhihong Xu [1,†], Kelin Li [1], Li Xie [1], Biao Han [2], Qiming Wang [2], Hainong Song [3,*] and Jian Zhang [1,3,*]

[1] Guangxi Key Laboratory of Clean Pulp & Papermaking and Pollution Control, School of Light Industry and Food Engineering, Guangxi University, Nanning 530004, China
[2] Scientific Research Academy of GuangXi Environmental Protection, Nanning 530022, China
[3] Guangxi Bossco Environmental Protection Technology Co., Ltd., Nanning 530007, China
[*] Correspondence: songhainong@163.com (H.S.); zhangjian_gx@gxu.edu.cn (J.Z.)
[†] These authors shared co-first authorship.

Abstract: The partial nitrification–anammox process offers a cost-effective, energy-efficient, and environmentally sustainable approach for nitrogen removal in wastewater treatment. However, its application under low ammonia nitrogen conditions faces operational challenges including prolonged start-up periods and excessive nitrite oxidation. This study employed a strategy combining polyurethane surface positive charge enhancement and zeolite loading to develop a carrier capable of microbial enrichment and inhibition of nitrate generation, aiming to initiate the partial nitrification-anammox process in a sequencing batch reactor. Operational results demonstrate that the modified carrier enabled the reactor to achieve a total nitrogen removal efficiency of 78%, with the effluent nitrate nitrogen reduced to 6.03 mg-N/L, successfully initiating the partial nitrification-anammox process. The modified carrier also exhibited accelerated biofilm proliferation (both suspended and attached biomass increased). Additionally, 16S rRNA revealed a higher relative abundance of typical anammox bacteria *Candidatus* Brocadia in the biofilm of the modified carrier compared to the original carrier, alongside a decline in nitrifying genera, such as *Nitrolancea*. These microbial shifts effectively suppressed excessive nitrite oxidation, limited nitrate accumulation, and sustained efficient nitrogen removal throughout the reactor's operation.

Keywords: partial nitrification–anammox; sequencing batch reactor; zeolite

1. Introduction

Biological nitrogen removal technology has a development history of nearly a century [1]. Conventional approaches relying on nitrification–denitrification processes remain energy- and resource-intensive, demanding significant aeration and external carbon supplementation while generating substantial operational costs [2]. Anaerobic ammonium oxidation (anammox) utilizes anaerobic ammonium-oxidizing bacteria (AnAOB) to directly oxidize ammonium into gaseous nitrogen (N_2), requiring only minimal aeration and no external carbon source [3]. AnAOB carry out the anaerobic ammonium oxidation reaction within the unique organelle (the anammoxosome), using ammonium nitrogen as the electron donor and nitrite nitrogen as the electron acceptor [4,5]. As an effective and environmentally friendly technology, this method has attracted extensive attention and is used to treat ammonia nitrogen-rich wastewater.

The anammox process requires co-existing ammonium (NH_4^+-N) and nitrite (NO_2^--N) substrates for its biochemical reactions, yet most wastewaters contain only ammonium. This fundamental limitation requires controlled partial ammonia oxidation to nitrite as the critical precursor for anammox-mediated nitrogen removal [6,7]. To address this substrate requirement, the partial nitrification–anammox (PN/A) process has been developed as an integrated solution for nitrogen-laden wastewater treatment [8]. The PN/A process is a two-step process where ammonia-oxidizing bacteria (AOB) first oxidize ammonia nitrogen to nitrite under aerobic conditions, followed by anaerobic ammonium-oxidizing bacteria (AnAOB) converting the remaining ammonia nitrogen and nitrite to N_2 under anaerobic conditions, thereby achieving nitrogen removal [9,10].

The PN/A process achieves efficient nitrogen removal through autotrophic ammonium conversion under limited aeration, offering significant economic and environmental advantages by eliminating external carbon requirements [11]. The prerequisite for successfully initiating the PN/A process is retaining a sufficient quantity of microorganisms in the reactor and constructing a microbial community with functionally compatible populations [12]. Key operational challenges include the following: (1) Members of the AnAOB grow slowly, with a long doubling time (12 days) [13]. Therefore, since AnAOB grow slowly, it is necessary to retain AnAOB in the reactor in the form of biofilms or granules to prevent biomass washout [14]. (2) The target product nitrite may be further oxidized by nitrite-oxidizing bacteria (NOB) to nitrate [10]. Nitrate cannot be removed in the absence of organic carbon sources in wastewater, leading to decreased total nitrogen removal efficiency. To address these challenges, the following two critical optimization strategies emerge: (1) enhancing microbial enrichment and improving the activity of anammox bacteria; (2) selectively inhibiting the activity of nitrite-oxidizing bacteria to prevent excessive nitrite oxidation [15].

The sequencing batch reactor (SBR) enables operational optimization through adjustable aeration periods for sedimentation control and hydraulic selection-based biomass retention [15]. However, hydraulic selection formed by controlling SBR settling duration cannot fundamentally regulate the AOB/NOB ratio. Improper control inevitably causes excessive ammonia oxidation, leading to effluent nitrate accumulation. Since AOB exhibit a stronger affinity for dissolved oxygen (DO) than NOB, intermittent aeration can be adopted to suppress NOB activity, preventing excessive nitrite oxidation [16]. Furthermore, since NOB are highly sensitive to free ammonia (FA) concentrations, establishing localized ammonia-rich zones can strategically inhibit NOB metabolic activity [17].

Sludge granulation and biofilm formation are the main means to meet the prolonged sludge age required for AnAOB enrichment. Under low ammonia nitrogen concentration conditions, granules are prone to disintegration, while introducing carriers to form biofilms provides higher stability, enabling tolerance to environmental fluctuations and hydraulic shear [18]. Polyurethane sponges are regarded as ideal microbial immobilization carriers due to their high porosity (facilitating substrate diffusion), large specific surface area, and fluidization capability. Furthermore, their interface design and modification can create interfacial microenvironments initiated by cell adhesion and microbial selection, accelerating PN/A process initiation. First, by introducing positively charged polyquaternium-10, the positive charge on the carrier surface is enhanced, promoting initial microbial attachment and biofilm formation [19]. Second, considering the sensitivity of NOB to free ammonia (FA), zeolite, with excellent ion adsorption and exchange capacity, can be incorporated onto the carrier surface [11]. By endowing the carrier with a dynamic buffering capacity for local ammonia nitrogen concentration, NOB-mediated side reactions are suppressed, improving the conversion efficiency of ammonia nitrogen to N_2 via the PN/A process pathway. Compared with the common polyurethane-modified carriers, the carrier developed in this study has higher stability, stronger resistance to environmental shocks, and due to its surface modification, facilitates easier biofilm formation and a faster start-up of the PN/A process.

This study employed a composite carrier with a positively charged surface modification and zeolite loading for microbial retention. By modifying the carrier surface with a positive charge potential, microbial adsorption was enhanced, promoting biofilm formation. Through the carrier modification of the local microenvironment, NOB growth was inhibited, while the activity of AOB and AnAOB was strengthened. The investigation focused on nitrogen removal efficiency during reactor start-up, morphological changes in sludge flocs and biofilms across operational phases, and shifts in the microbial community structure.

2. Materials and Methods

2.1. Material

2.1.1. Seed Sludge

The nitrifying activated sludge used in the experiment was sourced from the wastewater treatment reactor of Angel Yeast (Chongzuo) Co., Ltd., Chongzuo, China The retrieved granular sludge was mechanically disintegrated into flocculent sludge and mixed with anaerobic ammonium oxidation (anammox) flocculent sludge from the laboratory at a ratio of 2:1.

2.1.2. Synthetic Wastewater

To stimulate the activity of aerobic nitrifying bacteria, the chemical oxygen demand (COD) concentration was maintained at 100 mg per liter in the form of glucose during the initial ten days of the experiment [20].

Other influent substrates for the simulated effluent included 21.9 mg/L potassium dihydrogen phosphate (KH_2PO_4), 36 mg/L calcium chloride monohydrate ($CaCl_2{\cdot}H_2O$), and 25 mg/L magnesium sulfate hexahydrate ($MgSO_4{\cdot}6H_2O$). The pH value of the simulated wastewater was adjusted to 8.0–8.5 by adding appropriate amounts of potassium hydroxide (KOH) and hydrochloric acid (HCl), and 1 mL of trace element I and trace element II were added to each liter of wastewater [21]. The compositions of these elements are shown in Table 1.

Table 1. Trace element composition in the stock solution.

Trace Element I	Concentration (g/L)	Trace Element II	Concentration (g/L)
$2Na{\cdot}EDTA{\cdot}2H_2O$	15	$2Na{\cdot}EDTA{\cdot}2H_2O$	5
Na_2SeO_3	0.156	$FeSO_4{\cdot}7H_2O$	5
$Na_2MoO_4{\cdot}2H_2O$	0.22		
$NiCl_2{\cdot}6H_2O$	0.19		
$MnCl_2{\cdot}4H_2O$	0.99		
$CuSO_4{\cdot}5H_2O$	0.25		
$ZnSO_4{\cdot}7H_2O$	0.43		
$CoCl_2{\cdot}6H_2O$	0.24		
H_3BO_3	0.014		

2.2. Methods

2.2.1. Carrier Preparation and Reactor Description

The composite carrier was synthesized using the impregnation method. A predetermined amount of polyvinyl alcohol (PVA) was dissolved in deionized water at 95 °C to prepare a 10% (*w*/*v*) PVA solution. This solution was mixed with a 5% (*w*/*v*) polyquaternium-10 solution at a volume ratio of 10:1, stirred for 20 min to form a homogeneous solution. Zeolite powder and water-based polyurethane were blended at a weight ratio of 27:40, followed by the addition of the PVA-polyquaternium-10 solution. The mixture was stirred for 1 h to ensure thorough homogenization. A suitable number of 1 cm^3 polyurethane foam carriers were fully immersed in the homogeneous solution. After immersion, the carriers were rinsed three times with deionized water, dried, and stored in vacuum-sealed bags [22]. Each carrier contained approximately 0.06 g of zeolite. The experimental system employed a sequencing batch reactor (SBR), with its operational configuration schematically detailed in Figure 1. In the figure, R1 is the reactor with an unmodified carrier (polyurethane sponge not prepared by the impregnation method) and R2 is the reactor with a modified carrier (composite carrier prepared by the impregnation method).

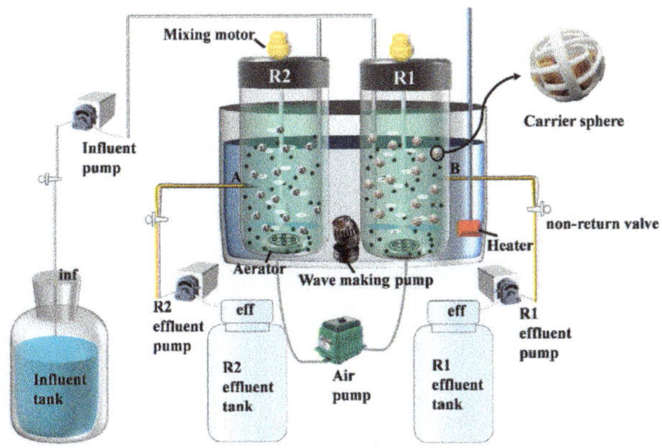

Figure 1. Schematic diagram of SBR reactors integrated with sponge carriers.

2.2.2. Reactor Operation and Control

The reactor was operated for 116 days. The whole operation period can be divided into three phases: the start-up period (Phase I: 1–20 d), the transition period (Phase II: 21–80 d), and the stabilization period (Phase III: 81–116 d). The specific parameters of the reactor are shown in Table 2.

Table 2. Phase-specific parameters for the operation of both SBR reactors.

Stage	Time (d)	NH4$^+$-N (mg N/L)	Air (L/min)	HRT (h)	NLR (kg-N/m^3·d^{-1})
I Start-up period	1~20	110	0.15	10	0.264
II Transition period	21~78	110	0.04~0.1	10	0.264
III Stable period	79~116	60	0.04	5	0.288

2.2.3. Sludge Activity Detection

Measurement of biofilm loading on the carrier [23]:

An appropriate number of carriers was removed from the reactor and placed in a 250 mL beaker. Then, 50 mL of deionized water was added. The mixture was stirred on a magnetic stirrer at 300 r/min for 1 h. The carriers were removed, squeezed to drain excess water, and returned to the reactor. The obtained suspension was tested for mixed liquor suspended solids (MLSS) and mixed liquor volatile suspended solids

(MLVSS) using standard methods. The biofilm loading on the carriers, expressed as attached growth biomass solids (AGBS) and volatile attached growth biomass solids (VAGBS), was calculated by dividing the MLSS and MLVSS by the number of carriers removed.

Measurement of activated sludge particle size [24]:

Three biofilm carriers were randomly selected from the reactor. After ultrasonic treatment to detach the sludge from the carriers, the carriers were returned to the reactor. The obtained suspension was dispersed using deionized water, and the particle size of the activated sludge was measured using a laser particle size analyzer. This procedure was repeated three times, and the average value was recorded.

2.2.4. Water Quality Detection

The reactor influent and effluent samples were monitored daily, including $NH_4{}^+$-N (ammonia nitrogen), $NO_2{}^-$-N (nitrite nitrogen), $NO_3{}^-$-N (nitrate nitrogen), and pH [25]. The ammonia nitrogen was detected by Nano reagent spectrophotometry, the nitrite nitrogen was detected by UV spectrophotometry, and the nitrate nitrogen by the thymol method [13]. The effluent parameters were tested daily during the first 10 days of operation and subsequently every 2 days (no statistical significance test was conducted).

The nitrogen loading rate (NLR, in kg-N/$m^3 \cdot d^{-1}$), the total nitrogen removal rate (TNRE, in %), the ammonia removal rate (ARE, in %), the ammonia removal rate (ARR, in kg-N/$m^3 \cdot d^{-1}$), the ammonia oxidation rate (AOR, in kg-N/$m^3 \cdot d^{-1}$), the nitrite oxidation rate (NOR, in kg-N/$m^3 \cdot d^{-1}$), and the total nitrogen removal rate (TNRR, unit: kg-N/$m^3 \cdot d^{-1}$) were calculated as shown in Equations (1)–(6), and the Δ $NO_3{}^-$-N/Δ $NH_4{}^+$-N in the PN/A was calculated as shown in Equation (7). The inf subscripts denote the influent water, and the eff subscripts denote the effluent water.

$$NLR = \left(\frac{C(TN)_{inf}}{HRT}\right) \times 24 \times 10^{-3} \quad (1)$$

$$TNRE = \frac{C(TN)_{inf-C(TN)_{eff}}}{C(TN)_{inf}} \times 100 \quad (2)$$

$$ARE = \frac{C(NH_4^+ - N)_{inf-}C(NH_4^+ - N)_{eff}}{C(NH_4^+ - N)_{eff}} \times 100 \quad (3)$$

$$AOR = \left(\frac{C(NH_4^+ - N)_{inf-}C(NH_4^+ - N)_{eff} - \frac{\Delta C(TN)}{2.04}}{HRT}\right) \times 24 \times 10^{-3} \quad (4)$$

$$NOR = \left(\frac{C(NO_3^- - N)_{eff} - \frac{0.26\Delta C(TN)}{2.04}}{HRT}\right) \times 24 \times 10^{-3} \quad (5)$$

$$TNRR = \frac{\Delta C(TN)}{HRT} \times 24 \times 10^{-3} \quad (6)$$

$$\frac{\Delta NO_3^-}{\Delta NH_4^+} = \frac{C(NO_3^- - N)_{eff-}C(NO_3^- - N)_{inf}}{C(NH_4^+ - N)_{inf-}C(NH_4^+ - N)_{eff}} \quad (7)$$

2.2.5. Extracellular Polymeric Substances (EPS) Extraction and Speciation

The improved total EPS extraction method was as follows: First, 5 mL of sludge was taken and washed three times with ultrapure water. The washed sludge was resuspended in 15 mL of 0.01 M PBS buffer (pH = 7.2) and centrifuged at 7000× g rpm for 20 min. NaOH was added to adjust the mixture to pH = 13. The supernatant was filtered through a 0.45 μm PES membrane to obtain the total EPS. The three subcomponents of the EPS were extracted using a modified heat method [26].

S-EPS: The washed 5 mL sludge was mixed with 15 mL PBS and centrifuged at 2000× *g* rpm for 1 min. The supernatant was discarded, and the remaining sludge was mixed with 15 mL of 0.05% NaCl solution, followed by centrifugation at 4000× *g* rpm for 5 min. The supernatant was filtered through a 0.45 μm PES membrane.

LB-EPS: Preheated 0.05% NaCl solution (60 °C) was added to the remaining sludge to adjust the total volume to 20 mL. After vortex mixing for 2 min, the mixture was centrifuged at 4000× *g* rpm for 10 min. The supernatant was collected by filtration through a 0.45 μm membrane.

TB-EPS: The remaining sludge was resuspended in 15 mL of 0.05% NaCl solution and heated in a water bath at 60 °C for 30 min. Following vortex mixing for 2 min and centrifugation at 4000 rpm for 15 min, the supernatant was collected through a 0.45 μm membrane [27].

The extracted EPS was characterized by three-dimensional excitation emission matrix (3D-EEM) fluorescence spectroscopy [28].

The protein content was analyzed using the modified Lowey method, and the polysaccharide content was determined using the phenol–sulfuric acid method [29].

2.2.6. Microbial Detection

Microbial community structure detection: Sludge samples were collected on days 20 and 100, respectively. Raw sludge and sludge samples from different reactors were labeled as Blank, R1, and R2. An appropriate amount of sludge was shipped under refrigeration to Majorbio BioPharm Technology Co., Ltd., Shanghai, China for high-throughput sequencing. The key steps of the high-throughput sequencing were as follows: DNA was extracted from the sludge using the E.Z.N.A.® Mag-Bind Soil DNA Kit (OMEGA), and bacterial and archaeal sequences were amplified using primers 52F10extF-Arch958RmodR and 338F-806R, respectively. PCR products were detected via 2% agarose gel electrophoresis, purified with the AxyPrep DNA Gel Extraction Kit, and sequenced on the Illumina MiSeq platform to generate a PE amplicon library. Operational taxonomic units (OTUs) were classified and analyzed [12].

3. Results and Discussion

3.1. Characterization of Modified Polyurethane Sponge Carrier

3.1.1. Surface Properties

Figure 2a–f present a comparative SEM characterization of the carrier surface morphology. Due to the zeolite loading, the composite carrier exhibited increased surface roughness, which aided in resisting hydraulic shear during the initial microbial adhesion stage, enriching microorganisms, and accelerating biofilm formation. Figure 2g displays the water contact angles of the carrier before and after modification. The modified carrier transitioned from hydrophobic to hydrophilic due to the incorporation of hydrophilic materials, polyvinyl alcohol (PVA) and zeolite, during fabrication [30]. Figure 2h illustrates the zeta potentials of the carrier before and after modification and the sludge biofilm. The composite carrier's positive potential increased from −7.45 mV to 1.79 mV owing to the doping of the positively charged material polyquaternium-10, while the anammox sludge showed a negative charge [31]. This charge complementarity enhanced electrostatic microbe–carrier interactions, achieving faster biofilm formation than unmodified carriers [23].

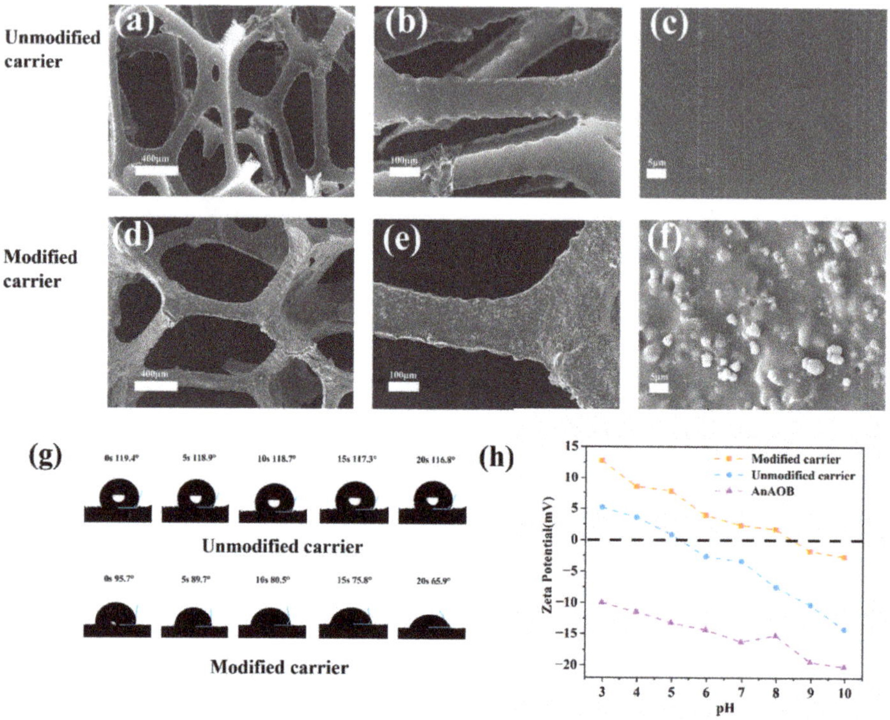

Figure 2. Modification of surface morphology (**a–f**), contact angle (**g**), and zeta potential (**h**).

3.1.2. Ammonia Nitrogen Adsorption Capacity

Figure 3a illustrates the NH_4^+-N adsorption performance of the modified carriers. The composite carrier exhibited an initial increase followed by stabilization in ammonia nitrogen removal, consistent across different temperatures, whereas the unmodified carrier showed no adsorption capacity. This is attributed to the zeolite-loaded modified carrier adsorption capability [32].

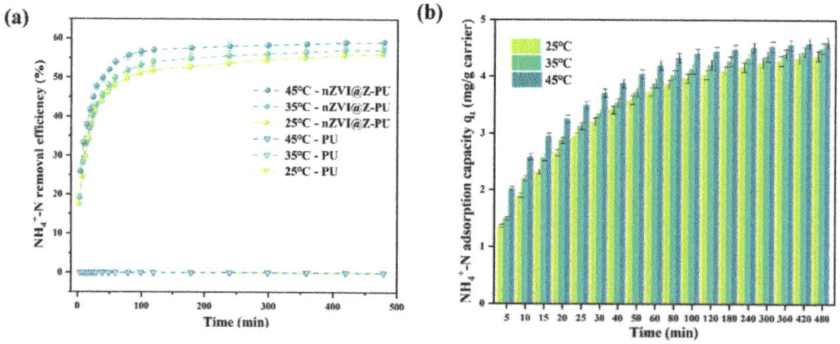

Figure 3. Kinetics of NH_4^+-N adsorption by the composite carrier at 35 °C (**a**); NH_4^+-N removal quantity before and after carrier modification (**b**).

As shown in Figure 3b, the NH_4^+-N adsorption capacity of the composite carrier displayed dynamic characteristics, with rapid initial adsorption followed by gradual stabilization. The maximum adsorption capacity reached 4.5028 mg/g carrier after 8 h.

3.2. Reactor Performance Enhanced by Carrier Modification

3.2.1. Nitrogen Removal

AOR, NOR, and TNRR indirectly reflect the theoretical in situ activities of the corresponding nitrifying microorganisms [33]. As shown in Figure 4, the AOR of both reactors gradually increased in Phase I, with the modified AOR being higher than the unmodified one. This is attributed to the modified carrier's superior initial microbial adsorption and ammonia-rich microenvironment construction capabilities, facilitating the enrichment of AOB.

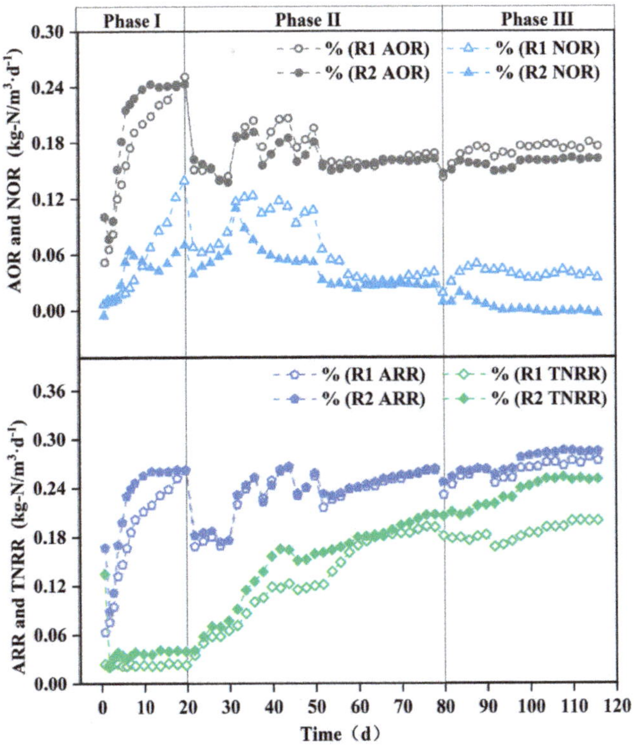

Figure 4. Enhancement of nitrogen removal in terms of AOR, NOR, ARR, and TNRR achieved by carrier modification. AOR: ammonia oxidation rate, NOR: nitrite oxidation rate, ARR: ammonia removal rate, TNRR: total nitrogen removal rate.

Sustained aeration and nitrite accumulation provided favorable conditions for NOB growth [34], leading to an increase in NOR and the recovery of NOB activity. However, the modified carrier achieved localized ammonia nitrogen inhibition, preventing a continuous NOR rise in the modified reactor.

The TNRE remained relatively low in both the modified and unmodified systems. In PN/A processes, total nitrogen removal primarily relies on AnAOB, which are inhibited under full aeration conditions [11]. Nevertheless, the TNRR of the reactor with the modified carrier was higher than that of the reactor with the unmodified carrier, due to the modified carrier's enhanced initial microbial adsorption.

Phase II initiation triggered abrupt ARR/AOR declines in both reactors due to the oxygen-limited suppression of AOB activity. Meanwhile, the gradual increase in the TNRR indicates that the low-oxygen environment led to AnAOB growth, though excessive ammonia accumulation limited total nitrogen removal. To sustain AOB and AnAOB activity, a "graded aeration reduction strategy" was adopted from day 30. ARR and AOR fluctuated upward during Phase II, demonstrating AOB's low-oxygen adaptability under fluctuating oxygen stress, while the steady rise in the TNRR reflects the continuous enhancement of AnAOB activity. Divergent NOR trends between the modified and unmodified carriers highlighted NOB as the primary factor influencing effluent nitrate differences [35]. The composite carrier's ammonia-rich microenvironment progressively eliminated NOB, exhibiting strong NOB suppression [36].

As shown in Figure 4, reactor R2 with modified carriers exhibited a significantly higher total nitrogen removal rate (TNRR) (0.25 kg-N/m^3·d^{-1} vs. 0.20 kg-N/m^3·d^{-1}) compared to the control reactor R1 in Phase III, while the ammonia removal rate (ARR) remained similar between the two reactors. This observation suggests that ammonia nitrogen in R1 may not have been sufficiently converted into nitrate for utilization in anaerobic ammonium oxidation (anammox). Since nitrate cannot be removed via denitrification in the absence of a carbon source, this limitation likely contributed to the lower TNRR in R1 compared to R2. These findings are corroborated by the reactor effluent data. As illustrated in Figure 5b, the ΔNO_3^--N/ΔNH_4^+-N ratio in the modified carrier reactor was significantly lower than that in the unmodified carrier reactor (0.11 vs. 0.22). This confirms that the modified carrier enhanced the selectivity of ammonia oxidation, effectively suppressing nitrate formation and accumulation. Consequently, the modified carrier facilitated efficient ammonia nitrogen removal via anammox, ensuring stable reactor performance.

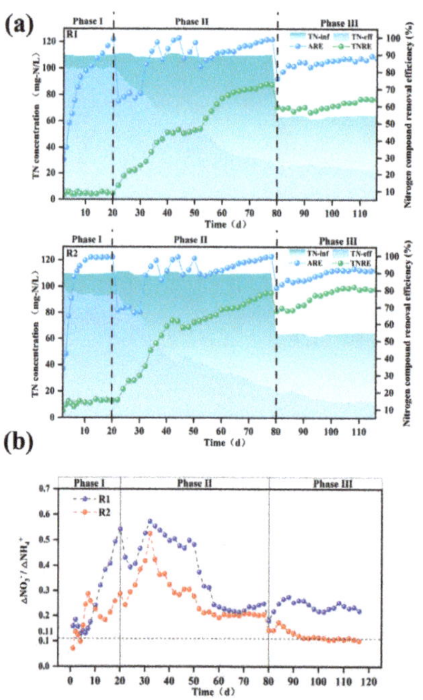

Figure 5. The TN, ARE, and TNRE of the reactor with unmodified (R1) and modified (R2) carriers (**a**); ΔNO_3^--N/ΔNH_4^+-N ratio throughout the operational period for both reactors (**b**). TN: total nitrogen, ARE:. ammonia removal efficiency, TNRE: total nitrogen removal efficiency.

3.2.2. Start-Up of PN/A Process

As shown in Figure 5a,b, it was observed that by the end of Phase I, both modified and unmodified systems achieved ammonia nitrogen removal efficiencies exceeding 95%, but the total nitrogen removal performance remained minimal. Upon reducing aeration in Phase II, ammonia removal rates experienced a trough, while the ΔNO_3^--N/ΔNH_4^+-N ratio gradually increased, surpassing the PN/A reaction's standard value of 0.11. After implementing the "graded aeration reduction" strategy, ammonia removal rates underwent repeated fluctuations of increase–decrease–increase. By day 52, the modified and unmodified systems reached ammonia removal rates of 81% and 89%, respectively, indicating restored AOB activity. Concurrently, the effluent total nitrogen in both reactors progressively declined, with total nitrogen removal efficiencies reaching 46.7% and 61.5% for R1 and R2. Moreover, the ΔNO_3^--N/ΔNH_4^+-N ratio of modified and unmodified systems decreased from peak values of 0.57 and 0.52 to 0.25 and 0.20 by the end of Phase II, demonstrating significant overall performance improvement.

In Phase III, fluctuations in the ammonia removal rates occurred due to a reduced influent ammonia concentration and shorter hydraulic retention time (HRT). As the reaction progressed, microbial synergy strengthened continuously. The modified carrier successfully enriched functional microorganisms while suppressing NOB, leading to the gradual washout of non-functional bacteria from the reactor. By day 98, the ΔNO_3^--N/ΔNH_4^+-N ratio of the modified system reached 0.12, approaching the PN/A standard value of 0.11. At this stage, the total nitrogen removal efficiency was 78%, the ammonia removal efficiency was 90%, and the TNRR was 0.24 kg-N·m^{-3}·d^{-1}, successfully achieving PN/A initiation under low ammonia nitrogen conditions. In contrast, the unmodified carrier failed to initiate PN/A.

3.3. *Biofilm and Microbial Community*

3.3.1. Biofilm and Biomass Growth on Carriers

Figure 6a illustrates the biofilm growth on the carriers. Sludge attachment was observed to follow the sponge skeleton framework, progressing gradually from the interior to the exterior. Upon reaction completion, the modified carrier formed a thick sludge biofilm, with significantly greater surface biomass accumulation compared to the unmodified carrier [37]. This was due to the electropositive surface of the modified carrier enhancing microbial adsorption. Sludge coverage reduced the oxygen penetration capacity in the modified carrier compared to the unmodified one. Owing to the porous structure of the polyurethane sponge, microorganisms continued to colonize the internal spaces of the carrier. Increased microbial attachment led to thicker biofilms, facilitating the development of anoxic zones within the carrier matrix, thereby promoting anammox and enhancing nitrogen removal [8]. This also explains the higher ammonia removal efficiency of the modified carrier.

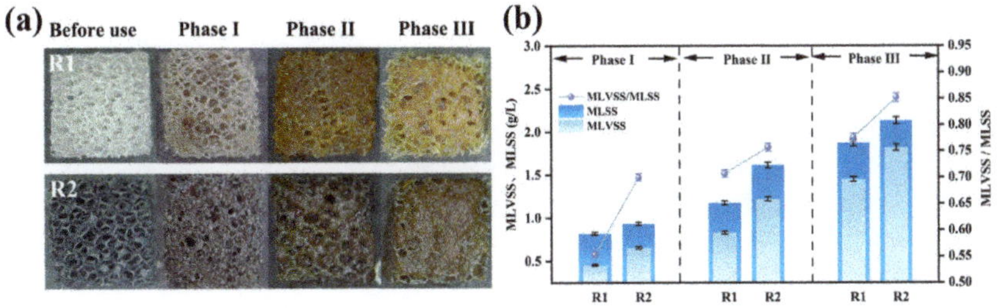

Figure 6. Biofilm growth on polyurethane sponge carrier: morphology of biofilm covering carrier surface (**a**); biofilm quantified by MLSS and MLVSS (**b**).

The MLVSS/MLSS ratio reflects the proportion of active biomass in the sludge biofilm [38]. As shown in Figure 6b, the MLSS and MLVSS in R2 (modified) remained higher than in R1 (unmodified). This phenomenon stems from the modified carrier's positive charge improving its initial microbial adsorption. During operation, suspended floccular sludge was continuously washed out, while microbial aggregates adhered to the carrier, resisting hydraulic shear and avoiding discharge, leading to gradual biomass accumulation.

The MLVSS/MLSS ratios for the modified and unmodified systems increased from 0.55 and 0.70 in Phase I to 0.77 and 0.85 in Phase III, respectively. The higher MLVSS/MLSS ratio in the modified system indicates greater biological activity in its biofilm. At the end of the reaction, the MLVSS reached 1.82 g/L for the modified system versus 1.45 g/L for the unmodified system. Due to oxygen concentration gradients, AnAOB preferentially thrive in biofilms; therefore, higher MLVSS confirm the enhanced nitrogen removal capacity of the modified carrier [14].

Post-reaction, the MLSS, MLVSS, attached granular biomass (AGBS), and volatile attached granular biomass (VAGBS) on individual carriers in the modified and unmodified reactors were quantified (Table 3).

Table 3. Suspended biomass and attached biomass on carriers.

	MLSS (g/L)	MLVSS (g/L)	AGBS (mg/Carrier)	VAGBS (mg/Carrier)
R1	0.82 ± 0.02	0.66 ± 0.05	6.83 ± 0.01	5.45 ± 0.01
R2	1.12 ± 0.06	0.95 ± 0.03	9.37 ± 0.01	7.92 ± 0.01
Growth	0.30 ± 0.04	0.29 ± 0.03	2.54 ± 0.01	2.47 ± 0.01

Both the MLSS and AGBS demonstrated superior performance in the modified reactor compared to the unmodified. The modified carrier exhibited a VAGBS of 7.92 ± 0.01 mg/carrier, 1.45-fold higher than the unmodified carrier's 5.45 ± 0.01 mg/carrier. The greater biomass in the modified reactor further verifies that the composite carrier promoted microbial adsorption more than the original carrier. The modified carrier effectively retained microorganisms under hydraulic scouring, enabling faster biofilm growth compared to the unmodified carrier [37]. Higher microbial abundance on the carrier accelerated metabolic rates, resulting in increased oxygen consumption.

3.3.2. EPS Composition and Speciation

Extracellular polymeric substances (EPS) are crucial components of biofilms, contributing to matrix structure integrity and stability and the protection of embedded microorganisms against toxic effects [39]. EPS exhibit distinct stratified structures in biofilms, including soluble EPS (S-EPS), which weakly associate with microbial aggregates or dissolve in the surrounding solution, forming a discrete overlayer with distinct boundaries outside the cell walls; loosely bound EPS (LB-EPS); and tightly bound EPS (TB-EPS), which collectively form a bilayer structure. All EPS fractions are enriched with negatively charged groups, with TB-EPS constituting the predominant proportion [40]. Figure 7a compares the concentrations and proportions of EPS fractions in the modified and unmodified reactors.

Experimental observations revealed minimal variation in the S-EPS and LB-EPS levels, with their contents in the sludge biofilms being significantly lower than TB-EPS. Therefore, EPS dynamics can primarily be attributed to TB-EPS changes. The TB-EPS correlated closely with the water quality and sludge morphology, serving as an indicator of microbial viability and aggregate structure. In the unmodified reactor, the TB-EPS increased from 27.03 mg/g VSS in Phase I to 70.34 mg/g VSS in Phase III, representing a 160% rise (43.31 mg/g VSS increment). In contrast, the modified reactor exhibited a more substantial TB-EPS increase from 35.44 mg/g VSS in Phase I to 111.83 mg/g VSS in Phase III, a 216% surge (76.39 mg/g VSS increment). EPS enhance biofilm adhesiveness, promote microbial aggregation and growth, and improve biofilm attachment [41]. The abundant EPS production accelerated biofilm formation on the modified carriers, increasing oxygen consumption and AnAOB activity, thereby enhancing ammonia removal efficiency [42].

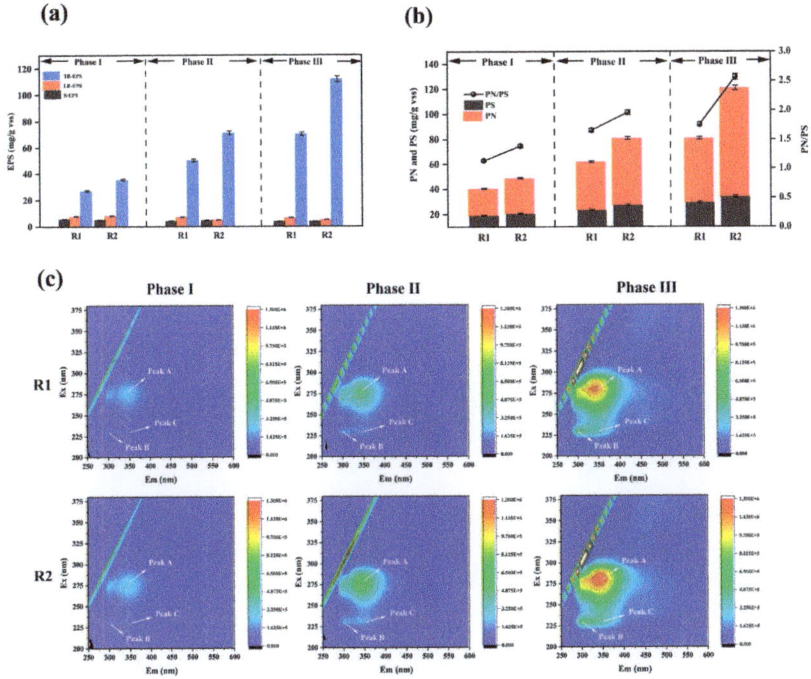

Figure 7. Growth of EPS on carriers during initiation of partial nitrification–anaerobic ammonia oxidation (**a**); PN and PS and their ratios in EPS on carriers during partial nitrification–anaerobic ammonia oxidation initiation at each phase (**b**); EEM spectra of TB-EPS on carriers during PN/A initiation (**c**).

EPS primarily comprise extracellular polysaccharides (PS) and proteins (PNs) [43]. The data of Figure 7b show that the PS content remained consistently lower than the PNs across all experimental phases in both the modified and unmodified systems. While the PS levels showed negligible differences between the modified and unmodified carriers, substantial PN disparities were observed, indicating that higher TB-EPS in the modified system predominantly resulted from increased PN accumulation. A higher PN/PS ratio reflects enhanced biofilm hydrophobicity. Increased sludge hydrophobicity is critical for microbial aggregation and biofilm development.

In Phase I, the PN/PS ratios for the modified and unmodified systems were 1.13 and 1.39, respectively, with PS values of 18.94 mg/g VSS and 23.32 mg/g VSS, showing minor differences. This is because during the initial adhesion stage, biofilms require increased PS secretion to form a buffer layer between the aqueous environment and microorganisms, protecting them from trace toxic substances and adapting to relatively harsh conditions. Meanwhile, PS enhance the compactness and stability of biofilm structures. EPS form gels via hydrogen bonds, and the PS content in EPS plays a critical role in the initial microbial attachment [44].

During Phases II and III, the PN/PS ratios in the modified reactor remained higher than those in the unmodified reactor. The increasing disparity reflects dynamic changes in the PN and PS compositions across reaction phases. The strong aggregation capacity of activated sludge is primarily attributed to PNs, which is dominated by hydrophobic amino acids. The hydrophobicity of the PNs correlates positively with EPS hydrophobicity, thereby promoting flocculation [45].

Figure 7c presents the EEM spectra of sludge TB-EPS from both reactors at different operational phase, revealing three distinct EPS-associated fluorescence peaks. Peak A (275 nm/340–360 nm) corresponds to tryptophan in proteins, Peak B (230 nm/300–330 nm) represents tyrosine, and Peak C (220 nm/335–355 nm) reflects aromatic protein-like substances. The peak positions show negligible shifts, indicating no significant changes in the EPS chemical composition. The fluorescence intensities gradually increased with the reaction's progression, aligning with the PN trends in the EPS. As the particle size and biofilm thickness increased during the operation, EPS secretion intensified. The fluorescence intensities of Peaks A and B rose significantly, with tryptophan becoming more prevalent during biofilm development, serving structural and stabilizing roles in EPS. Tryptophan fluorescence properties, monitored via EEM, aid in studying EPS composition and dynamics. The modified system exhibited higher intensity ratios of tryptophan-to-tyrosine peaks compared to the unmodified reactor (R1), reflecting enhanced protein synthesis and accumulation in the modified biofilms. The elevated PN/PS ratios increased the sludge surface's hydrophobicity, accelerating biofilm formation and improving structural robustness [46].

3.3.3. Microbial Community in Biofilm

To explore microbial community evolution during the start-up of a single-stage PN/A process, high-throughput sequencing was performed on sludge samples from a blank control (inoculated sludge) and reactors with modified or unmodified carriers. Figure 8a shows the microbial community composition at the phylum level. The dominant phyla included *Proteobacteria*, *Chloroflexi*, *Bacteroidota*, and *Planctomycetota*. *Proteobacteria* encompasses AOB and NOB [47]. *Chloroflexi* contains minor NOB populations, while *Planctomycetota* primarily harbors anaerobic AnAOB [48]. Figure 8b displays genus-level taxonomic profiles. *Denitratisoma* spp. represents denitrifying bacteria. *Candidatus* Brocadia spp. and *Candidatus* Kuenenia spp. are typical AnAOB genera in wastewater systems [49]. *Nitrosomonas* spp. (AOB) and *Nitrolancea* spp. (NOB) were also identified.

As shown in Figure 8a, *Planctomycetota* abundances in the modified and unmodified reactors were lower than in the blank control. This may be attributed to the glucose supplementation only during the first 10 days of Phase I, which likely eliminated heterotrophic microorganisms within *Planctomycetota* under unfavorable conditions. Since *Planctomycetota* hosts AnAOB, increased dissolved oxygen from aeration inhibited their growth [2]. Notably, the modified reactor exhibited higher *Planctomycetota* abundance than the unmodified reactor, possibly due to localized ammonia-rich microenvironments around the modified carrier, enhancing AnAOB activity and ammonia removal.

By the end of Phase III, decreased *Proteobacteria* and *Chloroflexi* abundances in both reactors likely resulted from NOB suppression and partial AOB washout from flocs. Bacteroidota enrichment correlated with biofilm development, as its associated microorganisms promoted denser biofilm structures. The increased *Planctomycetota* proportions confirm AnAOB proliferation. Notably, the *Planctomycetota* abundance in the modified reactor was 1.4-fold higher than in the unmodified system, demonstrating the modified carrier's superior ability to support AnAOB growth.

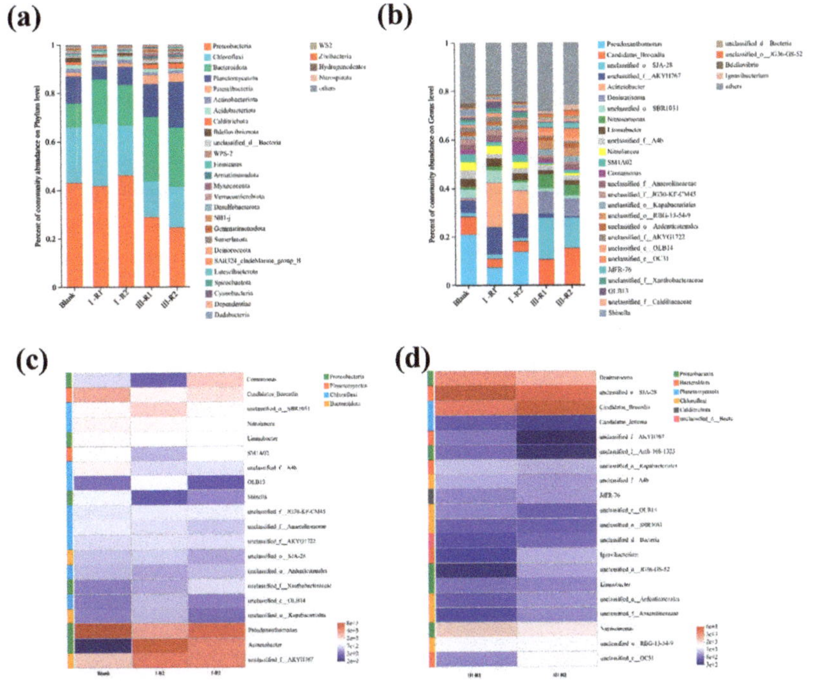

Figure 8. Evolutionary patterns at the phylum level of carrier microbial communities during PN/A initiation (**a**); evolutionary patterns at the genus level of vector microbial communities during PN/A initiation (**b**); heatmap depicting microbial abundance in seed sludge and various reactors during phase I (**c**); heatmap depicting microbial abundance in different reactors during phase III (**d**).

Figure 8b displays genus-level microbial community composition, revealing significant microbial shifts. The relative abundances of *Denitratisoma* spp. increased in both the modified and unmodified reactors during Phase III. This is likely because high aeration initially suppressed its activity, while reduced dissolved oxygen levels later provided favorable conditions for its recovery.

As shown in Figure 8c,d, the abundance of NOB was far lower than that of AOB, indicating that gradual aeration reduction successfully suppressed NOB activity, facilitated their washout, and effectively inhibited NOB proliferation. The enrichment of *Candidatus* Kuenenia spp. and *Nitrosomonas* spp. in the PN/A biofilm aligns with previous effluent results. This demonstrates that the gradual aeration reduction strategy maintained AOB activity while suppressing NOB activity and enhancing AnAOB activity. Higher proportions of *Candidatus* Kuenenia spp. and *Nitrosomonas* spp. in the modified reactor compared to the unmodified reactor suggest that the ammonia-rich microenvironment formed by the modified carrier promoted AnAOB activity improvement.

4. Conclusions

This study engineered a dual-functional polyurethane carrier incorporating surface charge modification (polyquaternium-10) and zeolite integration to optimize PN/A process initiation in SBR systems. By introducing the positively charged material polyquaternium-10, the carrier's surface charge increased from −7.45 mV to +1.79 mV, facilitating biofilm formation on the carrier surface and increasing both suspended biomass and attached biomass. Zeolite loading endowed the carrier with an ammonia nitrogen adsorption capacity of 4.50 mg/g carrier. The detection results of the microbial community structure reveal decreased NOB abundance and increased AnAOB abundance, suggesting that the zeolite-modified surfaces created an ammonia-rich microenvironment that suppressed NOB activity while enhancing AnAOB activity. The reactor operation results demonstrate that the composite-modified carrier successfully initiated the PN/A process, whereas reactors with unmodified carriers failed to establish the PN/A process. During the operation of the experimental conditions, the modified carrier did not break. However, due to the relatively loose pores of the polyurethane sponge, the material had a certain service life and needed to be replaced regularly. The modified carrier has the ability to rapidly start-up PN/A and has a certain resistance to environmental shocks. It has a promising future in larger-scale applications.

Author Contributions: Conceptualization, Z.L.; Methodology, Z.X., L.X. and Q.W.; Software, Z.L., L.X. and B.H.; Validation, Z.X., K.L., B.H., Q.W. and J.Z.; Formal analysis, Z.L., K.L. and H.S.; Investigation, Z.X., H.S. and J.Z.; Resources, H.S. and J.Z.; Data curation, K.L.; Writing—original draft, Z.X.; Writing—review & editing, Z.X. All authors have read and agreed to the published version of the manuscript.

Funding: This research was enthusiastically supported by the National Natural Science Foundation of China (222080666), Guangxi Key R&D Project (Guike AB23026119), the National Key Research and Development Program (2022YFC2105505), and the Guangxi Natural Science Foundation of China (2023GXNSFGA026001).

Institutional Review Board Statement: Not applicable.

Informed Consent Statement: Not applicable.

Data Availability Statement: The original contributions presented in this study are included in the article. Further inquiries can be directed to the corresponding authors.

Conflicts of Interest: Author Hainong Song and Jian Zhang were employed by the company Guangxi Bossco Environmental Protection Technology Co., Ltd. The remaining authors declare that the research was conducted in the absence of any commercial or financial relationships.

References

1. Liu, T.; Tian, R.; Li, Q.; Wu, N.; Quan, X. Strengthened Attachment of Anammox Bacteria on Iron-Based Modified Carrier and Its Effects on Anammox Performance in Integrated Floating-Film Activated Sludge (IFFAS) Process. *Sci. Total Environ.* **2021**, *787*, 147679. [CrossRef] [PubMed]
2. Huff Chester, A.L.; Eum, K.; Tsapatsis, M.; Hillmyer, M.A.; Novak, P.J. Enhanced Nitrogen Removal and Anammox Bacteria Retention with Zeolite-Coated Membrane in Simulated Mainstream Wastewater. *Environ. Sci. Technol. Lett.* **2021**, *8*, 468–473. [CrossRef]
3. Adams, M.; Xie, J.; Xie, J.; Guo, M.; Chen, C. Research Advances in Anammox Granular Sludge: A Review. *Crit. Rev. Environ. Sci. Technol.* **2022**, *52*, 631–674. [CrossRef]
4. Edwards, T.M. Ammonia and Aquatic Ecosystems—A Review of Global Sources, Biogeochemical Cycling, and Effects on Fish. *Sci. Total Environ.* **2024**, *907*, 167911. [CrossRef]
5. Cai, X. Remote Sensing Identification of Urban Water Pollution Source Types Using Hyperspectral Data. *J. Hazard. Mater.* **2023**, *459*, 132080. [CrossRef]
6. Qi, G. Ammonium Sulfite Pretreatment of Wheat Straw for Efficient Enzymatic Saccharification. *Sustain. Energy Technol. Assess.* **2018**, *29*, 12–18. [CrossRef]
7. Durán, U. Enhanced Ammonia Removal at Room Temperature by pH Controlled Partial Nitrification and Subsequent anaerobic ammonium oxidation. *Environ. Technol.* **2013**, *35*, 383–390. [CrossRef]
8. Bagchi, S.; Biswas, R.; Nandy, T. Alkalinity and Dissolved Oxygen as Controlling Parameters for Ammonia Removal through Partial Nitritation and ANAMMOX in a Single-Stage Bioreactor. *J. Ind. Microbiol. Biotechnol.* **2010**, *37*, 871–876. [CrossRef]
9. Tsushima, I.; Ogasawara, Y.; Kindaichi, T.; Satoh, H.; Okabe, S. Development of High-Rate Anaerobic Ammonium-Oxidizing (Anammox) Biofilm Reactors. *Water Res.* **2007**, *41*, 1623–1634. [CrossRef]
10. Wang, Z.; Zheng, M.; Duan, H.; Yuan, Z.; Hu, S. A 20-Year Journey of Partial Nitritation and Anammox (PN/A): From Sidestream toward Mainstream. *Environ. Sci. Technol.* **2022**, *56*, 7522–7531. [CrossRef]

11. Miao, Y.; Peng, Y.; Zhang, L.; Li, B.; Li, X.; Wu, L.; Wang, S. Partial Nitrification-Anammox (PNA) Treating Sewage with Intermittent Aeration Mode: Effect of Influent C/N Ratios. *Chem. Eng. J.* **2018**, *334*, 664–672. [CrossRef]
12. Si, Y. Rapid Pathogen Discovery in Diseased Turbot (Scophthalmus Maximus) Using 16S rRNA High Throughput Sequencing. *Aquac. Rep.* **2021**, *21*, 100835. [CrossRef]
13. Fang, A.; Feng, K.; Mei, X.; Liu, B.; Xie, G.; Ren, N.; Xing, D. The Simultaneous Recruitment of Anammox Granules and Biofilm by a Sequential Immobilization and Granulation Approach. *Chem. Eng. J.* **2021**, *417*, 128041. [CrossRef]
14. Qian, Y.; Guo, Y.; Shen, J.; Qin, Y.; Li, Y.-Y. Biofilm Growth Characterization and Treatment Performance in a Single Stage Partial Nitritation/Anammox Process with a Biofilm Carrier. *Water Res.* **2022**, *217*, 118437. [CrossRef]
15. Peng, Z.; Zhang, L.; Zhang, Q.; Li, X.; Peng, Y. Establishing a Two-Stage System to Efficiently Treat Real Domestic Sewage by Partial Nitrification-SBR and Air-Lift Anammox-UASB: Reactivating and Enhancing Anammox Bacteria to Optimize the Nitrogen Removal Performance. *Chem. Eng. J.* **2025**, *506*, 160333. [CrossRef]
16. Cui, H.; Zhang, L.; Peng, Y.; Zhang, Q.; Li, X. Achieving Stable Nitritation for Mainstream Anammox by Combining Nitrite Exposure Inhibition with High DO Reactivation. *J. Water Process Eng.* **2022**, *46*, 102589. [CrossRef]
17. Kent, T.R.; Sun, Y.; An, Z.; Bott, C.B.; Wang, Z.-W. Mechanistic Understanding of the NOB Suppression by Free Ammonia Inhibition in Continuous Flow Aerobic Granulation Bioreactors. *Environ. Int.* **2019**, *131*, 105005. [CrossRef]
18. Li, Z.; Li, D.; Zhang, J.; Wang, W.; Li, M.; Wang, S.; Zeng, H.; Zhang, J. Unravelling the Optimal Strategy on the Stable Mineralized Anammox Granular Systems in Treating Low-Strength Wastewater. *J. Water Process Eng.* **2024**, *59*, 105083. [CrossRef]
19. Yang, S.; Peng, Y.; Zhang, S.; Han, X.; Li, J.; Zhang, L. Carrier Type Induces Anammox Biofilm Structure and the Nitrogen Removal Pathway: Demonstration in a Full-Scale Partial Nitritation/Anammox Process. *Bioresour. Technol.* **2021**, *334*, 125249. [CrossRef]
20. Huang, T. Fast Start-up and Enhancement of Partial Nitritation and Anammox Process for Treating Synthetic Wastewater in a Sequencing Bath Biofilm Reactor: Strategy and Function of Nitric Oxide. *Bioresour. Technol.* **2021**, *335*, 125225. [CrossRef]
21. Wang, W. Dissolved Oxygen Microelectrode Measurements to Develop a More Sophisticated Intermittent Aeration Regime Control Strategy for Biofilm-Based CANON Systems. *Chem. Eng. J.* **2019**, *365*, 165–174. [CrossRef]
22. Liu, Z.; Chen, Y.; Xu, Z.; Lei, J.; Lian, H.; Zhang, J.; Wang, Z. Surface Modification of Polyurethane Sponge with Zeolite and Zero-Valent Iron Promotes Short-Cut Nitrification. *Polymers* **2024**, *16*, 1506. [CrossRef] [PubMed]
23. Sheng, X.; Ting, Y.P.; Pehkonen, S.O. Force Measurements of Bacterial Adhesion on Metals Using a Cell Probe Atomic Force Microscope. *J. Colloid Interface Sci.* **2007**, *310*, 661–669. [CrossRef] [PubMed]
24. Huang, J. Intermittent Aeration Improving Activated Granular Sludge Granulation for Nitrogen and Phosphorus Removal from Domestic Wastewater. *Bioresour. Technol. Rep.* **2021**, *15*, 100739. [CrossRef]
25. Zhang, S. Anammox Activity Improved Significantly by the Cross-Fed NO from Ammonia-Oxidizing Bacteria and Denitrifying Bacteria to Anammox Bacteria. *Water Res.* **2024**, *249*, 120986. [CrossRef]
26. Liu, T.; Xu, J.; Tian, R.; Quan, X. Enhanced Simultaneous Nitrification and Denitrification via Adding N-Acyl-Homoserine Lactones (AHLs) in Integrated Floating Fixed-Film Activated Sludge Process. *Biochem. Eng. J.* **2021**, *166*, 107884. [CrossRef]
27. Lei, J.; Xu, Z.; Chen, Y.; Yu, G.; Liu, Z.; Wang, S.; Zhang, J.; Li, K.; Xie, L. Methane Production Reduced by Lignin Derivatives in Pulping Wastewater: Inhibition of Free Hydrolase. *Fermentation* **2024**, *10*, 247. [CrossRef]
28. Yu, J. Spectroscopic Fingerprints Profiling the Polysaccharide/Protein/Humic Architecture of Stratified Extracellular Polymeric Substances (EPS) in Activated Sludge. *Water Res.* **2023**, *235*, 119866. [CrossRef]
29. Boleij, M. Solubilization and Characterization of Extracellular Proteins from Anammox Granular Sludge. *Water Res.* **2019**, *164*, 114952. [CrossRef]
30. Jiang, H. Modification of Polyurethane Sponge Filler Using Medical Stones and Application in a Moving Bed Biofilm Reactor for Ex Situ Remediation of Polluted Rivers. *J. Water Process Eng.* **2021**, *42*, 102189. [CrossRef]
31. Chen, R.; Takemura, Y.; Liu, Y.; Ji, J.; Sakuma, S.; Kubota, K.; Ma, H.; Li, Y.-Y. Using Partial Nitrification and Anammox To Remove Nitrogen from Low-Strength Wastewater by Co-Immobilizing Biofilm inside a Moving Bed Bioreactor. *ACS Sustain. Chem. Eng.* **2019**, *7*, 1353–1361. [CrossRef]
32. Chen, Z. Nitrite Accumulation Stability Evaluation for Low-Strength Ammonium Wastewater by Adsorption and Biological Desorption of Zeolite under Different Operational Temperature. *Sci. Total Environ.* **2020**, *704*, 135260. [CrossRef] [PubMed]
33. Yue, X.; Yu, G.; Lu, Y.; Liu, Z.; Li, Q.; Tang, J.; Liu, J. Community of the Completely Autotrophic Nitrogen Removal over Nitrite Process in a Submerged Aerated Biological FIlter. *Bioresour. Technol.* **2018**, *254*, 67–74. [CrossRef] [PubMed]
34. Chen, G. Optimizing of Operation Strategies of the Single-Stage Partial Nitrification-Anammox Process. *J. Clean. Prod.* **2020**, *256*, 120667. [CrossRef]
35. Mokete, R. Exploration of the Reactivity of Nanoscale Zero-Valent Iron (NZVI) Associated Nanoparticles in Diverse Experimental Conditions. *Chem. Eng.* **2020**, *150*, 107879. [CrossRef]
36. Liu, W.; Zhou, H.; Zhao, W.; Wang, C.; Wang, Q.; Wang, J.; Wu, P.; Shen, Y.; Ji, X.; Yang, D. Rapid Initiation of a Single-Stage Partial Nitritation-Anammox Process Treating Low-Strength Ammonia Wastewater: Novel Insights into Biofilm Development on Porous Polyurethane Hydrogel Carrier. *Bioresour. Technol.* **2022**, *357*, 127344. [CrossRef]

37. Sanchez-Huerta, C. Influence of Biofilm Thickness on the Removal of Thirteen Different Organic Micropollutants via a Membrane Aerated Biofilm Reactor (MABR). *J. Hazard. Mater.* **2022**, *432*, 128698. [CrossRef]
38. Guo, K. Potential Directions for Future Development of Mainstream Partial Nitrification-Anammox Processes: Ammonia-Oxidizing Archaea as Novel Functional Microorganisms Providing Nitrite. *Bioresour. Technol.* **2024**, *399*, 130605. [CrossRef]
39. Wang, B. Recovering Partial Nitritation in a PN/A System during Mainstream Wastewater Treatment by Reviving AOB Activity after Thoroughly Inhibiting AOB and NOB with Free Nitrous Acid. *Environ. Int.* **2020**, *139*, 105684. [CrossRef]
40. Sheng, G.-P.; Xu, J.; Li, W.-H.; Yu, H.-Q. Quantification of the Interactions between Ca^{2+}, Hg^{2+} and Extracellular Polymeric Substances (EPS) of Sludge. *Chemosphere* **2013**, *93*, 1436–1441. [CrossRef]
41. Hu, Y.-Q. Effect of Pure Oxygen Aeration on Extracellular Polymeric Substances (EPS) of Activated Sludge Treating Saline Wastewater. *Process Saf. Environ. Prot.* **2019**, *123*, 344–350. [CrossRef]
42. Vandana; Priyadarshanee, M.; Das, S. Bacterial Extracellular Polymeric Substances: Biosynthesis and Interaction with Environmental Pollutants. *Chemosphere* **2023**, *332*, 138876. [CrossRef] [PubMed]
43. Crovadore, J. Metatranscriptomic and Metagenomic Description of the Bacterial Nitrogen Metabolism in Waste Water Wet Oxidation Effluents. *Heliyon* **2017**, *3*, e00427. [CrossRef]
44. Benselfelt, T.; Cranston, E.D.; Ondaral, S.; Johansson, E.; Brumer, H.; Rutland, M.W.; Wag, L. Adsorption of Xyloglucan onto Cellulose Surfaces of Different Morphologies: An Entropy-Driven Process. *Biomacromolecules* **2016**, *17*, 2801–2811. [CrossRef]
45. Hudson, J.; Egan, S. Opportunistic Diseases in Marine Eukaryotes: Could Bacteroidota Be the next Threat to Ocean Life? *Environ. Microbiol.* **2022**, *24*, 4505–4518. [CrossRef]
46. Sheng, G.-P.; Yu, H.-Q. Characterization of Extracellular Polymeric Substances of Aerobic and Anaerobic Sludge Using Three-Dimensional Excitation and Emission Matrix Fluorescence Spectroscopy. *Water Res.* **2006**, *40*, 1233–1239. [CrossRef]
47. Guo, J. Comparative Study on the Treatment of Swine Wastewater by VFCW-MFC and VFCW: Pollutants Removal, Electricity Generation, Microorganism Community. *J. Environ. Manag.* **2023**, *342*, 118299. [CrossRef]
48. Brewer, T.E.; Albertsen, M.; Edwards, A.; Kirkegaard, R.H.; Rocha, E.P.C.; Fierer, N. Unlinked rRNA Genes Are Widespread among Bacteria and Archaea. *ISME J.* **2020**, *14*, 597–608. [CrossRef]
49. Luo, J.; Yang, J.; Li, S.; Li, X.; Chang, G.; Yang, Y. Initiating an Anaerobic Ammonium Oxidation Reactor by Inoculation with Starved Anaerobic Ammonium Oxidation Sludge and Modified Carriers. *Bioresour. Technol.* **2022**, *359*, 127438. [CrossRef]

polymers

MDPI

Article

Effect of Addition Amount of Ethylenediamine on Interlayer Nanochannels and the Separation Performance of Graphene Oxide Membranes

Na Meng *, Xin Sun, Jinxin Liu, Jialing Mi and Rong Rong

Jiangsu Key Laboratory of Industrial Pollution Control and Resource Reuse, School of Environment Engineering, Xuzhou University of Technology, Xuzhou 221018, China; sunxin2708@163.com (X.S.); ljx2667495913@163.com (J.L.); 17751282269@163.com (J.M.); rongrong114717@163.com (R.R.)
* Correspondence: mengna309@163.com

Abstract: In recent years, graphene oxide (GO)-based two-dimensional (2D) laminar membranes have attracted considerable attention because of their unique well-defined nanochannels and deliver a wide range of molecular separation properties and fundamentals. However, the practical application of 2D GO layered membranes suffers from instability in aqueous solutions as the interlayer *d*-spacing of GO membranes is prone to expansion caused by the hydration effect. In this study, the effects of the ethylenediamine (EDA) addition amount on the structure, crosslinking mechanism and separation performance of GO membranes were investigated systematically, and membrane performance was evaluated using water permeability and dye/salt rejection tests. The experimental results show that the amine groups of EDA chemically bond with the hydroxyl functional group (O=C–OH) of GO after intercalation, as evident from Fourier transform infrared spectroscopy (FTIR) and X-ray photoelectron spectroscopy (XPS). By further controlling the amount of the intercalated EDA, the as-prepared GO composite membranes show nanoscale-tuned *d*-spacing promising for downstream applications. In the demonstrated dye/salt nanofiltration scenario, the EDA intercalated and crosslinked GO membrane has enhanced permeability by over five times and a better dye rejection rate of over 96% compared with pure GO membranes. These findings highlight a facile strategy for controlling nanochannels by tuning the amounts of reactive intercalants.

Keywords: graphene oxide; D-spacing; crosslinking; EDA; destaining and desalting

1. Introduction

Nanofiltration membrane technology, which has developed rapidly over the past few years, shows significant advantages in the fields of seawater desalination, molecular separation and water filtration [1–3]. With a small pore size, nanofiltration membranes selectively screen separators mainly through sieve size and charge effect on membrane surfaces [4]. Meanwhile, they also show a good ability to remove the chrominance in water, which indicates their potential to treat dyes and desalting [5,6].

As a classic two-dimensional (2D) material, graphene oxide (GO) is abundant in oxygen-containing groups (carboxyl, hydroxyl and epoxide) on its surface and edge [7–9], which endows it with strong hydrophilicity and pollution resistance [10,11]. In addition, the membranes prepared by GO have a unique layered structure and nanoscale channels and are suitable for size-dependent molecular screening and water penetration. These properties make GO membranes suitable alternatives for conventional polymeric nanofiltration membranes [12–17]. However, the performance of GO membranes is affected by the solution diffusion mechanism [18]. When a GO membrane is placed in an aqueous solution, water molecules will first dissolve in the membrane material and diffuse through the membrane. This leads to the insertion of two–three layers of water molecules into the interlayer of GO channels, which increases the spacing between GO neighboring layers and thus the phenomenon of "expansion". This may give rise to the swelling of the layered structure of GO nanomembranes and destroy nanomass transfer channels, which thus limits the application of GO membranes in aqueous solutions [19,20].

To build a stable interlayer nanochannel of GO membranes and optimize the separation performance of membranes, some methods have also been tried to prepare GO membranes. At present, the main preparation methods include crosslinking, partial reduction and multi-material hybrid methods [21]. In general, the crosslinking method is employed to prepare GO membranes by connecting GO nanosheets with crosslinking agents to inhibit the hydration of GO membranes and improve their stability [22]. Commonly used crosslinkers include molecular polymers containing amino groups, such as ethylenediamine (EDA) [23], p-phenylenediamine [24] and polyethylenimine [25]. The amines contained in these polymers can be condensed with the carboxyl groups on the GO to produce a new C-N covalent bond to fix the GO membrane. For example, Hung et al. [23] selected three diamine monomers (EDA, butylenediamine and p-phenylenediamine) and successfully prepared GO composite nanofiltration membranes with different layer spacings via pressure-assisted self-assembly technology. Thus, the layer spacing stretching caused by membrane expansion was effectively inhibited. Wansuk et al. [26] coated GO multilayer membranes on the surface of polyamide membrane composite membranes by depositing GO nanosheets with opposite charges layer by layer. The results show that the prepared membranes have excellent anti-fouling and -chlorine properties.

In addition to using amine molecules with different structures to build interfacial molecular bridges to improve membrane performance, it has also become a direction to optimize the performance of GO membranes by embedding crosslinker molecules with different qualities. By adjusting the molecular weight of polyethyleneimine (PEI), Wang et al. [27] fine-tuned the interlayer structure, surface charge and hydrophilicity of a GO membrane, and the optimized membrane showed a high desalination rate and excellent mechanical stability. Kong et al. [28] found that the addition of more dopamine (DA) can weaken the hydration of GO nanosheets and overcome swelling in water, which effectively adjusts the interlayer spacing and crosslinking network structure. As a result, GO membranes can still maintain the performance of accurate separation in aqueous solutions. In previous studies [23,29,30], EDA has been reported to be an effective crosslinking agent for GO nanosheets, and EDA crosslinked GO composite membranes have lower interlayer spacing compared to other diamine monomers. Nevertheless, few studies have examined the effect of the amount of EDA on the material transport and interlayer structure of GO membranes. This research gap prompted researchers to systematically study the mass transfer process of GO membranes by adjusting different EDA amounts. This study aimed to study the feasibility and mechanism of EDA mediating GO membranes on nanofiltration structure and sieving performance by regulating the EDA amount.

In this study, a simple vacuum filtration method was used for synthesizing stable mixed cellulose ester (MCE)/GO/EDA membranes. By adding different amounts of EDA, an interfacial molecular bridge was constructed to regulate the layer spacing and screening performance of GO composite nanofiltration membranes. The structure and screening performance of EDA-mediated GO membranes were further studied.

2. Experimental Section

2.1. Materials

Single-layer graphene oxide (GO, Jiangsu Xianfeng Nanomaterials Technology Co., Ltd., Nanjing, China). Pure sodium chloride (NaCl, Sinopharm Chemical Reagent Co., Ltd., Shanghai, China), sodium sulfate (Na_2SO_4, Sinopharm Chemical Reagent Co., Ltd., Shanghai, China) and magnesium sulfate ($MgSO_4$, Sinopharm Chemical Reagent Co., Ltd., Shanghai, China). Ethylenediamine (EDA, Shanghai Maclin Biochemical Technology Co., Ltd., Shanghai, China). Hybrid Cellulose Ester Microporous Filter Membrane (MCE, Hangzhou Micropay Technology Co., Hangzhou, China) with a pore size of 0.22 μm.

Methylene Blue (MB, Shanghai Aladdin Biochemical Technology Co., Ltd., Shanghai, China). The experimental water used was deionized water.

2.2. Preparation of MCE/GO/EDA Composite Membranes

After 20 mg of GO powder was weighed and mixed with 500 mL of deionized water, 40 mg/L of GO dispersion was obtained by ultrasound at 30 °C for 30 min. After 40 mL and 40 mg/L of GO dispersion were evenly mixed with 100 mL of deionized water, different amounts of EDA were added, and the GO/EDA mixture was made using ultrasound at 30 °C for 30 min. Then, the mixture was filtered to the MCE membrane through a vacuum filtration device by the vacuum filtration method to obtain the MCE/GO/EDA membrane. The different types of MCE/GO/EDA membranes are named according to the amount of EDA added. For instance, MCE/GO/EDA-X indicates that the amount of EDA added to the graphene oxide composite membrane is "X" mL. Figure 1 shows the preparation process of the MCE/GO/EDA composite membrane.

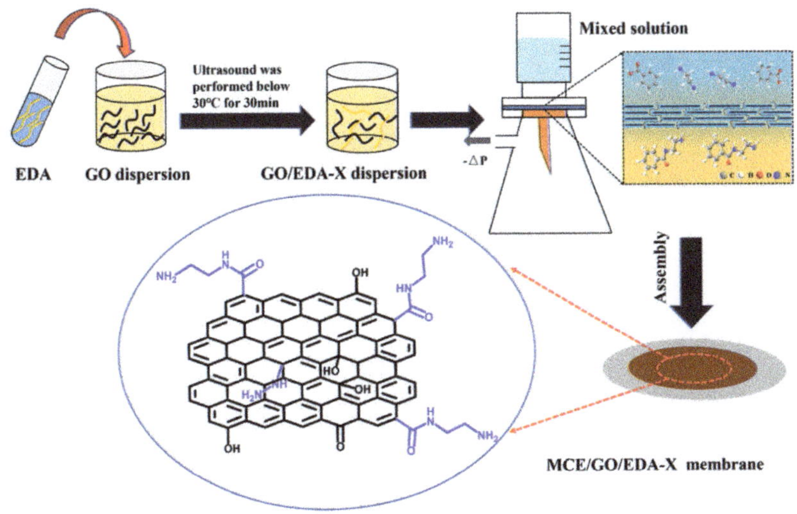

Figure 1. Preparation process of MCE/GO/EDA composite membranes.

2.3. Characterization of MCE/GO/EDA Membranes

Fourier transform infrared spectroscopy (FTIR, Nicolet iS10, Thermo Fisher Scientific, Waltham, MA, USA) and X-ray photoelectron spectroscopy (XPS, Thermo Scientific K-Alpha, Thermo Fisher Scientific Waltham, MA, USA) were utilized for analyzing the functional groups and structures of MCE/GO/EDA membranes. The surface and cross-section morphology of MCE/GO/EDA membranes were observed by scanning electron microscopy (SEM, Regulus 8100, Hitachi, Japan). The crystal structure of GO nanosheets was analyzed using an X-ray diffractometer (XRD, D8 ADVANCE, Bruker, Germany). The sample was scanned from 5 to 55°. The d-spacing (d) of the prepared GO membrane can be calculated by Bragg's law in Equation (1).

$$d = \frac{n\lambda}{2sin\theta} \quad (1)$$

where λ is X-ray wavelength; θ is the diffraction angle; and n is the diffraction order.

The hydrophilicity of the surface of the prepared membrane materials was tested using a contact angle tester (CA, JY-82C, China Chengde Dingsheng, China) at least twice for each membrane. The surface structure and roughness of membranes were observed using atomic force microscopy (AFM, Bruker Dimension ICON, Bruker, Germany). The surface charge of the GO membrane was observed using a zeta potential analyzer (Anton Paar surpass3, Anton Paar Austria Ltd., Graz, Austria).

2.4. Evaluation of Membrane Performance

The permeability of MCE/GO/EDA membranes was assessed by measuring membrane flux using deionized water as the feed solution. NaCl, Na_2SO_4, $MgSO_4$ and 10 ppm MB solutions with concentrations of 1000 ppm were used as feed solutions to test the rejection performance of membranes.

A dead-end filtration device was applied to test membrane performance, and a diagram of the device can be obtained from our previous study [31]. The filtration system consists of a nitrogen pressurizer, a sealable water tank, a digital balance, a 300 mL filter cup, a magnetic stirrer and computerized online measurement software. Regarding the measurement of rejection performance, the specific filtration process is as follows: (1) the membrane was compacted first for 2 h to achieve a stable flux, with a transmembrane pressure of 2 bar; (2) the transmembrane pressure was reduced to 1 bar, and the pure water flux was recorded every 30 s. At least 40 measurements were collected to obtain average flux values; (3) deionized water was replaced by salt and MB feed solution. Filter cups were stirred using a stirring rod at 600 rpm to minimize concentration polarization, and then the transmembrane pressure was restored to 1 bar. After filtering for a period of time, 15 mL of filtrate was gathered.

Membrane permeability (J, $Lm^{-2}\,h^{-1}\,bar^{-1}$) was determined according to Equation (2) by membrane effective area (A, m^2) per unit time (T, h) per unit pressure (P, bar) per volume of penetrant (V, L).

$$J = \frac{V}{T \times A \times P} \tag{2}$$

The conductivity meter was utilized to measure the conductivity of the filtrate and stock solution before filtration to evaluate the desalting performance of the membrane. The dye solution was measured using ultraviolet–visible spectrophotometry (UV-VIS) to evaluate the effect of the membrane on dye rejection.

The rejection of salt ions and dyes (R) was calculated using Equation (3). C_0 and C_1 corresponded to the concentrations of the feed solution and filtrate, respectively.

$$R = \frac{C_0 - C_1}{C_0} \times 100\% \tag{3}$$

3. Results and Discussion

3.1. FTIR Characterization of MCE/GO/EDA Membranes

The membrane was characterized by FTIR to illustrate the bond between EDA and GO. As shown in Figure 2, the original GO membrane is rich in oxygen-containing groups, among which absorption peaks can be observed at 3352, 1730, 1631 and 1274 cm^{-1} wave numbers. Corresponding to hydroxyl and carboxyl groups (–OH and –C=O), carbon–carbon double bond (–C=C–) and epoxy group (–C–O–C–), these data are consistent with the research results of Chen et al. [32]. After EDA treatment, the carboxyl group (–C=O) disappeared in the MCE/GO/EDA membrane; hydroxyl and epoxy groups (–OH and –C–O–C–) significantly decreased; and a new absorption peak (C–N, 1454 cm^{-1}) appeared, which indicated that the amine group of EDA would condense with the carboxyl group of GO. The nucleophilic substitution reaction with the epoxide group formed C–N covalent bonds [23], and the crosslinking reaction occurred successfully. Due to relatively mild crosslinking reaction conditions during the preparation of the composite membrane, the new absorption peaks generated were not significant. Hence, XPS was further used to analyze the composition of elements in the membrane and understand the reaction principle of GO and EDA.

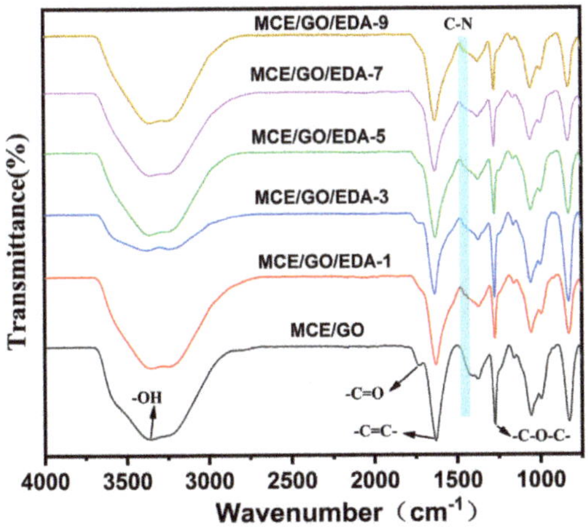

Figure 2. Fourier infrared spectra of MCE/GO and MCE/GO/EDA-X membranes.

3.2. XPS Characterization of MCE/GO/EDA Membranes

XPS was used to further analyze the crosslinking reaction between GO and EDA. As presented in Table 1, the increase in EDA amount from 1 to 9 mL was accompanied by a decrease in the O element of the EDA-loaded GO membrane from 27.1% to 21.2%, an increase in the N element from 4.79% to 5.74%, and a decrease in the O/C ratio from 0.40 to 0.29. The ratio (0.29) of O element to C element in the EDA-loaded GO membrane is significantly lower than that (0.49) of O element to C element in the original MCE/GO membrane. This is due to the reaction of EDA with the oxygen-containing functional groups of GO and the loss of oxygen-containing functional groups in the reaction process, which reduced the O/C ratio [33].

Table 1. XPS analysis of MCE/GO and MCE/GO/EDA-X membranes.

Sample	O/%	N/%	C/%	O/C
MCE/GO	32.14	-	67.86	0.47
MCE/GO/EDA-1	27.1	4.79	68.11	0.40
MCE/GO/EDA-3	24.72	4.9	70.38	0.35
MCE/GO/EDA-5	22.4	5.14	72.46	0.31
MCE/GO/EDA-7	21.89	5.63	72.48	0.30
MCE/GO/EDA-9	21.2	5.74	73.06	0.29

It can be seen from Figure 3 that the scanning spectra of the XPS C1s region of the original MCE/GO membrane and MCE/GO/EDA composite membrane with different amounts of EDA were decomposed into five peaks: C–C (284.8 eV), C–O/C–N (286 eV), C–O–C (286.85 eV), C=O (288.25 eV) and O–C=O (289.4 eV). Compared with the original MCE/GO membrane, the modified MCE/GO/EDA membrane exhibited a decrease in the C–O–C peak and an increase in the C–O/C–N peak. This is because GO contains active epoxy groups, and its exposure to the amine group leads to the active ternary epoxy ring opening reaction, which results in a new C–N bond [34]. This confirms that the crosslinking reaction between EDA and GO is successful. As illustrated in Figure 3b–f, the C–O–C peak of the MCE/GO/EDA membrane decreased, and the C–O/C–N peak increased more significantly with the increase in EDA amount. This also indicates that more oxygen-containing groups were consumed in the reaction to generate C–N covalent bonds, which greatly increased the stability of the membrane.

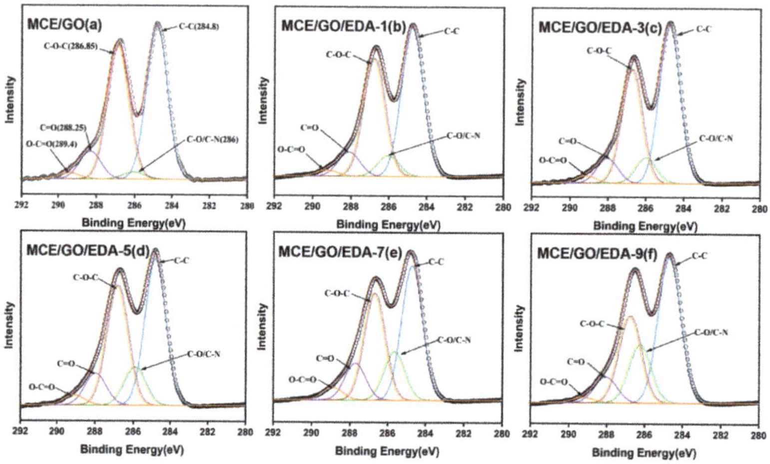

Figure 3. Analysis of XPS elements in the C1s region of the original MCE/GO membrane (**a**) and MCE/GO/EDA composite membrane (**b**–**f**).

3.3. Zeta Potential and Water CA Analysis

To investigate the effect of the embedding amount of EDA on the surface charge of the GO membrane, the zeta potentials of MCE/GO and MCE/GO/EDA-X membrane samples at Pondus hydrogenii (pH) = 7 were measured. Previous reports have demonstrated that GO membranes are negatively charged due to numerous carboxyl and hydroxyl functional groups on the surface [35,36]. As shown in Figure 4a, the surface potential of the original MCE/GO membrane is −46.02 mv, which is highly consistent with the research report of Zhang et al. [37]. In the meantime, it can be observed that the electronegativity of the membrane surface began to weaken with the increase in the embedding amount of EDA, but it still had a negative charge. This evolution of surface charge may be attributed to the removal of oxygen-containing functional groups by a large amount of EDA reduction. This indicates that surface charge can be properly regulated by changing the embedding amount of small amine molecules, which would play an important role in ion transport in the membrane.

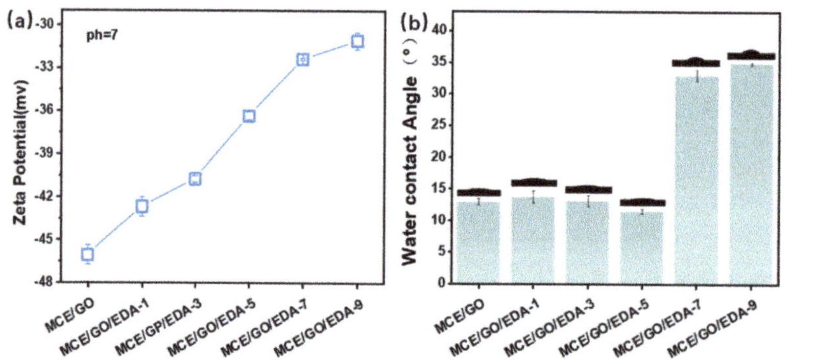

Figure 4. Surface zeta potentials (**a**) and water CAs (**b**) of MCE/GO and MCE/GO/EDA-X membranes.

It can be observed from Figure 4b that the CA of the MCE/GO/EDA membrane generally presented a trend of decrease with the increase in the EDA addition amount from 0 to 5 mL. Additionally, the hydrophilicity of the MCE/GO/EDA membrane was improved, which was related to the abundant oxygen-containing hydroxyl groups (–OH) on the membrane. The epoxide group (–C–O–C–) was associated with the amine group (–N–

H–) [28]. In addition, the change in the surface roughness of the membrane also changed its hydrophilicity [38,39]. However, the water CA of the MCE/GO/EDA membrane also began to increase, and its hydrophilicity decreased as the amount of EDA continued to increase. This indicates that the oxygen-containing groups on the surface of the MCE/GO/EDA membranes are reduced by EDA, and the hydrophilicity of the amines grafted onto the membranes is weaker compared to the hydrophilic hydroxyl groups, which greatly reduces the hydrophilicity of the membranes.

3.4. XRD Characterization of MCE/GO/EDA Membranes

According to the XRD pattern of the composite membrane, the Bragg formula was used to calculate the layer spacing of the composite membrane mediated by EDA with different amounts in dry and wet states (Figure 5a,c). In the dry state, the d-spacing corresponding to the original GO membrane is 0.747 nm, which is basically in line with the previously reported results [16,33]. With the increase in EDA amount, the layer spacing of the MCE/GO/EDA membrane increased. This shows that membrane spacing can be adjusted by adding different EDA amounts to different substances, which is expected to achieve the effect of accurately removing ions and molecules of different sizes. In addition, it can be seen from the MCE/GO and MCE/GO/EDA membranes tested in the wet state that the original MCE/GO membrane significantly expanded the size of the interlayer nanochannel due to the widespread swelling phenomenon (Figure 5d). Water molecules filled the membrane interlayer and stretched the interlayer spacing, which would seriously weaken the ion exclusion ability of the GO membrane [40]. Of note, the structure of the MCE/GO/EDA membrane did not change significantly with the increase in the embedding amount of EDA (Figure 5b). This is because the amine group of EDA formed effective C–N covalent bonds within the GO membrane to interlock GO nanosheets, which resisted the d-spacing stretching phenomenon [41]. The layer spacing of the MCE/GO/EDA-9 membrane cannot be accurately measured, possibly because excessive crosslinking would increase the disorder of the membrane in the water environment. Consequently, the intensity of the XRD peak was not observed.

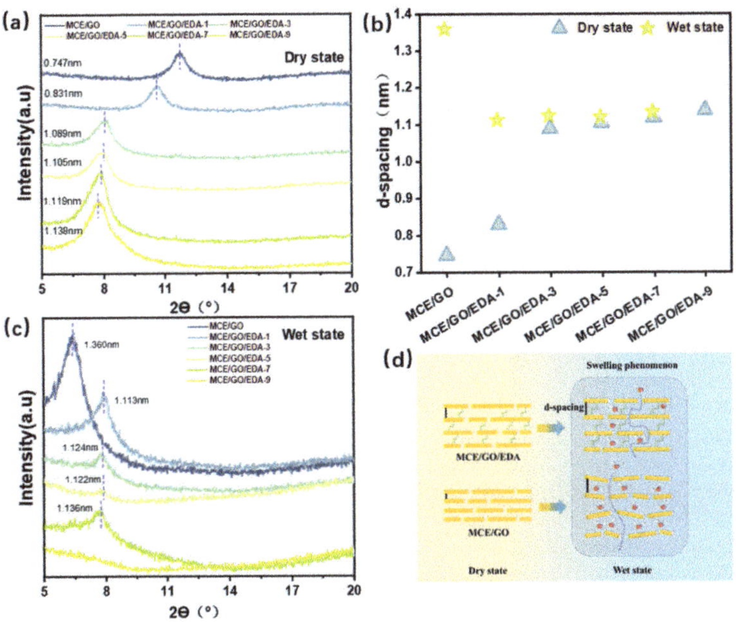

Figure 5. XRD patterns and d-spacing of MCE/GO and MCE/GO/EDA-X membranes: (**a**) dry state; (**b**) D-spacing; (**c**) wet state; (**d**) swelling phenomenon.

3.5. SEM Characterization of MCE, MCE/EDA and MCE/GO/EDA Membranes

The surface morphology and cross-sectional structure of the membrane were characterized by SEM (Figures 6 and 7). MCE and MCE/EDA membranes were characterized for their comparison after EDA treatment. As shown in Figure 6a,b, the surface aperture of the MCE/EDA membrane after EDA treatment did not change significantly compared with that of the original MCE membrane, which indicated that EDA did not change the surface morphology of the MCE base membrane. The surface morphology of MCE/GO/EDA membranes synthesized under different EDA loading loads is shown in Figure 6c–h. The surface of all MCE base membranes is covered by a uniform layer of GO nanosheets, with typical folded surface morphology and well-layered layered structure [42] owing to rich oxygen-containing groups on the surface of GO, which makes it form nanoscale folds in GO nanosheets and provides mass transfer channels for water molecules [43].

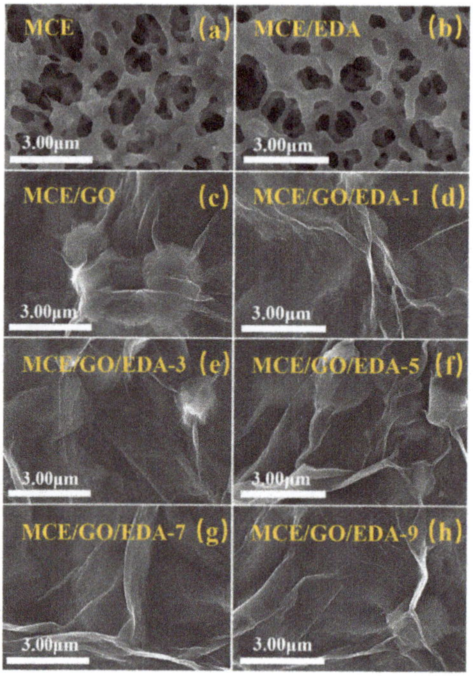

Figure 6. SEM surface images of MCE base (**a**), MCE/EDA (**b**), MCE/GO (**c**) and MCE/GO/EDA-X membranes (**d–h**).

As shown in Figure 7, no obvious gap existed between the GO layer and MCE base membrane in the MCE/GO/EDA membrane after EDA treatment compared with the MCE/EDA base membrane. This also indicates that GO and MCE membranes were closely combined through EDA crosslinking. Figure 7 shows the change in the thickness of the GO nanolayer under different EDA addition amounts, which is aligned with the expectation. With the increase in the EDA addition amount from 0 to 9 mL, the thickness of the GO/EDA layer increased from 273 to 605 nm. The change in the thickness of the GO layer meant that EDA embedded in the GO layer membrane could adjust the layer spacing of the GO layer membrane, which is consistent with the XRD test results.

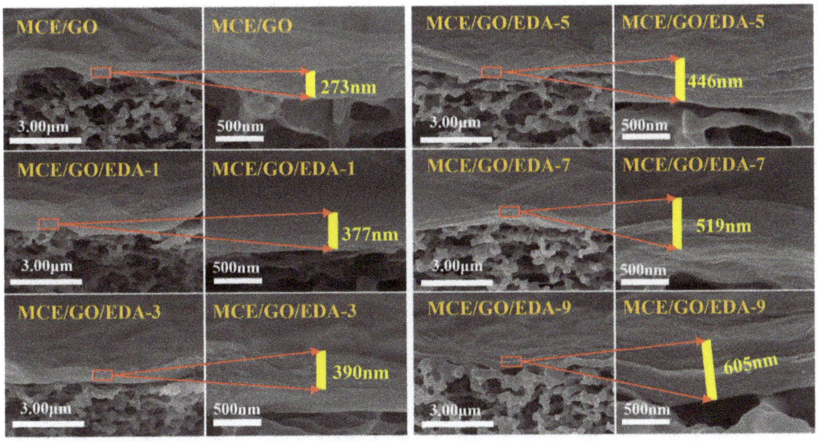

Figure 7. SEM cross-section images of MCE/GO and MCE/GO/EDA-X membranes.

3.6. AFM Characterization of MCE/GO/EDA Membranes

The effect of EDA on the surface roughness of the MCE/GO membrane was further studied by using AFM to characterize surface morphology. The 3D AFM images of the surface of the MCE/GO/EDA membrane under different EDA addition amounts are shown in Figure 8. It can be seen that the roughness of the membrane surface increases and then decreases with the amount of EDA. It was confirmed that the variation in the roughness of the MCE/GO/EDA membrane was linked to the crosslinking reaction of EDA and GO, and a higher crosslinking degree would lead to a more uniform deposition of GO on membrane surfaces, which resulted in a decrease in roughness [28]. When EDA was embedded into GO interlayer nanochannels, the roughness of the MCE/GO/EDA membrane became larger. Rough membranes can provide more water contact sites, which is conducive to the rapid flow of water through membranes and the increase in water flux [44].

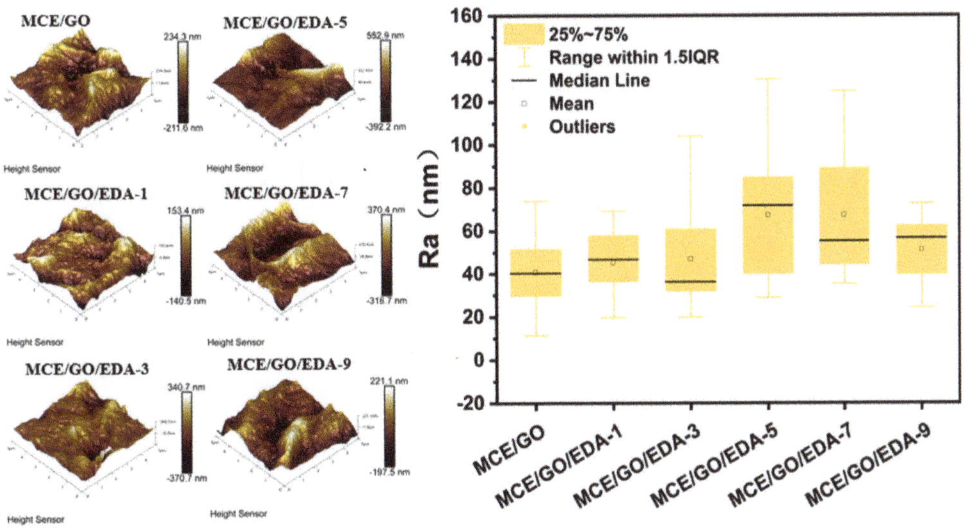

Figure 8. AFM images and surface roughness of MCE/GO and MCE/GO/EDA-X membranes.

3.7. Salt Rejection Performance

In the research process, a 1 bar terminal filtration device was used to test the water permeability, desalting performance and dye removal performance of the MCE/GO/EDA membrane under different EDA loads. As shown in Figure 9a, the MCE/GO membrane without EDA treatment exhibited low water flux, which resulted in the collapse of interlayer nanochannels because the GO membrane was easy to compact [45]. In general, it can be seen that EDA addition has a great influence on the water flux of the MCE/GO/EDA membrane. When the load value of EDA increased from 0 to 5 mL, the pure water flux of the MCE/GO/EDA membrane presented an increasing trend, which may be due to the increase in GO layer spacing and the expansion of interlayer nanochannels mediated by EDA into the GO layer. The rapid passage of water molecules was facilitated through the GO nanolayer. Nonetheless, the water flux of the MCE/GO/EDA membrane was reduced when the amount of EDA was further increased to 9 mL because of excessive EDA mediating the GO layer, which could easily cause a "blockage" phenomenon, thus hindering the flow of water molecules.

As shown in Figure 9b, the salt rejection rate of the GO membrane is Na_2SO_4 > $MgSO_4$ > NaCl, and the GO membrane has a higher rejection rate of Na_2SO_4 than NaCl. The reason is that high-valence co-ions need to overcome a larger interaction energy barrier than low-valence ones during ion transport [46]. Mg^{2+} transmission is inhibited, and $MgSO_4$ has a higher interception rate than NaCl since the hydration radius of Mg^{2+} is larger than that of Na^+. Moreover, the rejection rate of Na_2SO_4 is higher than that of $MgSO_4$, which concurs with previous research reports [47]. It can be observed from the figure that the rejection rate of Na_2SO_4 decreased slightly when the amount of EDA increased from 1 to 5 mL and began to rise when the amount of EDA increased to 9 mL. This is because excessive EDA made the structure of GO more dense [13], and the "blockage" phenomenon of interlayer nanochannels occurred. The rejection rate of Na_2SO_4 increased to 23.57%. By controlling the amount of EDA added, the layer spacing of the GO membrane can be adjusted, which thus affects its water flux and salt rejection performance. As demonstrated in Figure 9, the flux of the MCE/GO/EDA membrane reached the maximum value of 104.07 L/m^2·h·bar with the addition of 5 mL EDA, while the salt rejection rate decreased.

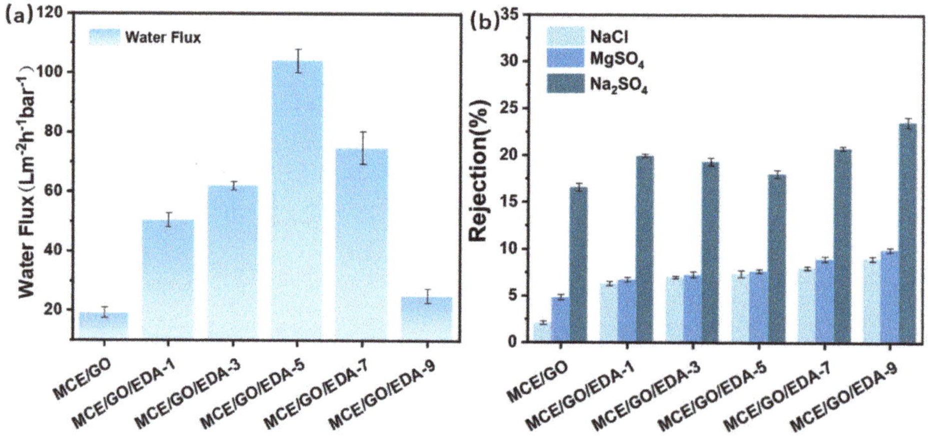

Figure 9. Effects of different EDA addition amounts on water flux (**a**) and salt rejection (**b**) of MCE/GO/EDA-X membranes.

3.8. Dye Rejection Performance

It is well known that the dye removal effect of nanofiltration membranes is closely related to the pore size of membranes and the charge property of membrane surfaces [48–50]. As illustrated in Figure 10a, the removal rate of MB dye for MCE/GO/EDA membranes with different EDA addition amounts is above 96%. This indicates that the negative electronegativity of the membrane surface is weakened by the embedding of EDA into the GO membrane, which reduces the attraction with the positively charged methylene blue molecules and attenuates the permeation of methylene blue molecules within the membrane. Despite the increase in interlayer nano-channels, the high rejection rate of MB was maintained.

Zhang et al. [33] prepared GO-UR/CA membranes by crosslinking graphene oxide (GO) with urea (UR), and the rejection rates of KCl, NaCl and $MgSO_4$ were 21.6%, 26.8% and 63.2%, respectively. Yuan et al. [51] used the covalent crosslinking of thiourea (TU) molecules with graphene oxide, which significantly increased the salt rejection rate (the rejection rate of NaCl was 95.6% and 90.2% for $MgCl_2$). Chandio et al. [52] prepared cGO/SAA membranes using serine amino acid (SAA)-modified GO with excellent separation efficiencies of 99% and 98% rejection of rhodamine B and methylene blue, respectively. Compared with previous studies, the composite membranes prepared by crosslinking EDA with GO in this paper showed lower rejection of salt ions but had relatively high water flux and high dye rejection, which can help to achieve dye/salt selective separation in practical applications.

To assess the stability of the membrane, it was immersed in a dye solution for 10 d, and the dye retention recovery rates of MCE/GO and MCE/GO/EDA-5 membranes were subsequently evaluated. As shown in Figure 10b, the initial dye retention rates of MCE/GO and MCE/GO/EDA-5 membranes were similar, and the dye retention recovery rate of the MCE/GO/EDA-5 membrane was 90.90% after 10 d of immersion, while that of the MCE/GO membrane was only 81.82% of the original value. This indicates that the MCE/GO/EDA-5 membrane has stronger stability than the MCE/GO one.

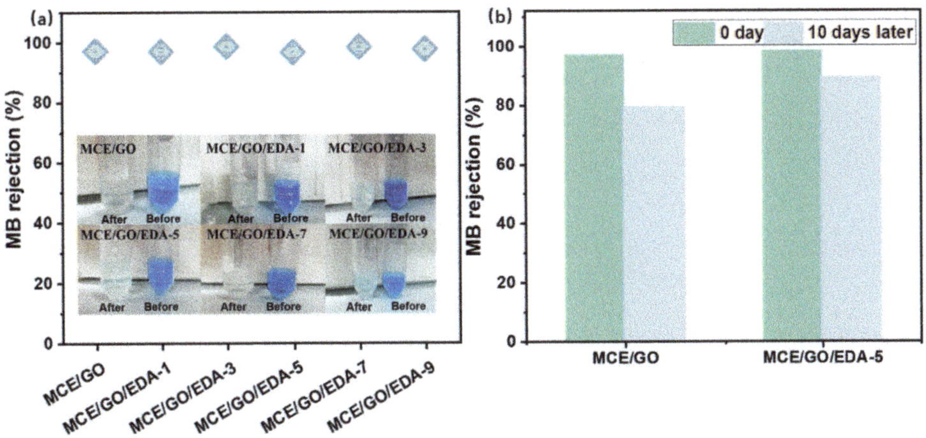

Figure 10. Effects of different EDA addition amounts on the MB dye removal (**a**) and stability (**b**) of the MCE/GO/EDA-X membrane.

The desalting and dye removal performance of the MCE/GO/EDA membrane was mainly related to the steric hindrance effect and electrostatic interaction [4]. As shown in Figure 11, the negatively charged ions would be repelled by electrostatic repulsion and could not penetrate the GO membrane because the membrane was negatively charged. Meanwhile, some ions whose hydration radius is larger than the aperture were also blocked outside the membrane. For the mass transport in pores, larger molecules or ions have a higher affinity for interlayer nanochannels owing to their weak interaction with bound water molecules, which resulted in the phenomenon of adhesion to transport channels [53].

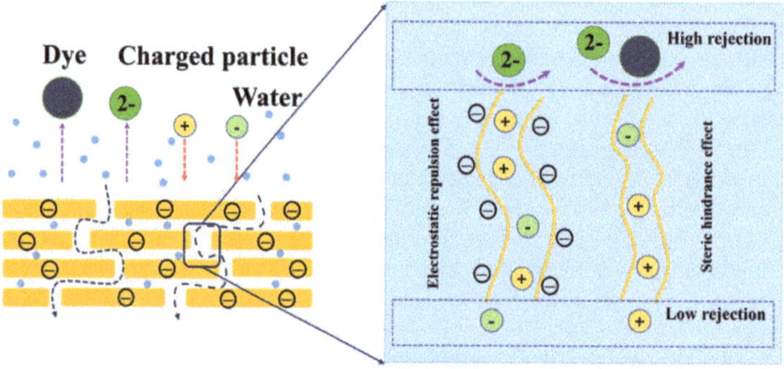

Figure 11. Ion transport mechanism of the MCE/GO/EDA-X membrane.

4. Conclusions

In this study, the effect of EDA amount on the structure and sieving performance of GO was studied by regulating interlayer nanochannels and adding different EDA amounts. EDA can form C–N covalent bonds via amidation with GO, which helps to inhibit the swelling of GO interlayer nanochannels. In the process of regulating layer spacing, it was found that an appropriate amount of EDA was beneficial to improving the mass transfer process of water molecules in interlayer nanochannels, whereas an excessive amount of EDA would lead to the dense structure of GO and improve the selectivity of Na_2SO_4 and NaCl. In addition, interlayer spacing was almost not increased. In conclusion, a stable interlayer nanomass transfer channel was constructed by embedding a high amount of EDA into the GO membrane, which is very important for GO to control ion transport as a nanofiltration membrane.

Author Contributions: Conceptualization, N.M.; methodology, X.S. and J.L.; experiment, X.S. and J.M.; data curation, R.R.; writing—original draft preparation, X.S. and N.M.; writing—review and editing, N.M.; supervision, N.M. All authors have read and agreed to the published version of the manuscript.

Funding: This research was funded by the Natural Science Foundation of the Jiangsu Higher Education Institutions of China (No. 23KJA610006).

Institutional Review Board Statement: Not applicable.

Informed Consent Statement: Not applicable.

Data Availability Statement: The data presented in this study are available on request from the corresponding author.

Conflicts of Interest: The authors declare that they have no conflicts of interest.

References

1. Cheng, X.; Lai, C.; Zhu, X.; Shao, S.; Xu, J.; Zhang, F.; Song, J.; Wu, D.; Liang, H.; Luo, X. Tailored ultra-low pressure nanofiltration membranes for advanced drinking water treatment. *Desalination* **2023**, *548*, 116264. [CrossRef]
2. Miao, J.; Hu, Q.; Zhang, Z.; Hu, Y. Bioinspired cell membrane-based nanofiltration membranes with hierarchical pore channels for fast molecular separation. *J. Membr. Sci.* **2024**, *697*, 122578. [CrossRef]
3. Figueira, M.; Rodríguez-Jiménez, D.; López, J.; Reig, M.; Cortina, J.L.; Valderrama, C. Evaluation of the nanofiltration of brines from seawater desalination plants as pre-treatment in a multimineral brine extraction process. *Sep. Purif. Technol.* **2023**, *322*, 124232. [CrossRef]
4. Zhou, S.; Guan, K.; Fang, S.; Wang, Z.; Li, Z.; Xu, P.; Nakagawa, K.; Takagi, R.; Matsuyama, H. Nanochannel characteristics contributing to ion/ion selectivity in two-dimensional graphene oxide membranes. *J. Membr. Sci.* **2024**, *689*, 122185. [CrossRef]

5. Aouni, A.; Fersi, C.; Cuartas-Uribe, B.; Bes-Pía, A.; Alcaina-Miranda, M.I.; Dhahbi, M. Reactive dyes rejection and textile effluent treatment study using ultrafiltration and nanofiltration processes. *Desalination* **2012**, *297*, 87–96. [CrossRef]
6. Nataraj, S.; Hosamani, K.; Aminabhavi, T. Nanofiltration and reverse osmosis thin film composite membrane module for the removal of dye and salts from the simulated mixtures. *Desalination* **2009**, *249*, 12–17. [CrossRef]
7. Lerf, A.; He, H.; Forster, M.; Klinowski, J. Structure of graphite oxide revisited. *J. Phys. Chem. B* **1998**, *102*, 4477–4482. [CrossRef]
8. Suk, J.W.; Piner, R.D.; An, J.; Ruoff, R.S. Mechanical properties of monolayer graphene oxide. *ACS Nano* **2010**, *4*, 6557–6564. [CrossRef]
9. Sun, X.; Shiraz, H.; Wong, R.; Zhang, J.; Liu, J.; Lu, J.; Meng, N. Enhancing the performance of PVDF/GO ultrafiltration membrane via improving the dispersion of GO with homogeniser. *Membranes* **2022**, *12*, 1268. [CrossRef]
10. Wang, F.; Fang, W.; Ming, X.; Liu, Y.; Xu, Z.; Gao, C. A review on graphene oxide: 2d colloidal molecule, fluid physics, and macroscopic materials. *Appl. Phys. Rev.* **2023**, *10*, 011311. [CrossRef]
11. Yu, H.; He, Y.; Xiao, G.; Fan, Y.; Ma, J.; Gao, Y.; Hou, R.; Yin, X.; Wang, Y.; Mei, X. The roles of oxygen-containing functional groups in modulating water purification performance of graphene oxide-based membrane. *Chem. Eng. J.* **2020**, *389*, 124375. [CrossRef]
12. Papageorgiou, D.G.; Kinloch, I.A.; Young, R.J. Mechanical properties of graphene and graphene-based nanocomposites. *Prog. Mater. Sci.* **2017**, *90*, 75–127. [CrossRef]
13. Hua, D.; Rai, R.K.; Zhang, Y.; Chung, T.-S. Aldehyde functionalized graphene oxide frameworks as robust membrane materials for pervaporative alcohol dehydration. *Chem. Eng. Sci.* **2017**, *161*, 341–349. [CrossRef]
14. Liu, G.; Jin, W.; Xu, N. Graphene-based membranes. *Chem. Soc. Rev.* **2015**, *44*, 5016–5030. [CrossRef]
15. Joshi, R.; Carbone, P.; Wang, F.-C.; Kravets, V.G.; Su, Y.; Grigorieva, I.V.; Wu, H.; Geim, A.K.; Nair, R.R. Precise and ultrafast molecular sieving through graphene oxide membranes. *Science* **2014**, *343*, 752–754. [CrossRef]
16. Huang, H.; Ying, Y.; Peng, X. Graphene oxide nanosheet: An emerging star material for novel separation membranes. *J. Mater. Chem. A* **2014**, *2*, 13772–13782. [CrossRef]
17. Koenig, S.P.; Wang, L.; Pellegrino, J.; Bunch, J.S. Selective molecular sieving through porous graphene. *Nat. Nanotechnol.* **2012**, *7*, 728–732. [CrossRef]
18. Wijmans, J.G.; Baker, R.W. The solution-diffusion model: A review. *J. Membr. Sci.* **1995**, *107*, 1–21. [CrossRef]
19. Ran, J.; Chu, C.; Pan, T.; Ding, L.; Cui, P.; Fu, C.-F.; Zhang, C.-L.; Xu, T. Non-covalent cross-linking to boost the stability and permeability of graphene-oxide-based membranes. *J. Mater. Chem. A* **2019**, *7*, 8085–8091. [CrossRef]
20. Wang, Z.; He, F.; Guo, J.; Peng, S.; Cheng, X.Q.; Zhang, Y.; Drioli, E.; Figoli, A.; Li, Y.; Shao, L. The stability of a graphene oxide (GO) nanofiltration (NF) membrane in an aqueous environment: Progress and challenges. *Mater. Adv.* **2020**, *1*, 554–568. [CrossRef]
21. Su, P.; Wang, F.; Li, Z.; Tang, C.Y.; Li, W. Graphene oxide membranes: Controlling their transport pathways. *J. Mater. Chem. A* **2020**, *8*, 15319–15340. [CrossRef]
22. Zhang, M.; Mao, Y.; Liu, G.; Liu, G.; Fan, Y.; Jin, W. Molecular bridges stabilize graphene oxide membranes in water. *Angew. Chem. Int. Ed.* **2020**, *59*, 1689–1695. [CrossRef] [PubMed]
23. Hung, W.-S.; Tsou, C.-H.; De Guzman, M.; An, Q.-F.; Liu, Y.-L.; Zhang, Y.-M.; Hu, C.-C.; Lee, K.-R.; Lai, J.-Y. Cross-linking with diamine monomers to prepare composite graphene oxide-framework membranes with varying d-spacing. *Chem. Mater.* **2014**, *26*, 2983–2990. [CrossRef]
24. Qian, Y.; Zhou, C.; Huang, A. Cross-linking modification with diamine monomers to enhance desalination performance of graphene oxide membranes. *Carbon* **2018**, *136*, 28–37. [CrossRef]
25. Halakoo, E.; Feng, X. Layer-by-layer assembled membranes from graphene oxide and polyethyleneimine for ethanol and isopropanol dehydration. *Chem. Eng. Sci.* **2020**, *216*, 115488. [CrossRef]
26. Choi, W.; Choi, J.; Bang, J.; Lee, J.-H. Layer-by-layer assembly of graphene oxide nanosheets on polyamide membranes for durable reverse-osmosis applications. *ACS Appl. Mater. Interfaces* **2013**, *5*, 12510–12519. [CrossRef]
27. Wang, Z.; Sun, J.; Li, N.; Qin, Y.; Qian, X.; Xie, Z. Tuning interlayer structure to construct steady dual-crosslinked graphene oxide membranes for desalination of hypersaline brine via pervaporation. *Sep. Purif. Technol.* **2022**, *286*, 120459. [CrossRef]
28. Kong, F.-X.; Liu, Q.; You, L.; Lu, P.; Liu, T.; Sun, G.-D.; Wang, Y.; Chen, J.-F. Facile preparation of dopamine mediated graphene oxide composite membranes with enhanced stability for nanofiltration: Structure, performance and stability. *Desalination* **2022**, *534*, 115778. [CrossRef]
29. Ju, H.; Duan, J.; Lu, H.; Xu, W. Cross-linking with diamine monomers to prepare graphene oxide composite membranes with varying d-spacing for enhanced desalination properties. *Front. Chem.* **2021**, *9*, 779304. [CrossRef]
30. Lou, Y.; Tan, F.J.; Zeng, R.; Wang, M.; Li, P.; Xia, S. Preparation of cross-linked graphene oxide on polyethersulfone membrane for pharmaceuticals and personal care products removal. *Polymers* **2020**, *12*, 1921. [CrossRef]
31. Meng, N.; Sun, X.; Liu, J.; Mi, J.; Chen, X.; Rong, R.J.P. Modulation of Interlayer Nanochannels via the Moderate Heat Treatment of Graphene Oxide Membranes. *Polymers* **2024**, *16*, 2200. [CrossRef] [PubMed]
32. Chen, X.; He, Y.; Zhang, Q.; Li, L.; Hu, D.; Yin, T. Fabrication of sandwich-structured ZnO/reduced graphite oxide composite and its photocatalytic properties. *J. Mater. Sci.* **2010**, *45*, 953–960. [CrossRef]
33. Zhang, Y.; Su, K.; Li, Z. Graphene oxide composite membranes cross-linked with urea for enhanced desalting properties. *J. Membr. Sci.* **2018**, *563*, 718–725. [CrossRef]
34. Park, S.; Dikin, D.A.; Nguyen, S.T.; Ruoff, R.S. Graphene oxide sheets chemically cross-linked by polyallylamine. *J. Phys. Chem. C* **2009**, *113*, 15801–15804. [CrossRef]

35. Ranjan, P.; Balakrishnan, J.; Thakur, A.D. Dye adsorption behavior of graphene oxide. *Mater. Today Proc.* **2019**, *11*, 833–836. [CrossRef]
36. Zhang, W.-H.; Yin, M.-J.; Zhao, Q.; Jin, C.-G.; Wang, N.; Ji, S.; Ritt, C.L.; Elimelech, C.; An, Q.-F. Graphene oxide membranes with stable porous structure for ultrafast water transport. *Nat. Nanotechnol.* **2021**, *16*, 337–343. [CrossRef]
37. Zhang, P.; Gong, J.-L.; Zeng, G.-M.; Deng, C.-H.; Yang, H.-C.; Liu, H.-Y.; Huan, S.-Y. Cross-linking to prepare composite graphene oxide-framework membranes with high-flux for dyes and heavy metal ions removal. *Chem. Eng. J.* **2017**, *322*, 657–666. [CrossRef]
38. Yu, D.; Sun, L.; Zhang, Y.; Song, Y.; Jia, C.; Wang, Y.; Wang, Y.; Kipper, M.J.; Tang, J.; Huang, L. Two-dimensional graphene oxide/MXene lamellar membrane cross-linked by urea with adjustable interlayer spacing for efficient dye rejection and ion sieving. *Chem. Eng. J.* **2024**, *480*, 148009. [CrossRef]
39. Hejazi, V.; Moghadam, A.D.; Rohatgi, P.; Nosonovsky, M. Beyond Wenzel and Cassie–Baxter: Second-order effects on the wetting of rough surfaces. *Langmuir* **2014**, *30*, 9423–9429. [CrossRef]
40. Zhang, M.; Zhao, P.; Li, P.; Ji, Y.; Liu, G.; Jin, W. Designing biomimic two-dimensional ionic transport channels for efficient ion sieving. *ACS Nano* **2021**, *15*, 5209–5220. [CrossRef]
41. Su, J.; Jia, M.; Wu, W.; Li, Z.; Li, W. Chemical vapor crosslinking of graphene oxide membranes for controlling nanochannels. *Environ. Sci. Nano* **2020**, *7*, 2924–2929. [CrossRef]
42. Jia, Z.; Wang, Y. Covalently crosslinked graphene oxide membranes by esterification reactions for ions separation. *J. Mater. Chem. A* **2015**, *3*, 4405–4412. [CrossRef]
43. Compton, O.C.; Nguyen, S.T. Graphene oxide, highly reduced graphene oxide, and graphene: Versatile building blocks for carbon-based materials. *Small* **2010**, *6*, 711–723. [CrossRef] [PubMed]
44. Dong, H.; Wu, L.; Zhang, L.; Chen, H.; Gao, C. Clay nanosheets as charged filler materials for high-performance and fouling-resistant thin film nanocomposite membranes. *J. Membr. Sci.* **2015**, *494*, 92–103. [CrossRef]
45. Zhang, M.; Guan, K.; Shen, J.; Liu, G.; Fan, Y.; Jin, W. Nanoparticles@ rGO membrane enabling highly enhanced water permeability and structural stability with preserved selectivity. *AIChE J.* **2017**, *63*, 5054–5063. [CrossRef]
46. Zhang, M.; Guan, K.; Ji, Y.; Liu, G.; Jin, W.; Xu, N. Controllable ion transport by surface-charged graphene oxide membrane. *Nat. Commun.* **2019**, *10*, 1253. [CrossRef]
47. Meng, N.; Zhao, W.; Shamsaei, E.; Wang, G.; Zeng, X.; Lin, X.; Xu, T.; Wang, H.; Zhang, X. A low-pressure GO nanofiltration membrane crosslinked via ethylenediamine. *J. Membr. Sci.* **2018**, *548*, 363–371. [CrossRef]
48. Yu, J.; He, Y.; Wang, Y.; Li, S.; Tian, S. Ethylenediamine-oxidized sodium alginate hydrogel cross-linked graphene oxide nanofiltration membrane with self-healing property for efficient dye separation. *J. Membr. Sci.* **2023**, *670*, 121366. [CrossRef]
49. Kim, H.; Kang, S.-O.; Park, S.; Park, H.S. Adsorption isotherms and kinetics of cationic and anionic dyes on three-dimensional reduced graphene oxide macrostructure. *J. Ind. Eng. Chem.* **2015**, *21*, 1191–1196. [CrossRef]
50. Soltanieh, M.; Mousavi, M. Application of charged membranes in water softening: Modeling and experiments in the presence of polyelectrolytes. *J. Membr. Sci.* **1999**, *154*, 53–60. [CrossRef]
51. Yuan, B.Q.; Wang, M.X.; Wang, B.; Yang, F.L.; Quan, X.; Tang, C.Y.; Dong, Y.C. Cross-linked graphene oxide framework membranes with robust nano-channels for enhanced sieving ability. *Environ. Sci. Technol.* **2020**, *54*, 15442–15453. [CrossRef] [PubMed]
52. Chandio, I.; Janjhi, F.A.; Memon, A.A.; Memon, S.; Ali, Z.; Thebo, K.H.; Pirzado, A.A.A.; Hakro, A.A.; Khan, W.S. Ultrafast ionic and molecular sieving through graphene oxide based composite membranes. *Desalination* **2021**, *500*, 114848. [CrossRef]
53. Shefer, I.; Lopez, K.; Straub, A.P.; Epsztein, R. Applying transition-state theory to explore transport and selectivity in salt-rejecting membranes: A critical review. *Environ. Sci. Technol.* **2022**, *56*, 7467–7483. [CrossRef] [PubMed]

polymers

Article

Methallylsulfonate Polymeric Antiscalants for Application in Thermal Desalination Processes

Ali A. Al-Hamzah [1,2], Christopher M. Fellows [1,2,*] and Osman A. Hamed [1]

1 Water Technologies Innovation Institute & Research Advancement, Saudi Water Authority, Al Jubail 31961, Saudi Arabia; aal-hamzah2@swcc.gov.sa (A.A.A.-H.); oahmed@yahoo.com (O.A.H.)
2 School of Science and Technology, The University of New England, Armidale, NSW 2351, Australia
* Correspondence: cfellows@une.edu.au; Tel.: +966-59-417-1150

Abstract: Nine copolymers of acrylic acid and sodium methallyl sulfonate were tested as scale inhibitors in thermal desalination. The nine antiscalants covered molar masses between 2000 and 9500 g.mol^{-1} and concentrations of sulfonated monomer ranging between 10 and 30 mole percent. A pressure measurement and control (P-MAC) unit and a high-temperature pressurized vessel were used to measure the effectiveness of the scale inhibitors in seawater, concentrated seawater, and model solutions at 125 °C. The effectiveness of the novel copolymers was comparable to commercial antiscalant at times up to 15 min and improved at longer times. Molar mass was a more important determinant of effectiveness than degree of sulfonation, with the greatest mitigation of calcium sulfate precipitation observed for antiscalants of molar mass 2000 to 2500 g.mol^{-1} regardless of sulfonate content. Antiscalants of molar mass 4500 to 5000 g.mol^{-1} showed a higher threshold effect than antiscalants of molar mass 7000 to 9500 g.mol^{-1}, with a 30% sulfonated polymer of molar mass 4500 g.mol^{-1} performing appreciably better than other polymers of a similar molar mass.

Keywords: thermal desalination; calcium sulfate; scale inhibitor; seawater desalination

1. Introduction

While the relative contribution of thermal desalination to global water production from seawater has declined significantly over the past few decades, in absolute terms it has only recently plateaued, and Multi-Stage Flash (MSF) and Multiple-Effect Distillation (MED) still provide the majority of the potable water for the Gulf Co-operation Council region, where the total dissolved solids content of seawater is significantly higher than that of standard seawater [1]. The factors that have led to the replacement of thermal desalination by membrane-based processes are purely economic [2], with thermal processes being significantly less vulnerable to negative variations in water quality, to defects in operation and maintenance, and to supply chain disruptions. The formation of inorganic scale, chiefly 'hard scale' composed of calcium sulfate, has a negative impact on the operation of thermal desalination plants by reducing heat exchanger effectiveness [1,3]. As calcium sulfate is an inverse solubility salt, its precipitation is the primary factor determining the top brine temperature of operations [4–9]. As the overall efficiency of thermal desalination increases with increasing top brine temperature, more effectively addressing the problem of inorganic scaling in thermal desalination is important. This is true not only for improving the efficiency of the operation of existing thermal desalination infrastructure, but for improving the economic competitiveness of potential future thermal desalination plants operating over an extended temperature range. The primary means by which hard scale formation is controlled is by the addition of scale inhibitors at the low-ppm level, typically polycarboxylate polymers of relatively low molar mass [10–16]. Carboxylates have been demonstrated to interact with forming calcium sulfate microcrystallites not only to inhibit precipitation but to enhance the formation of calcium sulfate hemihydrate at low temperatures and are not necessarily the ideal polymeric species for calcium sulfate control [17]. We have previously reported on the synthesis by Reversible Deactivation Radical Polymerization and testing of poly(acrylic acid) scale inhibitors incorporating various end groups on the formation of calcium sulfate scale under conditions relevant to thermal desalination [18]. The work reported on here, which was carried out previous to that work but has only become possible to publish at this time, deals with the testing of copolymers of acrylic acid and sodium methallyl sulfonate as scale inhibitors. In contrast to carboxylates and phosphonates, relatively little published work has appeared on the use of the sulfonate functional group in scale inhibition (e.g., [19–21]) and none to our knowledge on the prevention of hard scale under the conditions prevailing in thermal desalination. Due to the similarity in structure between the sulfonate group and the sulfate group, it is reasonable to expect relatively strong interactions between sulfonate and anion binding sites on the surface of forming calcium sulfate crystallites, impeding growth of such crystallites, and for this reasons sulfonate antiscalants have found application in the control of barium sulfate scaling in oilfields [19,20].

2. Materials and Methods

2.1. Copolymere

Nine copolymers of acrylic acid and sodium methallyl sulfonate (SMS) were synthesized in Degussa Laboratories in Krefeld, Germany, and supplied as viscous solutions of 29.9–31.5% solids. The major known characteristics of the copolymers as reported by Degussa Laboratories are given in Table 1.

Table 1. Composition and molar mass of antiscalants.

Sample Number	*w/w* % SMS	Molar Mass/10^3 (M_m) Degussa/RI/UV	Molar Mass/10^3 (M_n)
1	9	2.5/2.9/2.3	1.6/1.5/1.2
2	18	2.0/2.6/1.8	1.5/1.5/0.9
3	28	2.5//	1.6//
4	9	5.0/5.9/3.0	3.0/3.4/1.3
5	19	5.0/5.5/4.3	3.0/2.5/1.7
6	28	4.5/5.0/3.4	3.0/2.1/1.6
7	9	9.0/10.9/8.7	6.0/6.5/2.9
8	19	9.5//	6.0//
9	30	7.0/8.7/5.8	5.0/5.7/2.0

^{1}H (Figure S2) and ^{13}C (Figure S1) NMR of samples 2, 4–7, and 9 were collected on a Spinsolve 80 spectrometer (Magritek GmbH, Aachen, Germany) in aqueous solution and are shown in the Supplementary Information. Spectra showed a qualitative increase in peaks associated with SMS consistent with the reported composition, although the broad overlapping polymer peaks could not be integrated to give a quantitative composition.

Size Exclusion Chromatography was carried out on samples 1–2, 4–7, and 9 with a 1260 Infinity system (Agilent Technologies, Dubai, United Arab Emirates) using two PL Aquagel-OH 20 columns in series with both UV and RI detection (Figure S3). Samples of approximately 2 mg/mL were prepared in 0.34 M NaCl solution buffered with 1% *w/v* 1:1 $H_2PO_4^-$:HPO_4^{2-}. A calibration curve was prepared using five molar mass standards of poly(acrylic acid) (Agilent; M_p = 1550, 2925, 7500, 16,000, and 28,000 g/mol). The eluagrams obtained are shown in the Supplementary Information and poly(acrylic acid) equivalent molar masses are given in Table 1 for both RI and UV eluagrams.

2.2. Other Materials

Arabian Gulf Seawater of approximate composition given below (Table 2) and artificial seawater were employed. AR-grade sodium chloride, potassium chloride, magnesium sulfate, calcium chloride, and sodium bicarbonate were used to prepare artificial seawater.

Table 2. Representative Arabian Gulf seawater composition [6].

Parameter	Value	Unit
TDS	44,500	ppm
pH	8.1	
Chloride (Cl^-)	23,500	ppm
Sodium (Na^+)	12,400	ppm
Sulfate (SO_4^{2-})	3290	ppm
Magnesium (Mg^{2+})	1530	ppm
Calcium (Ca^{2+})	450	ppm
Potassium (K^+)	470	ppm
Hydrogen carbonate (HCO_3^-)	98	ppm

The supersaturation index (SI) for calcium sulfate hemihydrate was calculated for solutions of relatively low ionic strength using the expression derived by Shen et al. [22], and for ionic strengths greater than or equal to seawater the SI was calculated using the Pitzer model [23] for ion activities and the calcium sulfate hemihydrate solubility data of Zannoni et al. [24].

Total dissolved solids (TDS) was determined by gravimetry (ASTM D 1193), pH by potentiometry using a standard hydrogen electrode (ASTM D1293, AWWA Standard Method 4500-H+ B), chloride by titration with silver nitrate (ASTM D 512, AWWA Standard Method 4500 B), bromide by ion chromatography (ASTM D 6581, AWWA Standard Method #4110 B), calcium, magnesium, potassium, and sodium by inductively coupled plasma optical emission spectroscopy (AWWA Standard Method 3120B), and hydrogen carbonate by titration with hydrochloric acid (ASTM D 3875-15).

2.3. *Description of Experimental Units*

The bench-top tests were carried out in two separate units. Preliminary tests were first carried out in the pressure measurement and control (P-MAC) unit (P-MAC Systems, Aberdeen, United Kingdom). A high-temperature–high-pressure test unit constructed in the WTIIRA engineering workshop was then utilized for the screening tests.

2.3.1. Pressure Measurement and Control (P-MAC) Test Unit

Brine and scale inhibitor solution were separately pumped at constant flow rates of 10 mL/min with HPLC piston pumps into a mixing cell and the solution was then passed through a 1.016 mm ID stainless steel 316 capillary tube immersed in an oil bath heated to 125 °C (Figure 1). Any scale deposition reduces the bore of the tube and causes an increase in pressure drop across the cell. The rate of change in pressure across the coil was monitored via a ceramic pressure sensor inside the P-MAC unit. The resultant signal was conditioned and amplified to produce a buffered output suitable for displaying as a scaling curve on an external chart recorder. Each test was carried out until a differential pressure of 2.0 psi was obtained. This relatively low pressure, correlated with a low degree of scaling on the coil surface, was chosen to aid in cleaning of the coil. Cleaning of the coil after calcium sulfate deposition was carried out by sequential passage of 5% disodium ethylenediamine tetraacetic acid and 5% HCl solutions at room temperature.

Note that due to the requirement to obtain observable scaling effects for observation and comparison, this work was carried out at temperatures and concentrations above those used in commercial MSF plants, where the concentration of brine is rarely above 1.4 times that of seawater and temperatures are held well below 122 °C (where SI ~1.0 for a concentration factor of 1.4).

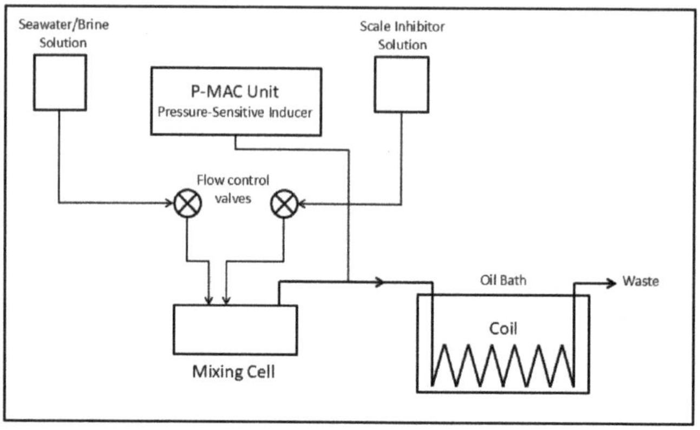

Figure 1. Schematic of pressure measurement and control (P-MAC) unit.

2.3.2. High Temperature-High Pressure Test Unit

A high temperature-high pressure test unit was constructed in order to evaluate the threshold effect of the antiscalants (Figure 2). The unit consisted of a glass reaction vessel with a heating coil molded inside its walls. The vessel was covered at the top by a stainlesssteel lid equipped with a stirrer, a thermocouple to sense the temperature, a pressure gauge, and a sampling point. The rate of heating and stirring was controlled by a separate control unit. The test unit is capable of handling chemical reactions at conditions up to 350 °C and 0.3 barg. An amount of 470 mL of test solution was poured into the reaction flask. The lid was tightly clamped to the flask in order to avoid any leakage. The solution was heated slowly with stirring, which was fixed at 300 rpm. Within around 20 min, the solution temperature reached 125 °C. The moment at which the temperature reached 125 °C was considered time zero. The rate of deposition was followed by monitoring the loss in calcium concentration with time. A series of 50 mL samples were collected at 5 min intervals for a period of 30 min for this purpose and filtered with a 0.45 µm filter. Calcium determination on the filtered samples was performed using a Metrohm Microprocessor (model 796).

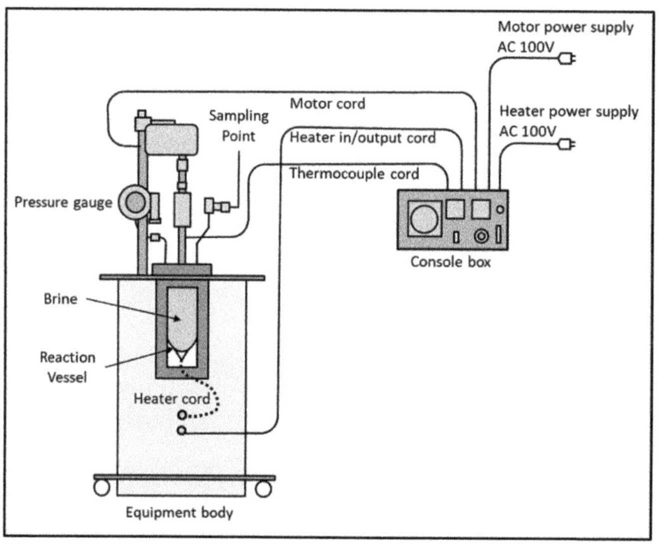

Figure 2. Schematic of high-temperature high-pressuretest unit.

3. Results

Three consecutive series of tests were carried out. In the first series, preliminary evaluation tests were carried out to identify the composition of the brine solution that was compatible with the response and performance of the pressure measurement and control (P-MAC) unit. In the second series, as a result of the technical problems which were experienced with the P-MAC unit, intensive evaluation and screening tests were performed on the high-temperature–high-pressure test unit available in the WTIIRA laboratories. In the third series, screening tests were carried out using the P-MAC unit based on the test results of the second series.

3.1. Series One: Preliminary Evaluation Tests on the P-MAC Unit

Preliminary evaluation tests were first performed on the P-MAC unit, with the primary objective of this series being to determine the concentration of the brine solution most appropriate for use in the P-MAC unit. In the first step, brine solutions prepared from seawater with twice (800 ppm Ca^{2+}, 6500 ppm SO_4^{2-}, SI = 2.34) and three times (1200 ppm Ca^{2+}, 9750 ppm SO_4^{2-}, SI = 6.16) the concentration of typical Arabian Gulf seawater were tested. It was observed that the formation of scale in the capillary tube was delayed for a period of more than 60 min in both cases. In the second step, synthetic brine solutions of different compositions and concentration ratios were tested. The time requirement for the P-MAC unit to reach the targeted pressure drop of 2.0 psi ranged between 90 and 16 min. However, at low scaling times which were obtained through the use of more concentrated solutions, calcium sulfate precipitates were observed forming in the preparation vessel.

To eliminate the possibility of salt precipitation during the preparation of highly concentrated brine solutions before they were pumped to the P-MAC unit, the cationic and anionic components of the selected brine solution were separated in two separate vessels into calcium chloride and sodium sulfate. The two solutions were then simultaneously pumped and mixed in the pipe system of the P-MAC well ahead of the heated capillary tube. Synthetic seawater of a concentration factor of two (SI = 6.16) was tested and scale precipitation occurred after around 40 min.

To induce more rapid precipitation, a solution consisting of calcium, sulfate, and bicarbonate with double the concentration of normal sea water and with a small quantity of sodium and chloride ions was tested (SI = 2.86). Precipitation of the scale occurred rapidly in two to three minutes. The more rapid scale formation at the lower SI was attributed to the role of co-precipitating $CaCO_3$ in the nucleation of crystal growth. It was concluded that this procedure was the most appropriate for use in the P-MAC unit due to the formation of scale in a reasonable time.

3.2. Series Two: Evaluation Tests on the High-Temperature–High-Pressure Unit

Four groups of experiments were carried out. In the first group of experiments, a direct comparison between the threshold effects of the nine antiscalants was carried out using a solution composed from calcium as calcium chloride and sulfate as sodium sulfate dissolved in distilled water. In the second group of experiments, the impact of additional sodium, chloride, bicarbonate, and magnesium ions on the threshold effects of the antiscalants was investigated. In the third group of this series, the best-performing antiscalants which were screened out in the first group of experiments were further tested using a synthetic brine solution composed of all constituents responsible for scale formation, and in the fourth group these antiscalants were tested using concentrated sea water.

3.2.1. Series 2, Group 1: Direct Comparison of the Performances of the Nine Antiscalants

A synthetic brine solution with only calcium and sulfate ions at close to double their concentrations in normal Arabian Gulf seawater (800 ppm Ca^{2+}, 6500 ppm SO_4^{2-}) gave good conditions for discrimination (SI = 2.98). Sodium, chloride, and magnesium ions were found to delay the precipitation of calcium sulfate significantly, as expected from theory, and for this reason they were excluded from test solutions [23]. A concentration of 10 ppm of the antiscalant solutions (4.0 to 4.5 ppm on solids basis) was used to provide discrimination between the performance of the different antiscalants. The results of the experiments are shown in Figure 3.

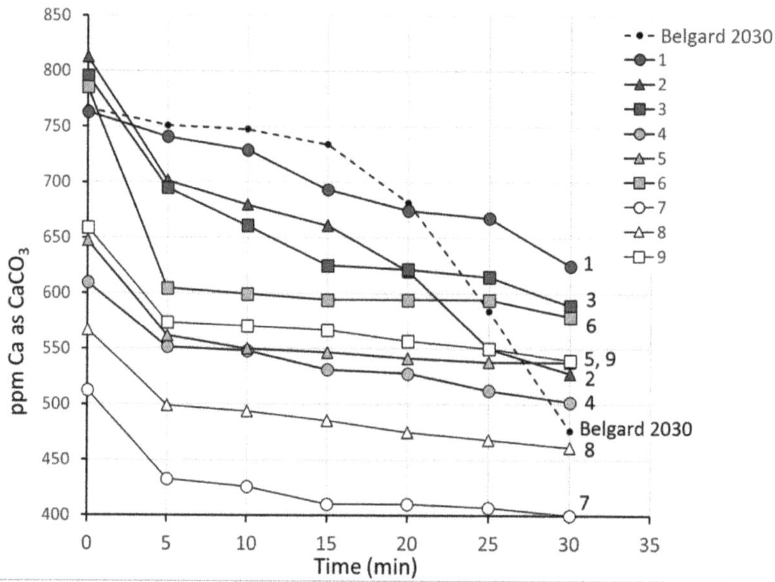

Figure 3. Reduction in calcium concentration in filtrate in the presence of copolymer antiscalants of average molar mass M_m 2000–2500 (1–3), 4500–5000 (4–6) and 7000–9500 (7–9) and commercial antiscalant Belgard 2030.

It can be seen in Figure 3 that the initial filtrable calcium concentrations vary greatly. Many of the antiscalants have a negative effect at room temperature, reducing the filterable calcium concentration relative to the reference. This effect scales with molar mass, being most significant for the highest molar mass samples (7–9) and absent for the lowest molar mass samples (1–3). Within a molar mass range, this effect is greatest for the polymers with the lowest sulfonated content (4, 7) and least for the polymers with the highest sulfonated content (6, 9). The cause of this reduction in filterable calcium is unknown but it is possibly due to the formation of insoluble calcium polyacrylate under transient conditions of high antiscalant concentration which can then nucleate the growth of calcium sulfate. This incompatibility of polyacrylic acids with calcium ions is well established at concentrations as low as 5 ppm for the polymer and is known to be more significant for homopolymers than copolymers [25]. However, such a phenomenon would only be expected to be significant in solutions of relatively low ionic strength [26].

In the first five minutes of treatment, all antiscalants can be seen to be equally effective (or ineffective), with drops in filtrable calcium of 22–181 ppm compared to 15 ppm for the control experiment. The calcium tolerance of polyacrylic acid has been reported to be inversely proportional to temperature over a relatively small temperature range (25–45 °C) [25,27], so it is possible that the differences between the incompatibilities of the antiscalants have been flattened over this time interval and they are all contributing to nucleation.

At times greater than five minutes, all antiscalants appear to show a positive effect. All but the lowest molar mass copolymers (1–3) have similar performance to the commercial antiscalant within the first 15 min, and all show better performance than the commercial product as tests continue beyond 15 min. The decline in filtrable calcium between 5 and 30 min was 275 ppm for the commercial antiscalant Belgard 2030, 105–175 ppm for the antiscalants of M_m 2000–2500, 25–50 ppm for the antiscalants of M_m 4500–5000, and 30–40 ppm for the antiscalants of M_m 7000–9500. Within a molar mass range, there was no clear trend in behavior with degree of sulfonation. This finding suggests that all species used were effective in retarding further growth of calcium sulfate nuclei by adsorbing to crystal growth sites [28], but that the lowest molar mass cohort of scale inhibitors was more susceptible to desorption under the conditions of the experiments. Considering all measured calcium loss, however, the set of lower molar mass antiscalants showed the best inhibition behavior.

3.2.2. Series 2, Group 2: Impact of Other Ions on Antiscalant Performance

In the second group of experiments, the role of other ions on the performance of a representative antiscalant was investigated under conditions where initial calcium was held constant at 1000 ppm and initial sulfate at 6500 ppm. Three conditions were investigated: (a) total concentration of the matrix with sodium chloride corresponding to approximately twice that of normal Arabian Gulf seawater, with 22,000 ppm sodium and 46,000 ppm chloride (SI = 2.0); (b) addition to (a) of sodium bicarbonate, giving a bicarbonate concentration of 310 ppm, approximately double that of standard Arabian Gulf seawater; and (SI = 2.0) (c) addition to (b) of magnesium as magnesium chloride to give a magnesium ion concentration of 3000 ppm, approximately double that found in standard Arabian Gulf seawater (Figure 4) (SI = 2.2). The pH of the tested solutions (b) and (c) was measured between 9.2 and 9.3.

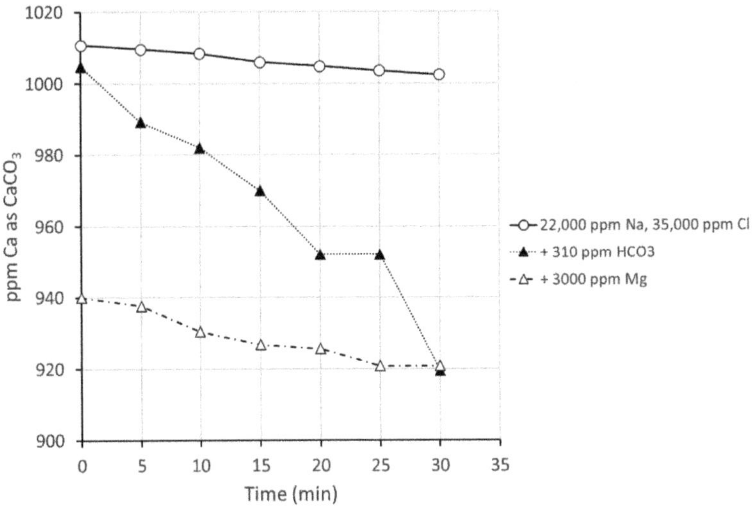

Figure 4. Reduction in calcium concentration in filtrate in the presence of low-molar-mass scale inhibitor 1 (9% SMS, M_m = 2500).

It can be seen that in comparison with the results for systems only containing the scale-forming ions (Figure 3), the performance of the inhibitor is much improved under the higher ionic strength conditions; this is not surprising, considering the strong dependence of the calcium sulfate solubility product on ionic strength [20]. The reduced performance in the presence of bicarbonate is consistent with the apparent accelerating effect of calcium carbonate in scale nucleation seen in the preliminary P-MAC investigations (Section 3.2.1). The initial drop in calcium concentration in the presence of a high amount of magnesium is unexpected; it is possible that the increase in overall hardness with the addition of such a large amount of magnesium (equivalent to over 12,000 ppm $CaCO_3$) leads to precipitation of insoluble polyacrylate salts. Disregarding the change in the calcium concentration at time = 0, it is clear that the performance of the scale inhibitor over the rest of the test is only slightly impaired by the addition of magnesium.

This synthetic brine containing calcium, sulfate, bicarbonate, magnesium, sodium, and chloride ions at concentrations double those of seawater was used as a medium for further discrimination of the low-molar-mass antiscalants 1–3 and 6.

3.2.3. Series 2, Group 3: Comparison of Antiscalants on Synthetic Concentrated Seawater

In this group of experiments, the four antiscalants 1–3 and 6 were tested at a 10 ppm concentration in the synthetic concentrated seawater prepared and a blank test was included for reference (Figure 5).

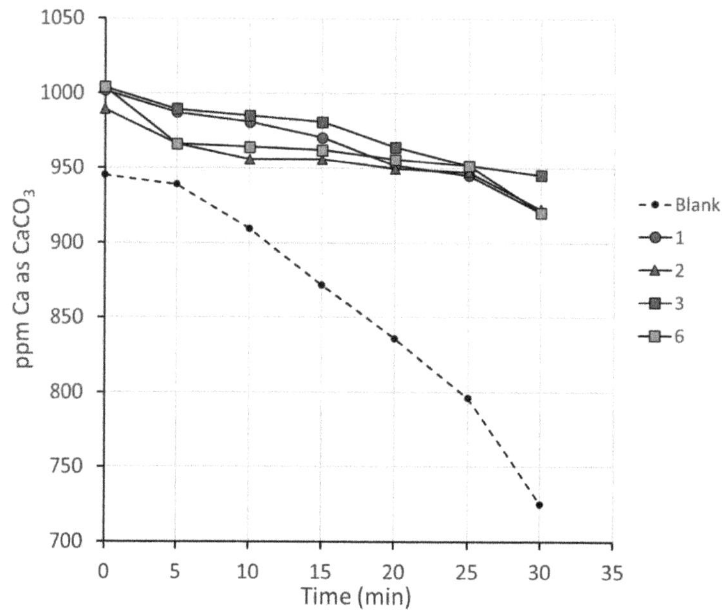

Figure 5. Reduction in calcium concentration in filtrate of concentrated artificial seawater in the presence of antiscalants 1–3 (M_m 2000–2500) and 6 (M_m 4500).

All four antiscalants were effective in delaying the precipitation of $CaSO_4$ compared to the blank solution and the drop before heating previously seen in the presence of magnesium (Figure 4) was not repeated, suggesting that result may have been artifactual. All four antiscalants showed a comparable threshold effect, with differences falling almost within experimental error. This group of experiments did not provide a clear guide for the discrimination between the four competing antiscalants. For 10 min, the approximate maximum residence time of brine in a large MSF system, the reduction in Ca concentration in the presence of the antiscalants was between 15 and 50% of that seen in the absence of any antiscalant.

3.2.4. Series 2, Group 4: Comparison of Antiscalants in Concentrated Seawater

As a last trial in the high-temperature–high-pressure test unit, real Arabian Gulf seawater concentrated in the laboratory to approximately double the initial concentration was used for further screening between the three lowest molar mass antiscalants (1–3). The pH of approximately 9 L of seawater was adjusted to 4.5 with diluted hydrochloric acid to destroy all the bicarbonate. The seawater was evaporated at 95 °C to half its original volume. Just before each experiment, the alkalinity of the concentrated brine was readjusted to double its concentration in seawater by the addition of sodium bicarbonate.

The threshold performances of the three antiscalants were to a large extent comparable to each other (Figure 6) and there was no clear trend with degree of sulfonate content, paralleling the results obtained in artificial seawater (Figure 5). Overall, the results were extremely similar to the results obtained in artificial seawater, with an average reduction in Ca concentration over 30 min of 6.8% in seawater compared to 6.9% in artificial seawater.

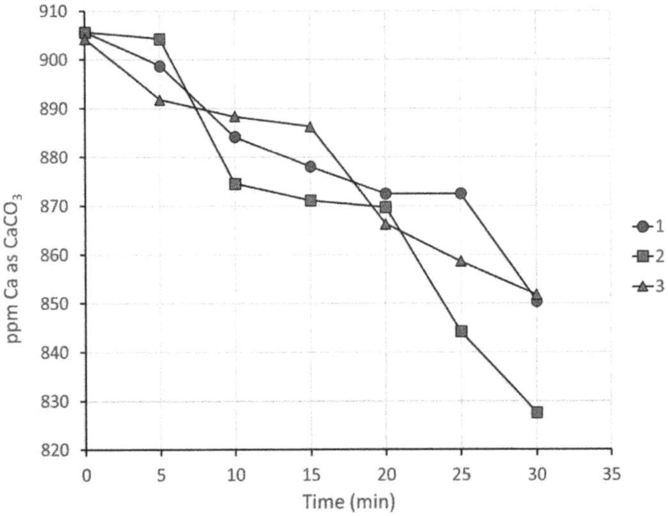

Figure 6. Reduction in calcium concentration in filtrate of concentrated seawater in the presence of antiscalants 1–3 (M_m 2000–2500).

The lack of any strong trend in inhibition behavior with sulfonate content at high-ionic-strength conditions suggests that the lower amount of sulfonate groups are sufficient to target the antiscalants to growth sites on crystallites and that the additional sulfonate content is superfluous.

3.3. Series Three: Performance of Low Molar Mass Antiscalants in the P-MAC Unit

Antiscalants 1–3 and 6 were then evaluated further in the P-MAC unit under aggressive operating conditions. A synthetic brine solution with concentrations of calcium, sulfate, and bicarbonate twice those of normal seawater was used as a feed to the P-MAC unit operating at a 125 °C temperature, as per Figure 1. Equal volumes of calcium and sulfate solutions were supplied to the instrument. All cations except sodium were fed as the chloride salts, and all anions except chloride (including the antiscalant) were fed as the sodium salts.

In an initial round of experiments, the optimum dose rate of antiscalant which would induce precipitation in the P-MAC unit in a reasonable time was determined. The effectiveness of antiscalant 1 was tested using dose rates of 10, 5.0, 2.0, and 1.0 ppm. At dose rates of 10 and 5.0 ppm, no scale formation was observed, while at a dose rate of 2.0 ppm, scale precipitation was observed after 26 min. Further reduction in antiscalant dose rate to 1.0 ppm resulted in rapid scale formation and was similar in behavior as the blank solution. A dose rate of 2.0 ppm was therefore selected for use in the P-MAC unit.

For each antiscalant, a wide variation was found in the time to reach the targeted pressure drop of 2.0 psi (Table 3). This variation is concerning and suggests that the method used could lead to the retention of nuclei for scale formation on the surface of the test coil.

While there is clearly a great degree of uncertainty in the P-MAC results, they are correlated positively with the drops in [Ca^{2+}] between 5 minutes and 30 min shown in Figures 3, 5, and 6.

Table 3. P-MAC results for antiscalants 1–3 and 6 for feed solution containing 2.0 ppm antiscalant solution, 1000 ppm Ca^{2+} 6500 ppm SO_4^{2-}, 22,000 ppm Na^+, 35,000 ppm Cl^-, 310 ppm HCO_3^-, and 3000 ppm Mg^{2+}.

Antiscalant	1	2	3	6
Time to reach pressure drop of 2 psi (min)	26.5	28.5	45	8.2
	14.37	28	7.03	3.02
	11.18	13	4.75	3
	5.5	18.5	4.02	3
	4.0	7.83	2.33	2.82
	3.37	7.78	2.03	2.23
	3.0	6.5		2.06
	2.93	3		2
	2.72	2.33		
	2.5			
	2.02			
Average	7.1	11.7	10.9	3.3
Standard Deviation	7.2	9.3	15.4	1.9
Maximum	26.5	28.5	45	8.2

4. Conclusions

A series of polycarboxylate antiscalants containing 10–30 weight% sulfonated monomer, methallyl sulfonate, were tested for their ability to retard calcium sulfate formation in a simple solution of calcium and sulfate ions at 125 °C and under conditions approximating those found in high-temperature Multi-Stage Flash (MSF) thermal desalination of seawater. At high-ionic-strength conditions, copolymers with M_m between 2000 and 4500 were shown to be most effective, inhibiting calcium sulfate formation with up to 85% effectiveness over the time scale required for MSF for realistic scale-forming solutions. No clear trend in effectiveness with sulfonate content was observed at high ionic strength, suggesting that extending the studies to copolymers with a lower fraction of allyl sulfonate comonomer may be valuable. At low-ionic-strength conditions, lower molar mass copolymers and copolymers with higher sulfonate content appeared to be more effective. Further work on the system is required in order to clarify the structure–property relationships and the role of the sulfonate group in the inhibition of calcium sulfate scaling.

Supplementary Materials: The following supporting information can be downloaded at: https://www.mdpi.com/article/10.3390/polym16192838/s1, Figure S1: 13C NMR of copolymers; Figure S2: 1H NMR of copolymers; Figure S3: Size Exclusion Chromatrography eluagrams; (a) RI detector; (b) UV detector.

Author Contributions: Conceptualization, A.A.A.-H. and O.A.H.; methodology, A.A.A.-H. and O.A.H.; formal analysis, A.A.A.-H. and C.M.F.; investigation, A.A.A.-H.; writing—original draft preparation, A.A.A.-H.; writing—review and editing, C.M.F.; project administration, A.A.A.-H. All authors have read and agreed to the published version of the manuscript.

Funding: This research received no external funding.

Institutional Review Board Statement: Not applicable.

Data Availability Statement: Data are available from the corresponding author upon request.

Acknowledgments: This work was carried out in collaboration with Jörg Issberner and Michael Voges of Stockhausen GmbH and would not have been possible without their invaluable support.

Conflicts of Interest: All authors were employed by the Saudi Water Authority. The authors declare no conflicts of interest.

References

1. Fellows, C.M.; Alhamzah, A.A.; East, C.P. Scale Control in Thermal Desalination. In *Water-Formed Deposits: Fundamentals and Mitigation Strategies*; Amjad, Z., Demadis, K., Eds.; Elsevier: London, UK, 2022; pp. 457–476.

2. Ihm, S.; Al-Najdi, O.Y.; Hamed, O.A.; Jun, G.; Chung, H. Energy cost comparison between MSF, MED and SWRO: Case studies for dual purpose plants. *Desalination* **2016**, *397*, 116–125. [CrossRef]
3. Zhou, L.; Jin, J.; Liu, Z. Controlling calcium sulfate scale in process of multi-stage distillation desalination of seawater. *Sea-Lake Salt Chem. Ind.* **2005**, *34*, 6–8.
4. Glater, J.; Dooly, R.L.; McCutchan, J.W. *Calcium Sulfate Scale Control in High Temperature Desalting Processes*; US Department of Interior: Washington DC, USA, 1973; Report UCAL-ENG 7364.
5. Sheikholeslami, R.; Ong, H.W.K. Kinetics and thermodynamics of calcium carbonate and calcium sulfate at salinities up to 1.5 M. *Desalination* **2003**, *157*, 217–234. [CrossRef]
6. Fellows, C.M.; Al-Hamzah, A. Thermal Desalination: Current Challenges. In *Mineral Scales and Deposits: Scientific and Technological Approaches*; Amjad, Z., Demadis, K., Eds.; Elsevier: London, UK, 2015; pp. 583–602.
7. Yu, B.; Miao, S.; Ding, M.; Ren, Y. Solubility and Physical Properties of α-Calcium Sulfate Hemihydrate in NaCl and Glycerol Aqueous Solutions at 303.15, 323.15, and 343.15 K. *J. Chem. Eng. Data* **2021**, *66*, 3686–3694. [CrossRef]
8. Taherdangkoo, R.; Tian, M.; Sadighi, A.; Meng, T.; Yang, H.; Butscher, C. Experimental Data on Solubility of the Two Calcium Sulfates Gypsum and Anhydrite in Aqueous Solutions. *Data* **2022**, *7*, 140. [CrossRef]
9. Reigl, S.; Van Driessche, A.E.S.; Mehringer, J.; Koltzenburg, S.; Kunz, W.; Kellermeier, M. Revisiting the Roles of Salinity, Temperature and Water Activity in Phase Selection during Calcium Sulfate Precipitation. *CrystEngComm* **2022**, *24*, 1529–1536. [CrossRef]
10. Solomon, D.H.; Rolfe, P.F. Polymers that inhibit the deposition of calcium sulfate. *Desalination* **1966**, *1*, 260–266. [CrossRef]
11. Amjad, Z. Calcium sulfate dihydrate (gypsum) scale formation on heat exchanger surfaces: The influence of scale inhibitors. *J. Coll. Interf. Sci.* **1988**, *123*, 523–536. [CrossRef]
12. Oner, M.; Dogan, O.G.; Oner, G. The influence of polyelectrolytes architecture on calcium sulfate dihydrate growth retardation. *J. Crystal Growth* **1998**, *186*, 427–437. [CrossRef]
13. Xue, X.; Change, F.; Li, N.; Zheng, F.; Yang, W.; Yang, X. Performance of a non-phosphorus antiscalant on inhibition of calcium-sulfate precipitation. *Water Sci. Technol.* **2012**, *66*, 193–200. [CrossRef] [PubMed]
14. Popov, K.; Rudakova, G.; Larchenko, V.; Tusheva, M.; Kamagurov, S.; Dikareva, J.; Kovaleva, N. A comparative performance evaluation of some novel (Green) and traditional antiscalants in calcium sulfate scaling. *Adv. Mater. Sci. Eng.* **2016**, *2016*, 7635329. [CrossRef]
15. Zhang, S.; Qu, H.; Yang, Z.; Fu, C.-E.; Tian, Z. Scale inhibition performance and mechanism of sulfamic/amino acids modified polyaspartic acid against calcium sulfate. *Desalination* **2017**, *419*, 152–159. [CrossRef]
16. Singh, Y.B.; Ng, K.C. Elucidation of dual-mode inhibition mechanism of a typical polymer-based antiscalant on Red seawater for thermal desalination at higher temperatures and higher concentration factors. *J. Petrol. Sci. Eng.* **2019**, *183*, 10638. [CrossRef]
17. Reigl, S.; Van Driessche, A.E.S.; Wagner, E.; Montes-Hernandez, G.; Mehringer, J.; Koltzenburg, S.; Kunz, W.; Kellermeier, M. Toward more sustainable hydraulic binders: Controlling calcium sulfate phase selection via specific additives. *ACS Sustain. Chem. Eng.* **2023**, *11*, 8450–8461. [CrossRef]
18. Al-Hamzah, A.; Wallace, A.G.; East, C.P.; Fellows, C.M.; Doherty, W.O.S. Inhibition by poly(acrylic acid) and morphological changes in calcium carbonate and calcium carbonate/calcium sulfate crystallization on silica fibers. *Ind. Eng. Chem. Res.* **2014**, *53*, 8793–8803. [CrossRef]
19. Emmons, D.H. Low Molecular Weight Polyvinyl Sulfonate for Low pH Barium Sulfate Scale Control. United States Patent US4710303A, 1 December 1987.
20. Falk, D.O.; Dormish, F.L.; Beazley, P.M.; Thompson, R.G. Polyvinyl Sulfonate Scale Inhibitor. United States Patent US5092404A, 3 March 1992.
21. Crossman, M.; Holt, S.P.R. Scale Control Composition for High Scaling Environments. United States Patent 6995120B2, 7 February 2006.
22. Shen, L.; Sippola, H.; Li, X.; Lindberg, D.; Taskinen, P. Thermodynamic modeling of calcium sulfate hydrates in the $CaSO_4$-H_2O system from 273.15 to 473.15 K with extension to 548.15 K. *J. Chem. Eng. Data* **2019**, *64*, 2697–2709. [CrossRef] [PubMed]
23. Pitzer, K.S. Theoretical considerations of solubility with emphasis on mixed aqueous electrolytes. *Pure Appl. Chem.* **1986**, *58*, 1599–1610. [CrossRef]
24. Zannoni, R.; Resini, I.; Liberti, L.; Santori, M.; Boari, G. Desulphation pretreatment for 138 °C (290 °F) operation and performance test of a 1 MGD plant at Doha East (Kuwait) power station. *Desalination* **1987**, *66*, 431–442. [CrossRef]
25. Amjad, Z.; Zuhl, R.W.; Zibrida, J.F. Factors influencing the precipitation of calcium-inhibitor salts in industrial water systems. In Proceedings of the Association of Water Technologies Annual Convention, Phoenix, AZ, USA, 3–6 November 2003.
26. Schweins, R.; Huber, K. Collapse of sodium polyacrylate chains in calcium salt solutions. *Eur. Phys. J. E* **2001**, *5*, 117–126. [CrossRef]
27. Amjad, Z.; Zuhl, R.W.; Zibrida, J.F. The impact of thermal stress on deposit control polymer performance. In Proceedings of the Association of Water Technologies Annual Convention & Exposition, Uncasville, CT, USA, 19–22 September 2013.
28. Senogles, E.; Doherty, W.O.S.; Crees, O.L.C. Scale Inhibitors. In *Encyclopedia of Polymer Science and Technology*, 2nd ed.; Mark, H.F., Bikales, N., Menges, G., Gaylord, N., Kroschwitz, J.I., Eds.; Wiley-Interscience: Hoboken, NJ, USA, 1996; pp. 7587–7594.

polymers

MDPI

Article

A Novel, Dual-Initiator, Continuous-Suspension Grafting Strategy for the Preparation of PP-g-AA-MAH Fibers to Remove of Indigo from Wastewater

Sijia Xie [1], Ziyang Fang [1], Zhouyang Lian [1,*], Zhengwei Luo [1], Xueying Zhang [1] and Shengxiu Ma [2]

[1] School of Environmental Science and Engineering, Nanjing Tech Univeristy, Nanjing 211816, China
[2] Karamay Zhiyuan Bochuang Environmental Protection Technology Co., Ltd., Karamay 834000, China
* Correspondence: lianzy@njtech.edu.cn

Abstract: The indigo dye found in wastewater from printing and dyeing processes is potentially carcinogenic, teratogenic, and mutagenic, making it a serious threat to the health of animals, plants, and humans. Motivated by the growing need to remove indigo from wastewater, this study prepared novel fiber absorbents using melt-blow polypropylene (PP) melt as a matrix, as well as acrylic acid (AA) and maleic anhydride (MAH) as functional monomers. The modification conditions were studied to optimize the double-initiation, continuous-suspension grafting process, and then functional fibers were prepared by melt-blown spinning the modified PP. The results showed that the optimum modification conditions were as follows: a 3.5 wt% interfacial agent, 8 mg/L of dispersant, 80% monomer content, a 0.8 mass ratio of AA to MAH, a 1000 r/min stir speed, 3.5 wt% initiator DBPH grafting at 130 °C for 3 h, and 1 wt% initiator BPO grafting at 90 °C for 2 h. The highest grafting rate of the PP-g-AA-MAH was 31.2%, and the infrared spectrum and nuclear magnetic resonance spectroscopic analysis showed that AA and MAH were successfully grafted onto PP fiber. This modification strategy also made the fibers more hydrophilic. The adsorption capacity of the PP-g-AA-MAH fibers was highly dependent on pH, and the highest indigo adsorption capacity was 110.43 mg/g at pH 7. The fiber adsorption capacity for indigo increased rapidly before plateauing with increasing time or indigo concentration, and the experimental data were well described in a pseudo-second-order kinetic model and a Langmuir isothermal adsorption model. Most impressively, the modified fiber adsorption capacity for indigo remained as high as 91.22 mg/g after eight regeneration and reuse cycles. In summary, the PP-g-AA-MAH fibers, with excellent adsorption-desorption characteristics, could be readily regenerated and reused, and they are a promising material for the removal of indigo from wastewater.

Keywords: suspension grafting; dual initiators; PP-based functional fiber; adsorption; indigo wastewater

1. Introduction

The textile industry produces large quantities of dyeing wastewater and is a major source of environmental pollution [1]. Compared with other wastewater, printing and dyeing wastewater has the characteristics of difficult biodegradable and complex composition. Generally, the pH of printing and dyeing wastewater is 6–10, COD_{Cr} is 400–1000 mg/L, BOD_5 is 100–400 mg/L, SS is 100–200 mg/L, and the chromaticity is 100–400 times, and it always contains a variety of dyes, which is difficult to treat [2,3]. Among dyes, indigo is widely used and has also been identified as a potential carcinogen, teratogen, and mutagen [4–6]. The enrichment of indigo in wastewater from the textile industry, therefore, poses a serious threat to the health of animals, plants, and human beings. Therefore, effective treatment strategies that can remove indigo from textile and printing waste are urgently needed.

At present, wastewater containing indigo is treated via coagulation precipitation [7] and electrochemical [8], biochemical [9,10], membrane separation [11], or adsorption methods [12–14]. Physical methods, such as adsorption, recycle dyes from wastewater without causing secondary pollution [15,16], making these strategies especially attractive. In particular, adsorption strategies are the treatment method of choice in sudden environmental accidents [17].

While polypropylene (PP) is the most common synthetic material produced on industrial scales with excellent chemical and physical properties [18,19], it is non-polar and, therefore, cannot be used in water treatment applications. To be able to extend the use of PP to this application area, polar functional groups must be added to the PP to make it more polar and more hydrophilic [20,21]. Stanislav Voronov [22] proposed the theory of using a free-radical mechanism to covalently graft polymers in order to modify polyolefins. Currently, most studies use graft polymerization to modify PP [23]. Among grafting polymerization methods, suspension grafting has the advantages of requiring low reaction temperatures, it uses an easily degraded matrix, it can be well controlled, and it can be used with simple post-treatment with water as suspension medium products [24], and as result, it has attracted more research attention in recent years.

Combining the advantages of the aforementioned materials and methods, novel PP-based materials were developed to remove indigo from wastewater in this study. Melt-blown PP resin was used as a matrix, as it is cheap and readily available, and acrylic acid (AA) and maleic anhydride (MAH) were used as functional monomers. The modification conditions were optimized to improve the suspension grafting rate with dual initiators containing 2,5-dimethyl-2,5-dihexane (DBPH) and benzoyl peroxide (BPO). The effects of the amount of initiator, interfacial agent, dispersant, monomer, reaction time, and temperature, as well as other process conditions, on the grafting reaction were investigated. The modified PP was then used to prepare fibers via melt-blown spinning and used to treat wastewater. The adsorption behavior of the fibers was also investigated.

2. Materials and Methods

2.1. Materials

PP melt-blown resin was purchased from Shanghai Expert in the Developing of New Material Co., Ltd. (Shanghai, China). AA, MAH, DBPH, BPO, anhydrous ethanol, sodium hydroxide (NaOH), hydrochloric acid (HCl), isopropyl alcohol, acetone, and all other reagents were obtained from commercial sources and were of analytical grade.

2.2. Preparation of PP-g-AA-MAH Fibers

A predetermined amount of PP resin and xylene, the interfacial agent, was added to a 10-L magnetic-driven reactor. The temperature was then raised to 140 °C to fully dissolve the PP resin in xylene. Subsequently, a predetermined amount of tap water was quickly pumped into the reactor to quench the reaction mixture and disperse the PP particles, and the desired amounts of AA and MAH monomers were added. The reactor was then sealed, it was heated to 130 °C, and the mixture was incubated for 60 min to ensure that the PP swelled. After the PP had swollen, the desired amount of DBPH initiator was added through the pressurized feeding port. After the desired time, the reaction was cooled to a set temperature, and the desired amount of BPO was added to continue the grafting reaction. The reaction products were then recovered via centrifugation and dried. The prepared PP-g-AA-MAH was then used to prepare fibers via melt-blown spinning.

2.3. Determination of Product Grafting Ratio and Monomer Grafting Efficiency

The PP-g-AA-MAH fibers were rinsed with pure water and then placed in a Soxhlet extractor and extracted with acetone for 16 h. After drying, 0.5 g of the fibers were removed and placed in a 250-mL round-bottomed flask. Xylene (100 mL) and a KOH/ethanol solution (70 mL) were then added to the round-bottom flask, the mixture was heated until it boiled, and the mixture was refluxed for 30 min. After the reaction, the mixture was cooled to room temperature, and then 2–3 drops of thymol blue indicator were added. The mixture was titrated with an HCl/isopropyl alcohol solution (0.05 mol/L), and the endpoint was defined as the drop-back volume at which the mixture turned egg-yellow. The un-grafted PP was used as the control. The grafting rate (Gr, %) and monomer grafting efficiency (Ge, %) were calculated according to Equations (1) and (2), respectively,

$$\mathrm{Gr} = \frac{(V_0 - V_1) \times \mathrm{c} \times (98.06 + 72.06)}{3 \times m} \times 100\% \tag{1}$$

$$\mathrm{Ge} = \frac{m'_{AA} + m'_{MAH}}{m_{AA} + m_{MAH}} \times 100\% \tag{2}$$

where V_0 (L) and V_1 (L) are the volumes of HCl/isopropyl alcohol solution consumed in the titration of the blank and sample, respectively, c (mol/L) is the concentration of the HCl/isopropyl alcohol solution, m (g) is the sample mass, $m_{AA} + m_{MAH}$ (g) is the amount of AA and MAH added to the grafting reaction, $m'AA + m'MAH$ (g) is the content of AA, and MAH in the product.

2.4. Characterization Methods

The functional groups present in the PP and PP-g-AA-MAH fibers were identified using Fourier-transform infrared spectroscopy (FT-IR) (Nicolet IS 10). FT-IR data were collected over a wavelength range of 400~4000 cm^{-1}. The micromorphology of PP and PP-g-AA-MAH fibers was observed via scanning electron microscopy (SU-8010, Hitachi (China) Ltd., Beijing, China). Proton nuclear magnetic resonance spectroscopy (^{1}H-NMR) data were also collected on the PP and PP-g-AA-MAH fibers using an AVANCE500 instrument. The wetting angle of the materials was measured via the capillary micropressure method and calculated using the Washburn equation [25]. The test device and contact angle correction are described in more detail in previous literature [26,27].

2.5. Adsorption and Regeneration of Indigo Using the PP-g-AA-MAH Fibers

The adsorption of indigo using the PP-g-AA-MAH fibers was investigated using static adsorption experiments. For the adsorption experiments, simulated indigo wastewater was placed in tapered bottles with plugs. A defined mass of PP-g-AA-MAH fiber was added to the flask, and the flask was sealed with plastic wrap and placed in a 25 °C constant-temperature oscillator. The fibers were removed from the solution at predetermined time points, and the concentration of indigo in the solution was measured at 610 nm in a UV-visible spectrophotometer [23]. The adsorption capacity (Q, mg/g) was calculated using Equation (3).

$$Q = \frac{(C_0 - C_1) \times V}{m} \tag{3}$$

where C_0 (mg/L) and C_1 (mg/L) are the concentration of indigo in the solution before and after the adsorption of the PP-g-AA-MAH fibers, V (L) is the volume of the solution, and m (g) is the mass of the fiber as an absorbent.

To study the desorption of the indigo dye from the PP-g-AA-MAH fibers, fibers saturated with dye were placed in an HCl (0.5 mol/L) solution and shaken for a predetermined amount of time. The indigo concentration in the solution was then determined, and the desorption rate (D, %) was calculated according to Equation (4).

$$D = \frac{V_2 \times C_2}{m \times e_e} \times 100\% \tag{4}$$

where C_2 (mg/L) is the concentration of indigo in the eluent, V_2 (L) is the volume of the eluent, m (g) is the mass of the fiber, and q_e (mg/g) is the adsorption capacity. All the experiments were performed in triplicate, and the presented results are the average values.

3. Results and Discussion

3.1. Mechanism of Suspension Grafting Polymerization

Figure 1 shows a schematic illustration of the suspension grafting polymerization reaction. Xylene and PP resin were added to a 10-L reactor, and the suspension was magnetically stirred. The mixture was heated until the PP dissolved, and then water was pumped into the reactor to quickly cool the PP. The quenching pretreatment step was performed to increase the amount of amorphous PP in the sample, as well as to increase the degree of swelling, and therefore the surface area of the sample, for the grafting polymerization reaction because the small-molecule initiators and monomers could more easily diffuse into voids on the surfaces of the swollen PP. Then, the PP particles, xylene, water, and grafted monomer were fully mixed at high stir speeds to form many "reaction bed" sites. The suspension–polymerization grafting reaction primarily occurred in the amorphous regions of the PP particles.

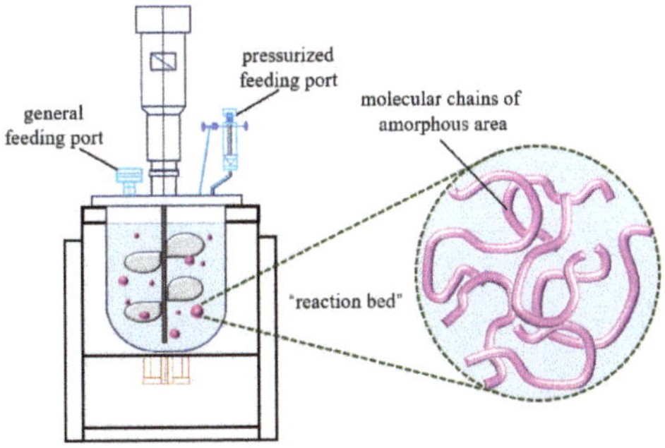

Figure 1. Illustration of the suspension-grafting reaction system.

The mechanism of the suspension-grafting polymerization reaction is shown in Figure 2. First, the initiator was decomposed into primary free radicals at high temperatures (reaction 1). These free radicals then grabbed α-H on the PP backbone to form a PP free radical (reaction 2). The PP free radical then reacted with monomers to initiate the grafting polymerization reaction (reaction 3). With strong reactivity and providing electrons for a reaction system, AA monomers were the first to react with the PP free radicals, and the grafted AA groups increased the intermolecular interactions between the polymer chains, thereby preventing the degradation of the PP and minimizing other side reactions. The stably grafted AA groups then reacted with other AA or MAH monomers as the reaction proceeded.

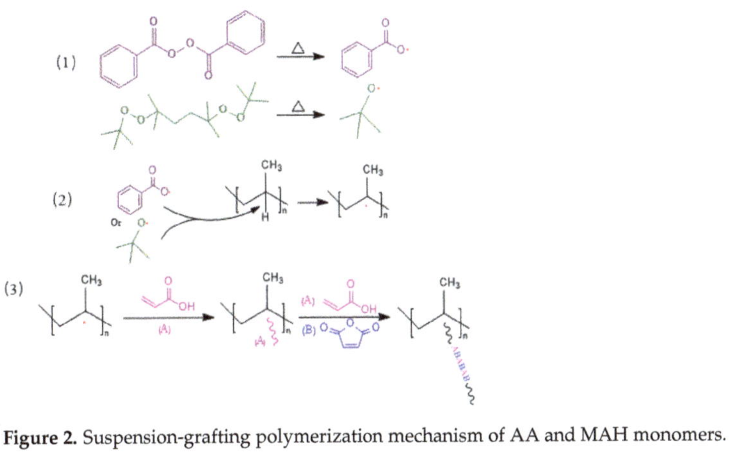

Figure 2. Suspension-grafting polymerization mechanism of AA and MAH monomers.

3.2. Influence of Reagent Concentrations on the Grafting Reaction

Reagent concentrations significantly influence the effectiveness of suspension-grafting polymerization. In this study, the relative amounts of the initiator, interfacial treatment agent, and dispersant, the total dosage of the monomer, and the monomer ratio were investigated as the main influencing factors. The results of these studies are shown in Figure 3.

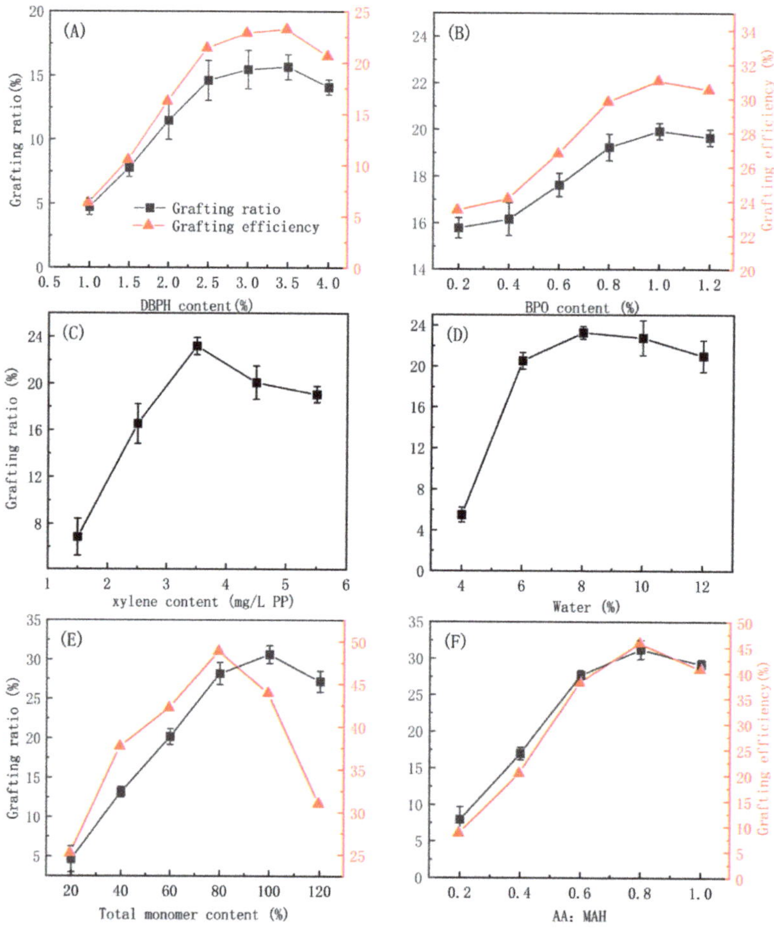

Figure 3. The effects of DBPH (**A**), BPO (**B**), xylene (**C**), water (**D**), the monomer amount relative to PP (**E**), and the ratio of AA to MAH monomers (**F**) on the grafting effectiveness.

It can be seen from Figure 3A,B that the concentration of both the DBPH and BPO initiators significantly influenced the grafting reaction. As the amount of initiator increased, the grafting ratio first increased rapidly and then increased more gradually until reaching a maximum value and then decreasing. The same overall trends were also seen in the monomer-grafting efficiency. Initially, increasing the initiator concentration increased the number of free radicals that formed at the reaction temperature, which in turn initiated the polymerization of more monomers and increased the grafting efficiency. However, as the concentration increased further, the concentration of free radicals was too high, and side reactions such as homopolymerization occurred [28,29]. The homopolymerization side reaction consumed a large amount of the monomer, resulting in less monomer grafted to the PP branch chain, which resulted in a decrease in the grafting rate and grafting efficiency.

While most studies use a single initiator in grafting reactions, here, two initiators were used to increase the grafting efficiency. In this unique strategy, the reaction temperature was increased to allow for an appropriate amount of the DBPH initiator to decompose, and then the temperature was lowered, and the BPO initiator was added. The cooling step in the reaction was similar to the quenching pretreatment used to swell the PP particles. After cooling, new active sites were generated, and the grafting polymerization continued, which in turn improved the grafting ratio and grafting efficiency.

Since both the initiator and grafted monomer used in this study were oil-soluble, xylene was chosen as the interfacial treatment agent. As can be seen from Figure 3C, the grafting ratio increased with the xylene dosage, and the maximum grafting rate was obtained using 3.5 wt% xylene relative to the PP. The increase in the grafting rate with the xylene concentration occurred because the added xylene swelled the amorphous regions of the PP, which promoted the diffusion of the initiator and monomer into the PP particles [30]. However, further increasing the xylene dosage above 3.5 wt% resulted in a slight decrease in the grafting rate because, at excessively high loading, the xylene coated the particle surface and prevented the diffusion of the monomers and initiators to the PP.

As can be seen from Figure 3D, the grafting ratio reached a maximum at 8 mL/g of dispersant to PP. A further increase in the amount of dispersant resulted in a slight decrease in the grafting rate. At low loadings, the dispersant could not ensure that the particles were well dispersed. During the reaction process, due to insufficient dispersion and adhesion, an independent suspension-grafting reaction system could scarcely be formed, and there were not enough sites for the reaction to occur. So, the grafting ratio was low at too-low loadings. At too high of a dispersant concentration, the number of individual particles, and therefore the reaction sites, was too high, and the relative concentration of the monomer to PP particles decreased. Therefore, the grafting ratio decreased at too high a dispersant concentration [31,32].

As can be seen from Figure 3E, both the grafting ratio and grafting efficiency increased with the initial monomer concentration. The highest grafting ratio was achieved when the total monomer dosage was 100% of the PP mass, while the highest grafting efficiency was achieved with a monomer dosage of 80% relative to the mass of PP. The likelihood of a monomer encountering a particle increased with the monomer concentration, which increased the grafting rate and efficiency. However, at too high concentration, the active sites on the main chain of PP that could participate in the grafting reaction were rapidly consumed due to a higher concentration of the monomer. On the other hand, the likelihood of a monomer encountering another monomer also increased, and therefore, the likelihood of side reactions also increased, and the grafting ratio and grafting efficiency decreased.

As can be seen from Figure 3F, as the AA-to-MAH mass ratio increased, the grafting ratio and grafting efficiency first increased and then slightly decreased. MAH struggles to self-polymerize because of its large-volume steric resistance, and AA self-polymerizes more easily. As the proportion of the AA monomer increased, the number of AA free radicals increased, as well as the likelihood of the AA free radicals coming into contact with the PP particles. The AA radical could also copolymerize with free MAH and then graft onto PP chain segments, which was also conducive to improving the grafting ratio and grafting efficiency. AA could also self-polymerize, and at too high an AA loading, the probability of an AA radical encountering another AA radical and a subsequent coupling termination reaction between AA radicals occurring also increased, resulting in an overall decrease in the grafting ratio and grafting efficiency. Moreover, the highly active AA monomers were more likely to react with the initiators at an excess AA loading, which reduced the probability of the initiators reacting with the PP and, thus, affected the grafting reaction.

3.3. Influence of Reaction Conditions on the Grafting Modification

In addition to the reagent dosages, the reaction conditions also significantly impacted the grafting reaction. Therefore, this study also explored the effects of the reaction time, reaction temperature, and stir rate on the grafting reaction, and the results are shown in Figure 4.

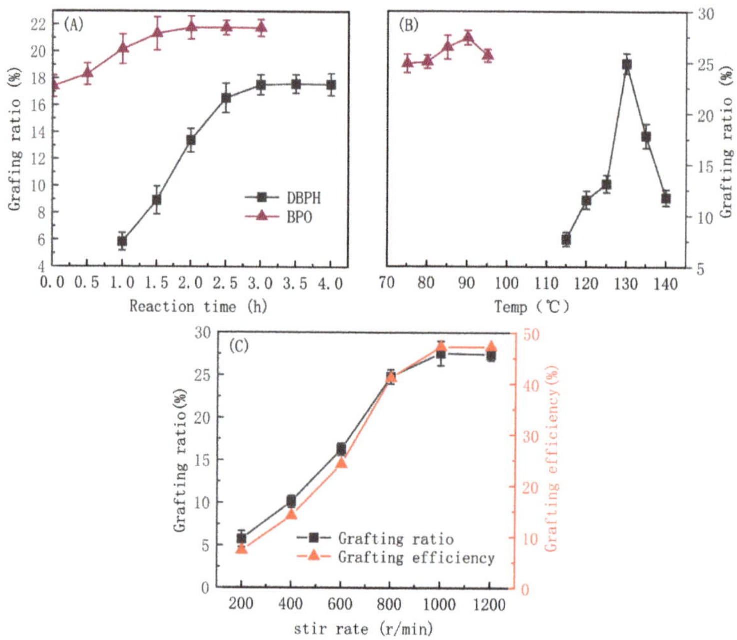

Figure 4. The effects of the reaction time (**A**), the reaction temperature (**B**), and the stir rate (**C**) on the grafting effect.

As can be seen from Figure 4A, the grafting ratio using either the DBPH or BPO initiators increased rapidly and then more slowly with an increasing reaction time, and it plateaued around 3 h and 2 h, respectively. As the reaction time increased, the number of free radicals formed via decomposition of the initiators increased, and the grafting reaction proceeded quickly. In addition, the relative monomer concentration was higher, which further increased the grafting ratio. The grafting ratio plateaued at long times because, once all of the initiators had decomposed and the concentration of active free radicals and monomers decreased, the reaction rate plateaued.

As seen in Figure 4B, the optimum reaction temperatures using DBPH and BPO are 130 °C and 90 °C, respectively. The grafting ratio decreased to varying degrees with a further increase or decrease in temperature. The nonmonotonic trend with temperature was due to competing effects on the reaction efficiency. On one hand, increasing the temperature promoted the swelling and dispersion of the PP particles, which promoted the suspension grafting reaction. On the other hand, the reaction temperature also determined the decomposition rate of the initiator [33]. The decomposition of the initiator was accelerated at higher temperatures, and as a result, more free radicals were produced, and the grafting reaction was faster. However, at too high a temperature, the initiator decomposed too rapidly, and the free-radical-coupling-termination and monomer-homopolymerization side reactions were more likely. Therefore, the grafting ratio decreased.

As can be seen from Figure 4C, the grafting ratio and grafting efficiency both increased rapidly with the stir rate before plateauing at high speeds. Stirring promoted the dispersion of the PP particles, and it also increased the degree of mixing and ensured even heat distribution, all of which promoted the grafting reaction. However, at very high stir speeds, there were limited gains with a further increasing rate, and the grafting ratio and grafting efficiency plateaued.

Based on the results presented above, the following were selected as optimum reaction conditions: 3.5 wt% interfacial agent, 8 mg/L of dispersant, 80% monomer content, a 0.8 mass ratio of AA to MAH, a 1000 r/min stir rate for the reactor, 3.5 wt% initiator DBPH grafting at 130 °C for 3 h, and 1 wt% initiator BPO grafting at 90 °C for 2 h, as summarized in Table 1. These conditions resulted in the highest PP-g-AA-MAH grafting rate of 31.2%.

Table 1. Optimum process parameters for the suspension grafting reaction.

PP	Monomers		Xylene	DBPH			BPO			Water	RPM	Grafting Ratio
	AA	MAH		Amount	Temp	Time	Amount	Temp	Time			
200 g	71 g	89 g	700 mL	7 g	130 °C	3 h	2 g	90 °C	2 h	1.6 L	1000 r/min	31.2%

3.4. *Characteristics of the PP-g-AA-MAH Fibers*

3.4.1. SEM Analysis of the Modified Fibers

SEM images of the fiber morphologies before and after the modification treatment are shown in Figure 5. The fibers prepared with the unmodified PP had smooth surfaces (Figure 5A), while those prepared with the modified PP-g-AA-MAH polymers had rough and uneven surfaces (Figure 5B). The fiber diameter had clearly increased, and it was the evidence of the grafted polymer on the modified fiber surfaces. It is observed that PP was grafted with AA and MAH after suspension grafting polymerization.

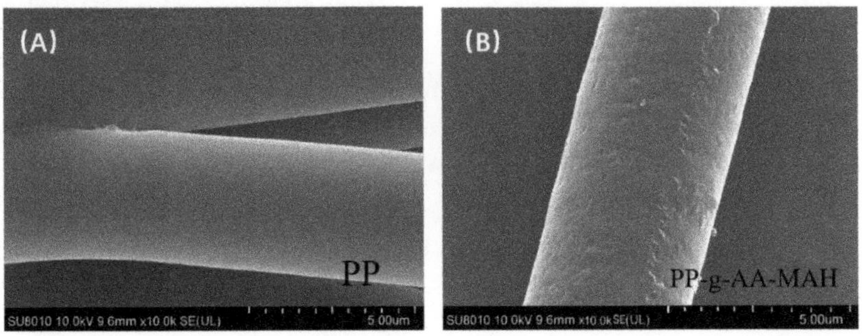

Figure 5. The micromorphology of the fibers before and after the suspension-grafting modification. (**A**) PP (**B**) PP-g-AA-MAH.

3.4.2. Fourier-Transform Infrared (FT-IR) Analysis of the Modified Fibers

Figure 6 shows the Fourier infrared spectra of the fibers prepared from either the neat PP or PP-g-AA-MAH. Note that the spectrum of PP-g-AA-MAH fibers contained the peaks characteristic of the PP fibers, as well as several new peaks. For example, the C-H symmetric and asymmetric -CH_3 stretching vibration peaks at 2964 cm^{-1} and 2838 cm^{-1}, respectively, the bending vibration absorption peaks from the CH_2 groups at 1376 cm^{-1}, and the absorption peaks characteristic of CH groups at 1460 cm^{-1} were present in both spectra. The new peak at approximately 1709 cm^{-1} in the spectrum of the PP-g-AA-MAH fibers was assigned as the stretching vibrations of the C=O in the carboxyl groups. The appearance of this peak confirmed that AA and MAH were successfully grafted to PP and that MAH was present as maleic acid.

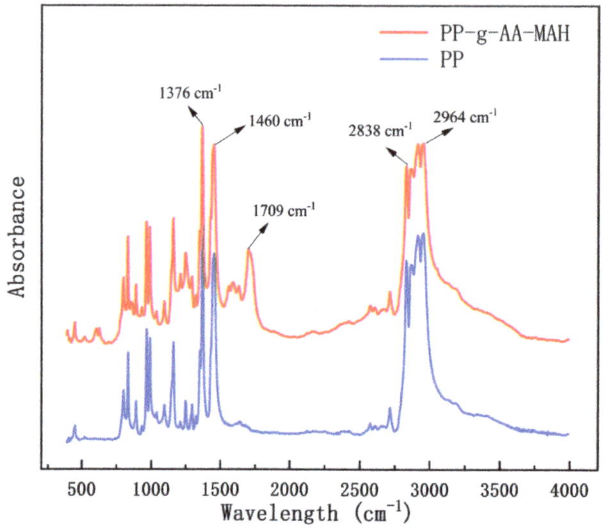

Figure 6. FT-IR spectra of the PP and PP-g-AA-MAH fibers.

3.4.3. Nuclear Magnetic Resonance (^{1}H-NMR) Analysis of the Modified Fibers

Figure 7 shows the ^{1}H-NMR spectra of the PP and PP-g-AA-MAH fibers. In the PP-g-AA-MAH fibers, the carboxyl group was expected to be connected to -CH, and the chemical shift of -CH containing carboxyl substituents should be seen at approximately 2.6 ppm. As expected, a peak at approximately 2.6 ppm [34] was seen in the spectrum of the PP-g-AA-MAH fibers but not the PP fibers, which confirmed that the carboxyl groups were successfully grafted onto PP.

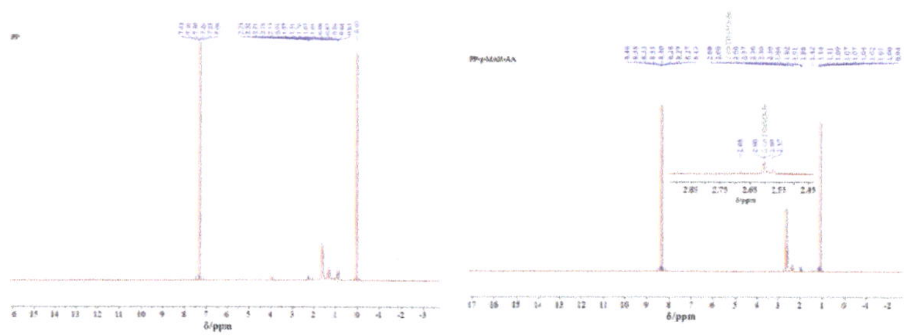

Figure 7. ^{1}H-NMR spectra of the PP and PP-g-AA-MAH fibers.

3.4.4. Surface-Property Analysis of the Modified Fibers

In this study, the capillary pressure method was used to test the hydrophilic properties of the product, first measuring the velocity of liquid penetration into the filling powder and then calculating the dynamic liquid contact angle using the Washburn [26] equation. The water contact angles were recorded using a micromanometer every 1 min. Curves of ΔP^2 versus t for PP, PP-g-AA-MAH, and insoluble starch were recorded, as shown in Figure 8, where the contact angle of starch was defined as 0°. Because PP was hydrophobic, water did not penetrate into the capillary tube containing the PP powder, and no change in pressure was recorded. Therefore, the measured ΔP^2 versus t was essentially constant. Meanwhile, both the starch and PP-g-AA-MAH fiber were hydrophilic, so water penetrated into the capillary, and the pressure changed over time. Based on these results, the water contact angle on the PP-g-AA-MAH fibers was determined to be 56.65°.

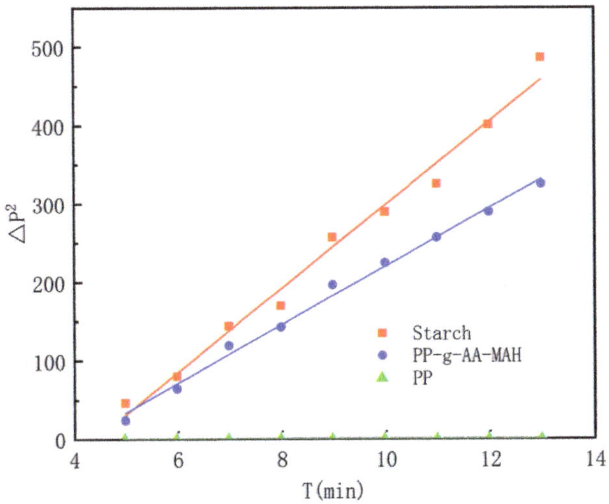

Figure 8. Measured $(\Delta P)^2 - t$ curves for the PP and PP-g-AA-MAH fibers.

3.4.5. Thermal Stability Analysis of the Modified Fibers

Figure 9 shows that the TG and DTG of PP and the graft product PP-g-AA-MAH fibers. With an increasing temperature, the grafted products lost weight before PP due to the exfoliation of AA and MAH copolymers on the product surface. Moreover, the maximum degradation temperature of PP was earlier than that of PP-g-AA-MAH, indicating that the grafting process had some cross-linking effect on the PP macromolecular chain. The maximum temperature of the melt-blown spinning process was 230 ° C, while PP-g-AA-MAH had no obvious loss at 230 °C, indicating that the product had good thermal stability and would not produce decomposition during the melt-blown spinning process.

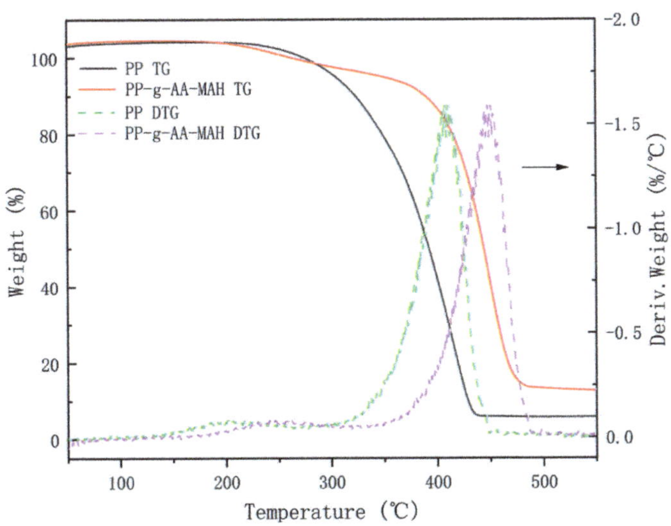

Figure 9. TG and DTG of PP and PP-g-AA-MAH fibers.

3.5. Adsorption and Regeneration of Indigo Dye Using the PP-g-AA-MAH Fibers

3.5.1. Effect of pH on the Fiber Adsorption Capacity for Indigo

The pH of the solution can affect the surface charge and morphology of an adsorbent, and therefore, it is an important factor that influences the adsorption performance of a material. Therefore, the effects of the solution pH on the adsorption capacity of the PP-g-AA-MAH fibers at a constant adsorption temperature and adsorption time, as well as the initial concentration of the indigo solution of 25 °C, 2 h, and 200 mg/L, respectively, were studied. The results are shown in Figure 10.

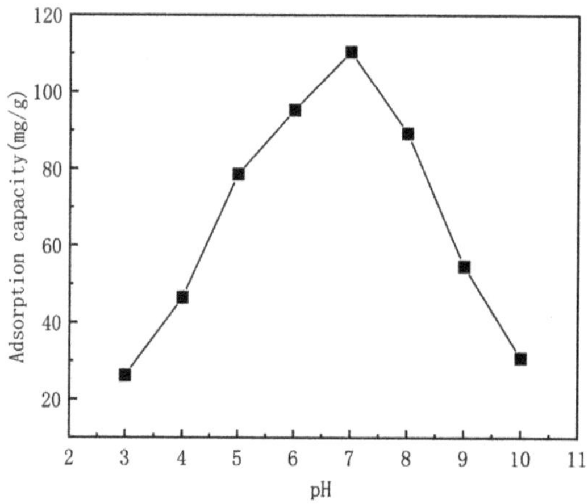

Figure 10. Effect of solution pH on the fiber adsorption capacity for indigo.

As can be seen from Figure 10, the fiber adsorption capacity first increased and then decreased with an increasing pH. The highest adsorption capacity of 110.43 mg/g was measured for a solution pH of 7. The nonmonotonic trend with pH was because the charge state of the fibers and dyes changed with the pH. At a low-solution pH, the fibers were positively charged because of the high concentration of free H^+ in the solution, and the indigo was only slightly soluble; therefore, the adsorption was not strong. The surface charge of the fibers gradually decreased with an increasing pH, and therefore, the number of -COO^- on the surface increased. These groups then interacted with the dye via electrostatic and hydrogen-bonding interactions, which increased the adsorption capacity of the fibers for the dye. At pH values greater than 7, the number of -COO^- groups on the fiber surface increased, and the indigo was either singly or doubly charged gradually under the action of insurance powder. The resulting electrostatic repulsion between the negatively charged fibers and dyes, therefore, decreased the adsorption capacity of the fibers at high pH values [26,35].

3.5.2. Adsorption Kinetics

At fixed conditions of an indigo solution concentration of 200 mg/L, a solution pH of 7, and a temperature of 25 °C, the adsorption kinetics were investigated. The measured adsorption capacities over time were fit with both pseudo-first-order and pseudo-second-order kinetic models [36], and the results are shown in Figure 11 and Table 2.

Figure 11 shows that the PP-g-AA-MAH fiber adsorption capacity for indigo first increased rapidly before plateauing at long times. At early time points, there were many active sites on the PP-g-AA-MAH fiber surface that could interact with the dye, and therefore, the adsorption rate was fast. The number of available binding sites on the fiber surfaces, as well as the concentration of free dye in the solution, decreased over time, and therefore, the adsorption rate gradually plateaued at long times. At long times, the active site on the surface of the fiber was saturated, and no additional dye could adsorb to the fibers.

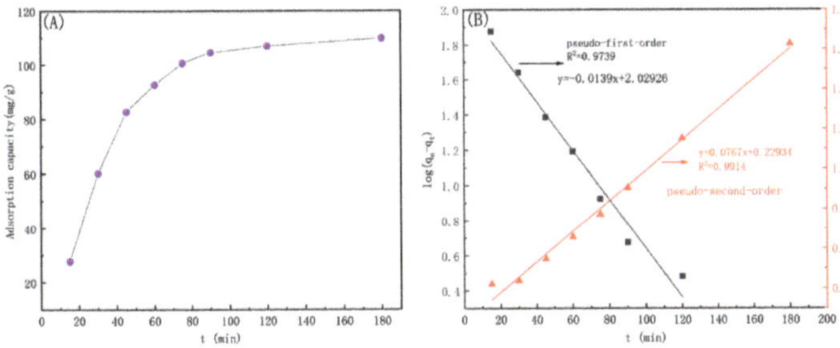

Figure 11. Adsorption capacity over time and (**A**) corresponding fits to pseudo-first-order and pseudo-second-order kinetic models (**B**).

Table 2. Fit parameters for pseudo-first-order and pseudo-second-order kinetic models.

Kinetic Model	q_e (mg/g)	K	R^2
Pseudo-first-order kinetic model	92.45	0.062	0.9739
Pseudo-second-order kinetic model	127.66	0.00078	0.9914

The R^2 of the fit of the pseudo-first-order kinetic model to the experimental data was 0.9739, while that for the pseudo-second-order kinetic model was 0.9914, indicating that the pseudo-second-order model better described the data. The better agreement with the pseudo-second-order model suggests that the indigo dye adsorbed to the PP-g-AA-MAH fibers via chemisorption, and the number of available active sites controlled the adsorption rate.

3.5.3. Adsorption Isotherms

The effects of indigo concentration on the fiber adsorption capacity were investigated at 25 °C, pH = 7 with an adsorption time of 2 h. The measured isothermal adsorption isotherms were then fit with the Langmuir [37], Freundlich [38], and Temkin [39] adsorption models, and the results are shown in Figure 12 and Table 3.

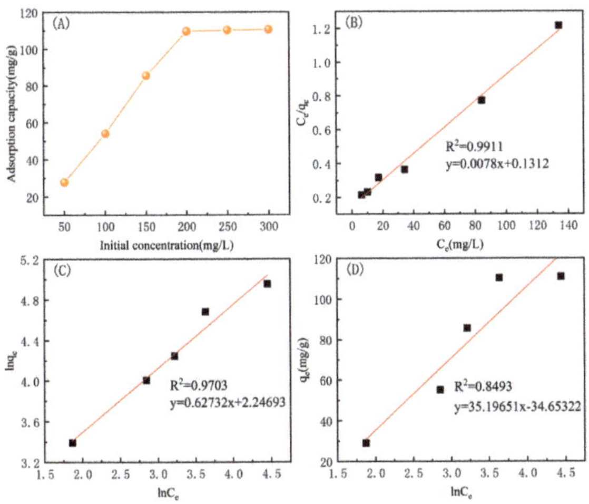

Figure 12. The effect of initial indigo concentration on the fiber adsorption capacity (**A**) and corresponding fits of the data to the Langmuir (**B**), Freundlich (**C**), and Temkin (**D**) models.

As can be seen from the Figure 12, the adsorption capacity of the PP-g-AA-MAH fiber for indigo increased with the initial dye concentration in the solution. The adsorption capacity increased rapidly from 25.03 mg/g to 110.43 mg/g when the initial dye concentration increased from 50 mg/L and 200 mg/L. However, because there was a finite number of adsorption sites on the PP-g-AA-MAH fibers, the amount of adsorbed dye increased until the adsorption sites were saturated, at which point the amount did not increase further.

Table 3. Extracted parameters from fits to the Langmuir, Freundlich, and Temkin isotherm models.

Langmuir			Freundlich			Temkin		
Q_m/mg/g	K_L	R^2	n	K_F	R^2	B_T	K_T	R^2
111.74	1.72	0.9911	2.0253	5.2069	0.9703	28.9843	0.5369	0.8493

The linear correlation coefficient of the fit to the Langmuir isothermal adsorption model (R^2 = 0.9911) was closer to 1 than that of the Freundlich model (R^2 = 0.9703) and the Temkin model (R^2 = 0.8493). The best fit with the Langmuir isothermal adsorption model suggested that the PP-g-AA-MAH fibers had a uniform adsorption surface and all active sites had the same affinity for indigo. Moreover, the adsorption likely proceeded via monolayer adsorption. The maximum theoretical adsorption capacity from the fit to the data of 111.74 mg/g was also very close to the experimental value of 110.43 mg/g.

3.5.4. Adsorption Performance of Regenerated PP-g-AA-MAH Fibers

An ideal adsorbent has a high adsorption capacity for contaminants, but the contaminants can also be easily released (desorbed) from the material to reduce the material costs. Therefore, the desorption of indigo and regeneration of PP-g-AA-MAH fibers were investigated, and the results are shown in Figure 13.

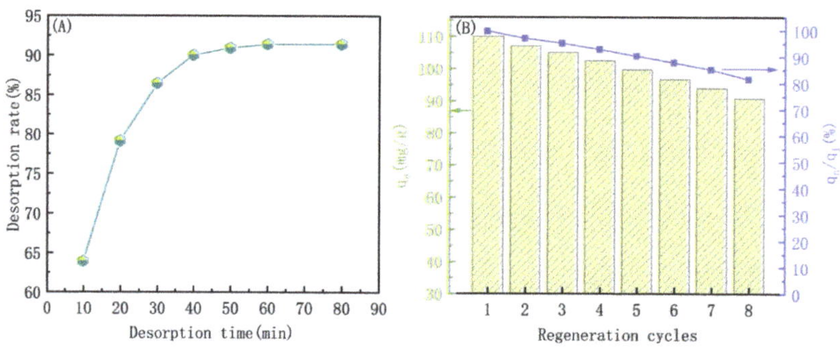

Figure 13. Effect of desorption time on the desorption rate of indigo (**A**); relationship between adsorption capacity and the regeneration cycle (**B**).

As can be seen from Figure 13A, the amount of indigo released from the fibers increased over time before plateauing at 90.05% after 40 min. These results showed that the addition of a 0.5-mol/L HCl solution to the dye-saturated PP-g-AA-MAH fibers weakened the electrostatic and hydrogen-bonding interactions with the dye, and the dye was readily released back into the solution. Moreover, the PP-g-AA-MAH fibers were also highly acid-resistant and could be treated with a strongly acidic solution to remove the dye.

Figure 13B shows the adsorption capacity of PP-g-AA-MAH fiber for indigo after multiple adsorption/regeneration cycles. While the adsorption capacity decreased slightly after multiple use cycles, the adsorption capacity remained as high as 91.22 mg/g, or 81.7% of the original adsorption capacity, after eight use cycles. These results clearly show that the PP-g-AA-MAH fibers could be readily regenerated and reused to treat indigo wastewater.

3.5.5. Comparison of Adsorption Capacity with Other Adsorbents Comparisons between the maximum adsorption capacities (q_{max}) of PP-g-AA-MAH fibers and other adsorbents for indigo reported in the literature are presented in Table 4. The results show that PP-g-AA-MAH fibers exhibit a reasonable capacity for indigo adsorption.

Table 4. Adsorption capacity of various adsorbents for indigo.

Adsorbents	q_{max}/mg·g^{-1}	Reference
Chitosan-modified flamboyant pods (CMFP)	22.45	[40]
Coriander seeds	124	[41]
Nanocelluloses using acid hydrolysis and oxidizing agents	39.7	[42]
A composite hydrogel (DMCHA)	306.08	[43]
Zizyphus joazeiro Mart peel (ZJP)	50	[44]
Natural clay (NC)	57	[45]
CaO nanoparticles made from eggshells	260	[46]
Carbonized sugarcane bagasse	8.88	[47]
Spirulina platensis	91	[48]
PP-g-AA-MAH fibers	110.43	This work

4. Conclusions

In the study presented here, modified PP-g-AA-MAH was prepared from PP melt-blow resin using a continuous suspension grafting strategy with AA and MAH functional monomers and a combination of DBPH and BPO initiators. Systematic experiments showed that the optimum modification conditions were as follows: 3.5 wt% xylene as an interfacial agent, 8 mg/L of water as a dispersant, 80% monomer content, a 0.8 mass ratio of AA to MAH, a 1000 r/min stir speed, 3.5 wt% initiator DBPH grafting at 130 °C for 3 h, and 1 wt% initiator BPO grafting at 90 °C for 2 h. The highest PP-g-AA-MAH grafting rate was 31.2%, as confirmed with FTIR and ^{1}H NMR. Importantly, the modification increased the hydrophilicity of the fibers such that the water contact angle of the PP-g-AA-MAH fibers was 56.58°.

The prepared PP-g-AA-MAH fibers were then used to remove indigo from simulated wastewater from printing and dyeing processes. The solution pH value significantly influenced the PP-g-AA-MAH fiber adsorption capacity, and the highest adsorption capacity of 110.43 mg/g was measured at pH 7. The adsorption capacity increased rapidly with time before plateauing, and the indigo adsorption to the fibers was well described with a pseudo-second-order kinetic model and Langmuir isothermal adsorption model. Most notably, the dye could be readily desorbed from PP-g-AA-MAH fibers, and the adsorption capacity of the regenerated fibers reached 91.22 mg/g after 8 use cycles, which was 81.7% of the original adsorption capacity. Together, these results highlight that the PP-g-AA-MAH fibers prepared here, with excellent adsorption, desorption, regeneration, and reuse properties, effectively and efficiently removed indigo from wastewater.

Author Contributions: Conceptualization, Z.L. (Zhouyang Lian) and X.Z.; data curation, S.X., Z.F. and Z.L. (Zhengwei Luo); formal analysis, S.X. and Z.L. (Zhengwei Luo); investigation, S.X. and S.M.; methodology, Z.L. (Zhouyang Lian), S.X., Z.F. and X.Z.; writing—original draft, S.X., Z.F., Z.L. (Zhouyang Lian), S.M. and X.Z.; writing—review and editing, Z.L. (Zhouyang Lian). All authors have read and agreed to the published version of the manuscript.

Funding: This work was supported by the Key Research and Development Program of Karamay (Grant Number 2024zdyf0009).

Institutional Review Board Statement: Not applicable.

Data Availability Statement: The original contributions presented in the study are included in the article, further inquiries can be directed to the corresponding author.

Acknowledgments: Z.L. (Zhouyang Lian) would like to thank the support from the Key Research and Development Program of Karamay (Grant Number 2024zdyf0009).

Conflicts of Interest: Author Shengxiu Ma was employed by the company Karamay Zhiyuan Bochuang Environmental Protection Technology Co., LTD. The remaining authors declare that the research was conducted in the absence of any commercial or financial relationships that could be construed as a potential conflict of interest.

References

1. Chen, Y.; Zhang, Y.; Yuan, Y.; Li, H.; Dong, K.; Li, X. Progress Research on Treatment Methods of Printing and Dyeing Wastewater. *IOP Conf. Ser. Earth Environ. Sci.* **2019**, *330*, 032073. [CrossRef]
2. Yang, Q.; Ren, S.; Zhao, Q.; Lu, R.; Hang, C.; Chen, Z.; Zheng, H. Selective separation of methyl orange from water using magnetic ZIF-67 composites. *Chem. Eng. J.* **2018**, *333*, 49–57. [CrossRef]
3. Ren, Z.; Yang, X.; Zhang, W.; Zhao, Z. Preparation, characterization and performance of a novel magnetic Fe-Zn activated carbon for efficient removal of dyes from wastewater. *J. Mol. Struct.* **2023**, *1274*, 134407. [CrossRef]
4. Shukla, A.K.; Alam, J.; Rahaman, M.; Alrehaili, A.; Alhoshan, M.; Aldalbahi, A. A Facile Approach for Elimination of Electroneutral/Anionic Organic Dyes from Water Using a Developed Carbon-Based Polymer Nanocomposite Membrane. *Water Air Soil Pollut.* **2020**, *231*, 104. [CrossRef]
5. Olivo-Alanís, D.; Atilano-Camino, M.M.; García-González, A.; Humberto-Álvarez, L.; García-Reyes, R.B. Chlorophyll-sensitized phenolic resins for the photocatalytic degradation of methylene blue and synthetic blue wastewater. *J. Sol.-Gel Sci. Technol.* **2021**, *100*, 538–554. [CrossRef]
6. Chowdhury, M.F.; Khandaker, S.; Sarker, F.; Islam, A.; Rahman, M.T.; Awual, M.R. Current treatment technologies and mechanisms for removal of indigo carmine dyes from wastewater: A review. *J. Mol. Liq.* **2020**, *318*, 114061. [CrossRef]
7. Gao, J.Y.; Gao, F.Z.; Zhu, F.; Luo, X.H.; Jiang, J.; Feng, L. Synergistic coagulation of bauxite residue-based polyaluminum ferric chloride for dyeing wastewater treatment. *J. Cent. South Univ.* **2019**, *26*, 449–457. [CrossRef]
8. Wang, Z.; Chu, Y.; Zhao, G.; Yin, Z.; Kuang, T.; Yan, F.; Zhang, L.; Zhang, L. Study of surface wettability of mineral rock particles by an improved Washburn method. *ACS Omega* **2023**, *8*, 15721–15729. [CrossRef]
9. Yang, Q.; Zhang, W.; Zhang, H.; Li, Y.; Li, C. Wastewater treatment by alkali bacteria and dynamics of microbial communities in two bioreactors. *Bioresour. Technol.* **2011**, *102*, 3790–3798. [CrossRef]
10. Pan, Z.; Tao, D.; Ren, M.; Cheng, L. A Combinational Optimization Method for Efficient Production of Indigo by the Recombinant Escherichia coli with Expression of Monooxygenase and Malate Dehydrogenase. *Foods* **2023**, *12*, 502. [CrossRef]
11. Dhas, P.G.T.N.; Gulyas, H.; Otterpohl, R. Impact of powdered activated carbon and anion exchange resin on photocatalytic treatment of textile wastewater. *J. Environ. Prot.* **2015**, *6*, 54421. [CrossRef]
12. Hashemian, S.; Hidarian, M. Synthesize and characterization of sawdust/$MnFe_2O_4$ nano composite for removal of indigo carmine from aqueous solutions. *Orient. J. Chem.* **2014**, *30*, 1753. [CrossRef]
13. Qiu, M.; Huang, C. Removal of dyes from aqueous solution by activated carbon from sewage sludge of the municipal wastewater treatment plant. *Desalin. Water Treat.* **2015**, *53*, 3641–3648. [CrossRef]
14. Peng, S.; Zhang, D.; Huang, H.; Jin, Z.; Peng, X. Ionic polyacrylamide hydrogel improved by graphene oxide for efficient adsorption of methylene blue. *Res. Chem. Intermed.* **2018**, *45*, 1545–1563. [CrossRef]
15. Mostafa, M.M.H.; Nkudede, E.; Sulemana, H.; Zhang, B.; Zhu, K.; Hu, S.; Gblinwon, R.T.; Kosiba, A.A.; Manickam, S. Treatment of simulated printing and dyeing wastewater using ozone microbubble. *E3S Web Conf.* **2021**, *261*, 04005. [CrossRef]
16. Benhalima, T.; Sadi, A.; Dairi, N.; Ferfera-Harrar, H. Multifunctional carboxymethyl cellulose-dextran sulfate/AgNPs@ zeolite hydrogel beads for basic red 46 and methylene blue dyes removal and water disinfection control. *Sep. Purif. Technol.* **2024**, *342*, 127001. [CrossRef]
17. Qian, J.; Han, X.; Yang, S.; Kuang, L.; Hua, D. A strategy for effective cesium adsorption from aqueous solution by polypentacyanoferrate-grafted polypropylene fabric under γ-ray irradiation. *J. Taiwan Inst. Chem. Eng.* **2018**, *89*, 162–168. [CrossRef]
18. Mahboubizadeh, S.; Hasanzadeh, M.; Panahi, F.; Arzaqi, Z. A Review of Processing Methods and Modification of Reinforcements Used in Polypropylene Composites. *J. Iran. Chem. Eng.* **2021**, *20*, 36–63.
19. Correa-Aguirre, J.P.; Luna-Vera, F.; Caicedo, C.; Vera-Mondrag6n, B.; Hidalgo-Salazar, M.A. The Effects of Reprocessing and Fiber Treatments on the Properties of Polypropylene-Sugarcane Bagasse Biocomposites. *Polymers* **2020**, *12*, 1440. [CrossRef]
20. Mandal, D.K.; Bhunia, H.; Bajpai, P.K.; Chaudhari, C.V.; Dubey, K.A.; Varshney, L. Radiation-induced grafting of acrylic acid onto polypropylene film and its biodegradability. *Radiat. Phys. Chem.* **2016**, *123*, 37–45. [CrossRef]
21. Abdel Ghaffar, A.; El-Arnaouty, M.; Aboulfotouh, M.E.; Taher, N.; Taha, A.A. Radiation graft copolymerization of butyl methacrylate and acrylamide onto low density polyethylene and polypropylene films, and its application in wastewater treatment. *Radiat. Eff. Defects Solids* **2014**, *169*, 741–753. [CrossRef]
22. Nosova, N.; Roiter, Y.; Samaryk, V.; Varvarenko, S.; Stetsyshyn, Y.; Minko, S.; Stamm, M.; Voronov, S. Polypropylene surface peroxidation with heterofunctional polyperoxides. *Macromol. Symp.* **2004**, *210*, 339–348. . [CrossRef]
23. Basuki, B.; Suyitno, S.; Kristiawan, B. Absorbance and electrochemical properties of natural indigo dye. In *AIP Conference Proceedings*; AIP Publishing: Long Island, NY, USA, 2018; Volume 1931.

24. Li, Z.; Ma, Y.; Yang, W. A facile, green, versatile protocol to prepare polypropylene-g-poly(methyl methacrylate) copolymer by water-solid phase suspension grafting polymerization using the surface of reactor granule technology polypropylene granules as reaction loci. *J. Appl. Polym. Sci.* **2013**, *129*, 3170–3177. [CrossRef]
25. Ji, L.; Shi, B. A novel method for determining surface free energy of powders using Washburn's equation without calculating capillary factor and contact angle. *Powder Technol.* **2015**, *271*, 88–92. [CrossRef]
26. Qing, C.; Bigui, W.; Yingdong, H. Capillary pressure method for measuring lipophilic hydrophilic ratio of filter media. *Chem. Eng. J.* **2009**, *150*, 323–327. [CrossRef]
27. Lian, Z.; Xu, Y.; Zuo, J.; Qian, H.; Luo, Z.; Wei, W. Preparation of PP-g-(AA-MAH) Fibers Using Suspension Grafting and Melt-Blown Spinning and its Adsorption for Aniline. *Polymers* **2020**, *12*, 2157. [CrossRef]
28. Bora, D.; Dutta, H.; Saha, B.; Reddy, Y.A.K.; Patel, R.; Verma, S.K.; Sellamuthu, P.S.; Sadiku, R.; Jayaramudu, J. A review on modification of polypropylene carbonate (PPC) using different types of blends/composites and its advanced uses. *Mater. Today Commun.* **2023**, *37*, 107304. [CrossRef]
29. Yang, M.; Sun, L.; Liu, Q.; Deng, Y.; Liu, C.; Li, N.; Xu, J.; Chen, Y.; Jian, X. Ultra high density carbon fiber derived from isotactic polypropylene fiber without skin-core structure. *Polym. Adv. Technol.* **2024**, *35*, e6300. [CrossRef]
30. George, G.; Joseph, K.; Saritha, A.; Nagarajan, E.R. Influence of fiber content and chemical modifications on the transport properties of PP/jute commingled biocomposites. *Polym. Compos.* **2017**, *39*, E250–E260. [CrossRef]
31. Jiao, W.; Liu, W.; Yang, F.; Jiang, L.; Jiao, W.; Wang, R. Improving the interfacial property of carbon fiber/vinyl ester resin composite by grafting modification of sizing agent on carbon fiber surface. *J. Mater. Sci.* **2017**, *52*, 13812–13828. [CrossRef]
32. Liang, F.; Yuan, H.; Shao, Q.; Song, W. Study of suspension grafting process of polypropylene. *Des. Monomers Polym.* **2018**, *21*, 130–136. [CrossRef]
33. Wang, S.; Muiruri, J.K.; Soo, X.Y.D.; Liu, S.; Thitsartarn, W.; Tan, B.H.; Suwardi, A.; Li, Z.; Zhu, Q.; Loh, X.J. Bio-Polypropylene and Polypropylene-based Biocomposites: Solutions for a Sustainable Future. *Chem. Asian J.* **2023**, *18*, e202200972. [CrossRef] [PubMed]
34. Badertscher, M.; Bühlmann, P.; Pretsch, E. *Structure Determination of Organic Compounds: Tables of Spectral Data*; Springer: Berlin/Heidelberg, Germany, 2009; Volume 443.
35. Wang, J.; Sun, C.; Xia, W.; Cao, Z.; Sheng, G.; Xie, X. Pd/BN catalysts for highly efficient hydrogenation of maleic anhydride to succinic anhydride. *Appl. Catal. A Gen.* **2022**, *630*, 118471. [CrossRef]
36. Musah, M.; Azeh, Y.; Mathew, J.T.; Umar, M.T.; Abdulhamid, Z.; Muhammad, A.I. Adsorption kinetics and isotherm models: A review. *CaJoST* **2022**, *4*, 20–26. [CrossRef]
37. Yuan, H.; Shao, Q.; Liang, F.; Shi, H.; Song, W. Mechanism of crosslinking in benzoyl peroxide-initiated functionalization of vinyltriethoxysilane onto polypropylene in the water medium. *J. Appl. Polym. Sci.* **2020**, *137*, 49534. [CrossRef]
38. Zhao, D.; Wang, L.; Zhang, R.; Ma, Y.; Chen, D.; Zhao, C.; Yang, W. Preparation of toughened polypropylene-g-poly (butyl acrylate-co-acrylated castor oil) by suspension grafting polymerization. *Polym. Eng. Sci.* **2018**, *58*, 86–93. [CrossRef]
39. Zhong, J.; Yang, B.; Feng, Y.; Chen, Y.; Wang, L.G.; You, W.D.; Ying, G.G. Enhanced photo–Fenton removal efficiency with core-shell magnetic resin catalyst for textile dyeing wastewater treatment. *Water* **2021**, *13*, 968. [CrossRef]
40. Okoya, A.A.; Diisu, D.O.; Olaiya, O.O.; Adegbaju, O.S. Adsorption Efficiency of Flamboyant Pods for Indigo Dye Removal from Textile Industrial Wastewater. *Environ. Pollut.* **2020**, *9*, 224371109. [CrossRef]
41. Kadiri, L.; Ouass, A.; Hsissou, R.; Safi, Z.; Wazzan, N.; Essaadaoui, Y.; Lebkiri, I.; El Khattabi, O.; Housseine Rifi, E.; Lebkiri, A. Adsorption properties of coriander seeds: Spectroscopic kinetic thermodynamic and computational approaches. *J. Mol. Liq.* **2021**, *343*, 116971. [CrossRef]
42. Basta, A.H.; Lotfy, V.F. Impact of pulping routes of rice straw on cellulose nanoarchitectonics and their behavior toward Indigo dye. *Appl. Nanosci.* **2022**, *13*, 4455–4469. [CrossRef]
43. Lv, A.; Lv, X.; Xu, X.; Chen, Y.; Zhang, J.; Shao, Z.-B. Tailored multifunctional composite hydrogel based on chitosan and quaternary ammonium ionic liquids@montmorillonite with different chain lengths for effective removal of dyes and 4-nitrophenol. *Sep. Purif. Technol.* **2024**, *342*, 127019. [CrossRef]
44. Ribeiro, J. Study of the Ziziphus Joazeiro Peel for Indigo Blue Adsorption. *Int. J. Adv. Res.* **2019**, *7*, 171–178. [CrossRef] [PubMed]
45. Missaoui, M.; Tahri, N.; Daramola, M.O.; Duplay, J.; Schäfer, G.; Ben Amar, R. Comparative investigation of indigo blue dye removal efficiency of activated carbon and natural clay in adsorption/ultrafiltration system. *Desalin. Water Treat.* **2019**, *164*, 326–338. [CrossRef]
46. Radhi, I.K.; Hussein, M.A.; Kadhim, Z.N. Investigation of ni grosine, alizarin, indigo and acid fuchsin removal by modification of CaO derived from eggshell with AgI: Adsorption, kinetic and photocatalytic studies. *Eur. J. Chem.* **2019**, *10*, 64–71. [CrossRef]
47. Sanda, O.; Amoo, K.I.; Taiwo, E.A. Studies on the adsorption of indigo dye from aqueous media onto carbonised sugarcane bagasse. *Niger. J. Basic Appl. Sci.* **2018**, *25*, 40–47. [CrossRef]
48. Robledo-Padilla, F.; Aquines, O.; Silva-Núñez, A.; Alemán-Nava, G.S.; Castillo-Zacarías, C.; Ramirez-Mendoza, R.A.; Zavala-Yoe, R.; Iqbal, H.M.N.; Parra-Saldívar, R. Evaluation and Predictive Modeling of Removal Condition for Bioadsorption of Indigo Blue Dye by *Spirulina platensis*. *Microorganisms* **2020**, *8*, 82. [CrossRef] [PubMed]

polymers

Article

Coaxial Electrospun Porous Core–Shell Nanofibrous Membranes for Photodegradation of Organic Dyes

Yaoyao Yang *, Shengwei Zhou, Xianyang Cao, He Lv, Zhiyuan Liang, Rui Zhang, Fujia Ye and Dengguang Yu *

School of Materials and Chemistry, University of Shanghai for Science and Technology, 516 Jungong Road, Shanghai 200093, China; 213353182@st.usst.edu.cn (S.Z.); 202342936@st.usst.edu.cn (X.C.); 201850158@st.usst.edu.cn (H.L.); 212203084@st.usst.edu.cn (Z.L.); 212203081@st.usst.edu.cn (R.Z.); 1935041028@st.usst.edu.cn (F.Y.)

* Correspondence: yyyang@usst.edu.cn (Y.Y.); ydg017@usst.edu.cn (D.Y.)

Abstract: In this study, a series of AgCl/ZnO-loaded nanofibrous membranes were prepared using coaxial electrospinning. Their physical and chemical characteristics were evaluated by SEM, TEM, XRD, XPS, IR, PL, and UV–visible spectrometer, and the photocatalytic experiments using methylene blue (MB) as a model pollutant. The formation of AgCl/ZnO heterojunction and the structure of core–shell nanofibers with porous shell layer were confirmed. AgCl/ZnO photocatalysts were also effectively loaded onto the surfaces of the porous core–shell nanofibers. The results of photocatalytic experiments revealed that the AgCl/ZnO (M_{AgCl}:M_{ZnO} = 5:5)-loaded nanofibrous membrane achieved a degradation efficiency of 98% in just 70 min and maintained a photocatalytic efficiency exceeding 95% over the first five experimental cycles, which successfully addressed the issues of photocatalytic efficiency loss during the photodegradation of MB with AgCl/ZnO nanoparticles as photocatalyst. The photodegradation mechanism was also researched and proposed.

Keywords: electrospinning; core–shell nanofibers; porous membranes; AgCl/ZnO heterojunction; photodegradation

1. Introduction

Due to the swift advancement of contemporary society, organic pollutants [1–3] like dyes, pesticides, and fertilizers have emerged as the principal factors contributing to water pollution. The rapid and efficient removal of water pollutants has become a popular topic and a focus of research in the environmental field. Organic dyes, as common pollutants, can lead to reduced oxygen levels, increased toxicity, and excessive nutrient contents in water bodies. Photocatalysis induces oxidation–reduction reactions in semiconductor materials [4–9] to degrade organic pollutants and is now a highly regarded green and sustainable development strategy. Heterostructures formed by two or more photocatalytic materials often exhibit a wider range of light absorption wavelengths compared with single photocatalysts [10–13]. By effectively suppressing the recombination of electron–hole pairs, photocatalytic performance can be enhanced.

Photocatalytic nanoparticles are often in the form of a powder, which has disadvantages, such as easy aggregation and difficult recovery after a reaction, resulting in a loss of photocatalytic activity and decreased recyclability [14,15]. Photocatalytic nanoparticles can be loaded on nanofiber skeletons by combining electrospinning technology with photocatalysis [16,17], which can effectively improve recycling performance while ensuring high catalytic efficiency. In recent years, the rapid advancement of multifluid electrospinning technology [18,19] has resulted in a diverse range of internal structural designs for electrospun nanofibers. Additionally, the controllable manufacturing of multilevel complex structures [20,21] has significantly enhanced the functionality of fiber materials.

Porous nanofibers have high specific surface areas, making them ideal carriers for photocatalysts [22–26]. However, such porous structures may reduce the mechanical properties of fibers. Nanofibers that are prepared using coaxial electrospinning technology, featuring both a core–sheath structure and a porous sheath layer, exhibit outstanding mechanical properties and a high specific surface area, effectively overcoming the limitations of conventional porous nanofibers.

In this study, silver chloride (AgCl) nanoparticles (NPs) were prepared using a water bath method and then combined with zinc oxide (ZnO) to form AgCl/ZnO composite photocatalytic particles. Porous nanofibers were prepared using cellulose acetate (CA) and polycaprolactone (PCL) as polymer matrices, and AgCl/ZnO was loaded onto their surfaces. A series of tests were carried out on the prepared functional fiber membranes to examine their surface morphology and physical properties. The performance and mechanism of the photocatalytic degradation of methylene blue by porous nanofiber membranes loaded with photocatalysts were investigated through transient photocurrent response, electrochemical impedance, free radical scavenging, and cyclic degradation experiments.

2. Materials and Methods

2.1. Reagents

The materials used in this study were a silver nitrate standard solution (0.1000 mol/L, Macklin), zinc oxide nanoparticles (30 ± 10 nm, Shanghai Macklin Biochemical Technology Co., Ltd., Shanghai, China), hydrochloric acid (HCl, 36–38%, analytical-grade, China Pharmaceutical, Shanghai, China), polyvinylpyrrolidone (PVP K10, China Pharmaceutical), anhydrous ethanol (EtOH, 99.7%, analytical-grade, Macklin, Shanghai, China), deionized water (H_2O), methylene blue (MB, >98%, Macklin), cellulose acetate (CA, 39.3–40.3 wt.% acetyl, Sigma-Aldrich, St. Louis, MO, USA), polycaprolactone (PCL, 99%, Mw = 80,000, Sigma-Aldrich), hexafluoroisopropanol (HFIP, 99.5%, analytical-grade, Mw = 168.04, Macklin), trichloromethane (CF, ≥99.0%, China Pharmaceutical), ethylenediaminetetraacetic acid disodium salt (Na_2-EDTA, 99.96%, Macklin), isopropanol (IPA, ≥99.9%, Macklin), and benzoquinone (P-BQ, ≥99.5%, Macklin). These regents were used directly after purchase.

2.2. Sample Preparation

2.2.1. Photocatalytic NPs

First, 20 mL of a 0.1 mol/L $AgNO_3$ standard solution was transferred to 10 mL of deionized water using a pipette and stirred thoroughly. Next, 0.073 g of HCl was dissolved in 10 mL of deionized water and added dropwise to the aforementioned solution. To obtain a milky white suspension, the solution was stirred with a magnetic stirrer at room temperature for 2 h and on a heated stirrer at 60 °C for 1 h. The thoroughly mixed suspension was centrifuged at a rate of 4000 r/min. Then, the precipitate was washed three times with water and ethanol. Afterwards, the sediment was transferred into a vacuum drying oven and heated to 60 °C for 12 h. After cooling, the dried precipitate was ground to obtain AgCl nanoparticles.

Next, 0.1112 g of PVP was dissolved in 20 mL of anhydrous ethanol to form a 0.05 mol/L PVP solution as the cementing agent for ZnO and AgCl. Afterward, 30 mL of a 0.1 mol/L $AgNO_3$ standard solution was transferred into a beaker that held 10 mL of deionized water. The solution was mixed thoroughly until it was clear and colorless. After that, 10 mL of a 0.3 mol/L HCl solution was slowly introduced into the beaker, and stirring continued with a magnetic stirrer for 2 h at room temperature. To the acidified $AgNO_3$ aqueous solution, ZnO nanoparticles (0.243 g, 0.567 g, and 2.187 g) were added, and the solution was stirred at 60 °C for 3 h. Finally, 10 mL of a 0.05 mol/L PVP solution was added to the mixture and stirred continuously at 60 °C for 3 h. The thoroughly mixed suspension was centrifuged at a rate of 4000 r/min, and the precipitate was washed three times with water and ethanol. Afterwards, the sediment was transferred into a vacuum drying oven and heated to 60 °C for 12 h. After cooling, the dried precipitate was ground to obtain composite nanoparticles of AgCl/ZnO with different molar ratios (M_{AgCl}:M_{ZnO} = 1:9, 3:7, and 5:5). A schematic diagram illustrating the process of preparation is shown in Figure 1.

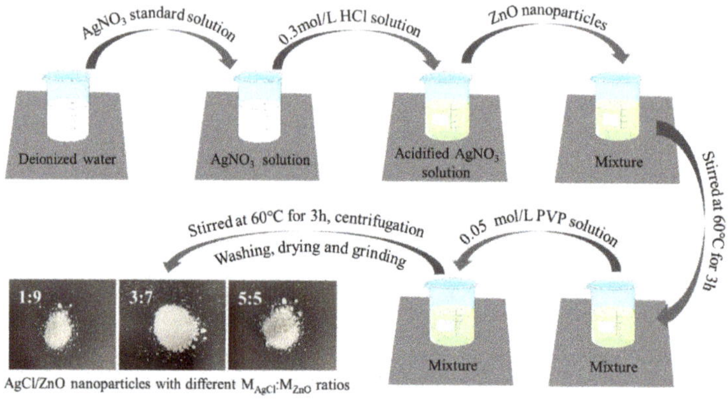

Figure 1. Schematic diagram of the process for preparing AgCl/ZnO nanoparticles.

2.2.2. Nanofiber Membranes Supported by Photocatalytic NPs

Next, 0.8 g of CA was dissolved in 10 mL of HFIP, and 0.08 g of photocatalytic NPs were added to the solution. The mixture was stirred with a magnetic stirrer for 24 h until a uniform white suspension was obtained as the sheath fluid. Photocatalytic NPs of ZnO, AgCl, AgCl/ZnO (M_{AgCl}:M_{ZnO} = 1:9), AgCl/ZnO (M_{AgCl}:M_{ZnO} = 3:7), and AgCl/ZnO (M_{AgCl}:M_{ZnO} = 5:5) were used to fabricate F1, F2, F3, F4, and F5, respectively. Then, 0.8 g of PCL was dissolved in 10 mL of CF and stirred with a magnetic stirrer for 12 h until a uniform, colorless, and transparent solution was obtained as the core fluid. The obtained sheath and core fluids were used to execute coaxial electrospinning. The collector distance was set to 15 cm, the flow rate of the core and sheath solutions was maintained at 1.0 mL/h, and the voltage was set to 10 kV, as shown in Figure 2.

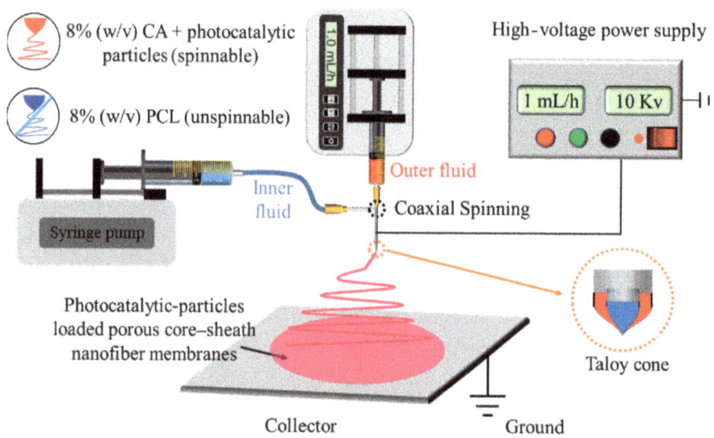

Figure 2. Schematic diagram of the preparation process of electrospun nanofibers.

2.3. *Characterization*

2.3.1. Morphology and Structure

The surface morphology was examined under an accelerated voltage of 30 kV using a field emission scanning electron microscope (FESEM, Quanta FEG 450, FEI Corporation, USA). First, the sample stage was cleaned thoroughly, and a conductive adhesive was applied to its surface. The samples were placed on the conductive adhesive, and a Au coating was sprayed onto the surface using a coating instrument to provide conductivity to the samples.

The photocatalytic particles and the fiber membrane loaded with photocatalytic particles were placed on a copper grid, and the internal structure was examined using a transmission electron microscope (TEM, Tecnai G2 F30, FEI Corporation, USA) at an accelerating voltage of 100 kV.

The specific surface area, pore volume, and pore size distribution of the electrospun nanofiber membranes were analyzed using a gas adsorption analyzer (ASAP2020, micromeritics, USA) providing nitrogen adsorption–desorption isotherms. The samples were heated and subjected to vacuum degassing to remove adsorbed impurities on their surfaces.

2.3.2. Physical and Chemical State

The synthesized samples were characterized using an X-ray diffractometer (XRD, Karlsruhe, Germany) employing CuK α (λ = 1.541 Å) radiation as the X-ray source, with a scanning range of 10–70°.

X-ray photoelectron spectroscopy (XPS, ESCALAB Xi+, ThermoFischer, USA) was used to obtain XPS data.

The functional groups present in the coaxial electrospun nanofiber membranes were analyzed using Fourier transform infrared spectroscopy (FTIR, PerkinElmer, Billerica, MA, USA). KBr was ground together with the measured sample. The instrument's scanning range was from 500 cm^{-1} to 4000 cm^{-1} with a resolution of 2 cm^{-1}.

The UV absorption performance of the samples was characterized using a evaluate (Lambda 750, PerkinElmer, Billerica, USA). $BaSO_4$ was pressed and treated as a background. Subsequently, a small amount of the photocatalyst was placed on its surface and pressed. The scan range was 300–800 nm.

2.3.3. Photocatalytic Performance Testing

The photocatalytic performance of the samples was tested using a 300 W xenon lamp (PLS-SXE300D/300DUV, PerfectLight, China) to simulate sunlight. First, 10 mg of a sample was added to 100 mL of an MB solution and stirred for 30 min under dark conditions using a magnetic stirrer to achieve adsorption equilibrium between the functional particles and the organic pollutant solution. Second, the xenon lamp was positioned at a height of 15 cm above the MB solution. After turning on the xenon lamp, 3 mL of the solution was extracted and centrifuged every 10 min. Then, the absorbance of the solution was measured using a UV–visible spectrometer. The photodegradation rate of the organic pollutants was determined using the formula $Q = (C_0 - Ct)/C_0 \times 100\%$, where C_0 and Ct represent the absorbance values of the solution at time *t* before and after irradiation with a xenon lamp, respectively, and Q is the degradation rate. After undergoing each photocatalytic degradation process, the sample was thoroughly rinsed with ethanol and deionized water multiple times and then dried at 60 °C for 24 h. Afterwards, the cyclic stability of the sample was examined. The error bar was calculated from three independent experiments.

2.3.4. Analysis of Photocatalytic Mechanism

The photoluminescence (PL) spectrum of the functional fiber membrane was measured using an Edinburgh steady-state/transient fluorescence spectrometer (FLS1000, Edinburgh Instruments Ltd., UK) with an excitation wavelength of λ = 254 nm.

First, 5 mg of the photocatalyst-supported nanofiber membranes were crushed, ground, and dispersed in 1 mL of ethanol (containing 20 μL of Nafion) to prepare a slurry. The slurry was coated onto a fluorine-doped tin oxide transparent glass substrate and dried to create a working electrode. A carbon rod and a Ag/AgCl electrode were used as the counter electrode and the reference electrode, respectively. The photocatalyst's photocurrent response (I-t) was measured under a switchable illumination cycle of 0.4 V and 100 s using a 300 W xenon lamp as a simulated solar light source. The distance between the photocatalyst and the xenon lamp was maintained at around 16 cm (with the xenon lamp operating at 14 A with an approximate light output power density of 200 $mW \cdot cm^{-2}$). Electrochemical impedance spectroscopy and Mott–Schottky curves were measured under open-circuit conditions with a disturbance voltage of 5 mV in the frequency range of 0.01 Hz to 100 kHz.

Next, 1 mmol of Na_2-EDTA, IPA, and P-BQ were separately added to the MB solution as free radical scavengers. During the photocatalytic process, Na_2-EDTA, IPA, and P-BQ captured holes (h^+), hydroxyl radicals (·OH), and superoxide radicals ($\cdot O_2^-$), respectively. This condition allowed us to determine which type of free radical was the main factor affecting the photocatalytic degradation efficiency of MB.

3. Results and Discussion

3.1. Photocatalytic Nanoparticles

It is widely recognized that AgCl, ZnO, and their AgCl/ZnO complexes can be utilized as photocatalysts. The ZnO utilized in this article was commercially available, and AgCl and AgCl/ZnO were synthesized following the method described in Section 2.2.

The morphology and structure of these photocatalytic nanoparticles are shown in Figure 3. The prepared AgCl NPs exhibited a uniform size, an irregular cubic structure, and excellent dispersibility. The ZnO NPs had a size range of 20–30 nm and exhibited severe aggregation, which was due to the high surface activation energy or alterations in the surface charge state of the nanoparticles. The series of AgCl/ZnO NPs shared similar morphologies. As a representative of the series materials, an SEM image of AgCl/ZnO (M_{AgCl}:M_{ZnO} = 5:5) is shown in Figure 3c. The surface morphology of the composite nanoparticles changed relative to individual AgCl and ZnO, and it effectively improved the agglomeration of ZnO. Considerable intermixing of the two types of particles was also observed, and a heterostructure was formed. The TEM image in Figure 3d provides further evidence and clearly shows the interface formed by ZnO and AgCl nanoparticles with lattice spacings of 0.260 and 0.196 nm, respectively [27].

Figure 3. SEM images of (**a**) AgCl NPs, (**b**) ZnO NPs, and (**c**) AgCl/ZnO NPs (M_{AgCl}: M_{ZnO} = 5:5) and (**d**) TEM image of AgCl/ZnO NPs (M_{AgCl}: M_{ZnO} = 5:5).

Figure 4 presents the XRD patterns of the photocatalytic nanoparticles. The peaks labeled with red triangles "▼" and gray circles "●" match the standard data of ZnO (JCPDS card No. 36-1451) and AgCl (JCPDS card No. 31-1238), respectively [27]. The characteristic peaks of AgCl and ZnO can be identified in the series of AgCl/ZnO NPs, with no other crystallization peaks present, indicating the high purity of the AgCl/ZnO NPs. This is because the AgCl and AgCl/ZnO NPs in this work were prepared under dark conditions, so there was no reduction or a minimal reduction of AgCl. As shown in Supplementary Figure S1, the XRD spectrum of zinc oxide remained largely unchanged after the photocatalytic reaction. However, a crystalline peak with a position at 38.8, which corresponds to the standard data of silver (JCPDS card No. 04-0783), was identified within the AgCl and AgCl/ZnO NPs after the photocatalytic reaction. These data demonstrate the decomposition of silver chloride during the photocatalytic process.

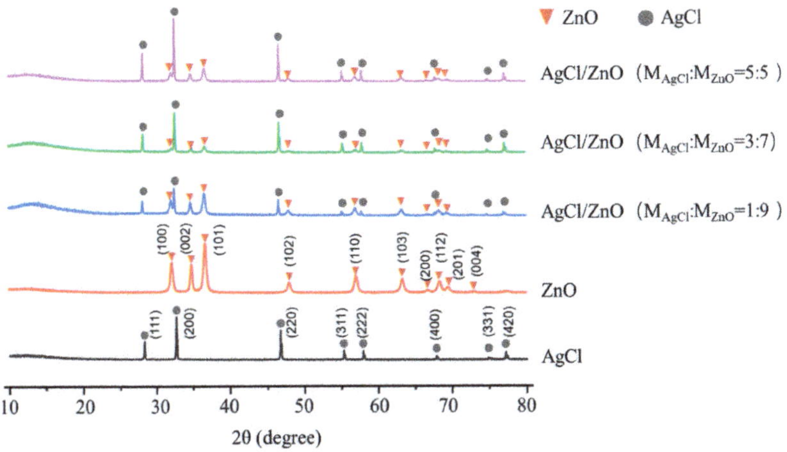

Figure 4. XRD patterns of ZnO NPs, AgCl NPs, and AgCl/ZnO NPs with different molar ratios.

Furthermore, the details of the material composition and element valance state of AgCl/ZnO NPs (MAgCl: MZnO = 5:5) were analyzed using energy dispersive X-ray photoelectron spectroscopy (XPS). As indicated in the wide scan survey spectra (Figure 5a), the characteristic peaks of Zn *2p*, O *1s*, Ag *3d*, C *1s*, and Cl *2p* are observed. In Figure 5b, the C 1 s spectrum consists of deconvoluted peaks at approximately 288.5 eV, 286.5 eV, and 284.8 eV, which are signatures of C = O, C-N, and C-C carbon species, respectively [28]. In the magnified XPS spectrum of Zn *2p* (Figure 5c), the two spin orbital peaks at 1045.3 eV and 1022.2 eV belong to Zn *2p* 1/2 and Zn *2p* 3/2, respectively, reflecting the stable Zn^{2+} state [29]. The O 1 s spectrum shown in Figure 5d reveals oxygen components situated at around 530 eV, and 532 eV, which are typically related to ZnO, and C = O bonds, respectively [29]. In the Ag 3d spectrum in Figure 5e, the clear peaks at 372.7 eV and 366.7 eV are associated with two satellite peaks corresponding to Ag *3d* 3/2 and Ag *3d* 5/2, which is ascribed to the Ag in AgCl [30]. Shown in Figure 5f are peaks at 198.7 eV and 197.1 eV, which are signatures of the Cl^- in AgCl, and peaks at 200.2 eV and 198.6 eV, which are signatures of the Cl^0 [15,30].

The operation of the photocatalytic experiment is detailed in Section 2.3.3. As shown in Figure 6a, the photocatalytic efficiency of AgCl/ZnO NPs was much higher than those of the individual AgCl NPs and ZnO NPs. As the content of AgCl increased, so too did the photocatalytic efficiency of the AgCl/ZnO NPs. The kinetic analysis of photocatalytic MB degradation is illustrated in Figure 6b. The fitting correlation coefficients are 0.96717, 0.99187, 0.96742, 0.98773, and 0.98041, respectively, indicating that the photocatalytic processes of MB degradation by the five different photocatalysts all follow the first-order kinetic model. The lower photocatalytic efficiency of AgCl and ZnO may be due to the easier recombination of electrons in the valence band (VB) with defects (holes) in the conduction band (CB). The formation of a heterojunction between AgCl and ZnO can effectively suppress the recombination of electrons and holes. O_2 can be reduced to $\cdot O_2^-$ by the electrons on the CB, whereas H_2O can be oxidized to ·OH by the holes in the VB. When the molar ratio of AgCl to ZnO is 1:1, the heterojunction between AgCl and ZnO becomes highly enriched, effectively suppressing electron–hole recombination and attaining the utmost photocatalytic efficiency.

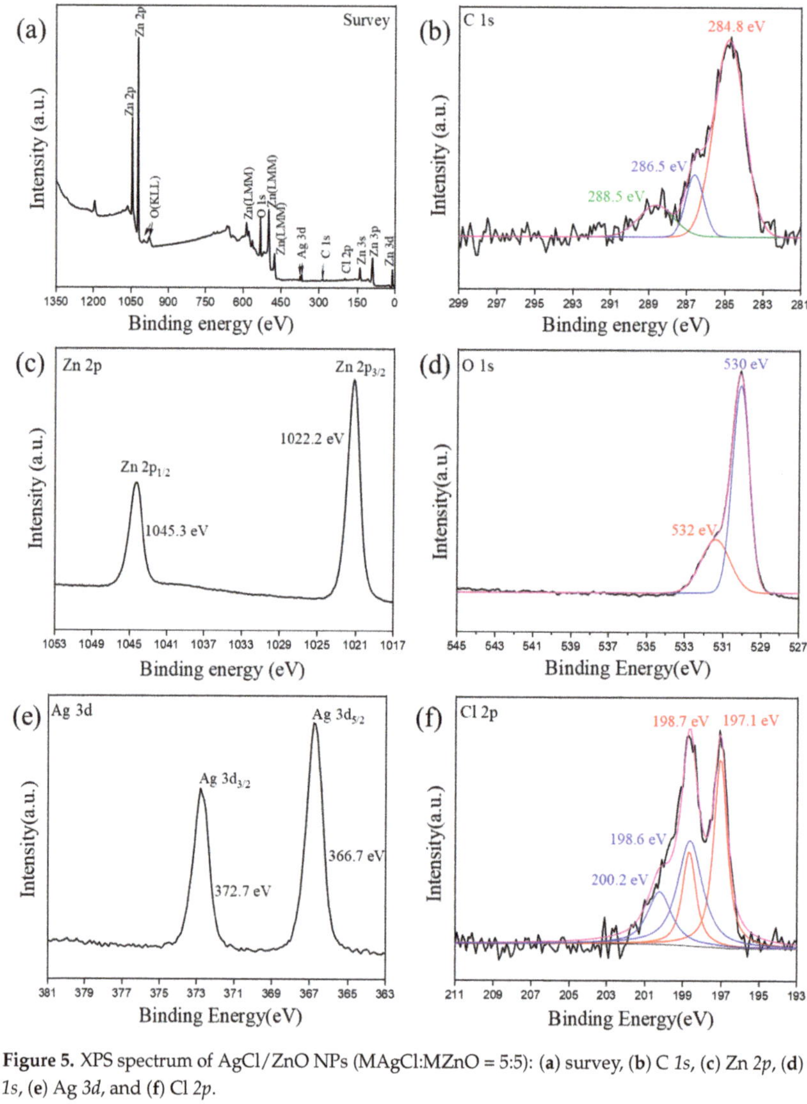

Figure 5. XPS spectrum of AgCl/ZnO NPs (MAgCl:MZnO = 5:5): (**a**) survey, (**b**) C *1s*, (**c**) Zn *2p*, (**d**) O *1s*, (**e**) Ag *3d*, and (**f**) Cl *2p*.

Figure 6c shows that after four photocatalytic experiments, the mass of the photocatalyst decreased from 10 mg to 7.2 mg. Figure 6d indicates that as the cyclic experiments progressed, the photocatalytic degradation efficiency of MB decreased. The degradation rates for the first, second, third, and fourth cycles were 99.70%, 93.76%, 88.56%, and 79.64%, respectively. This was mainly due to the loss of photocatalytic particles during the centrifugation and drying processes of the photocatalytic experiments.

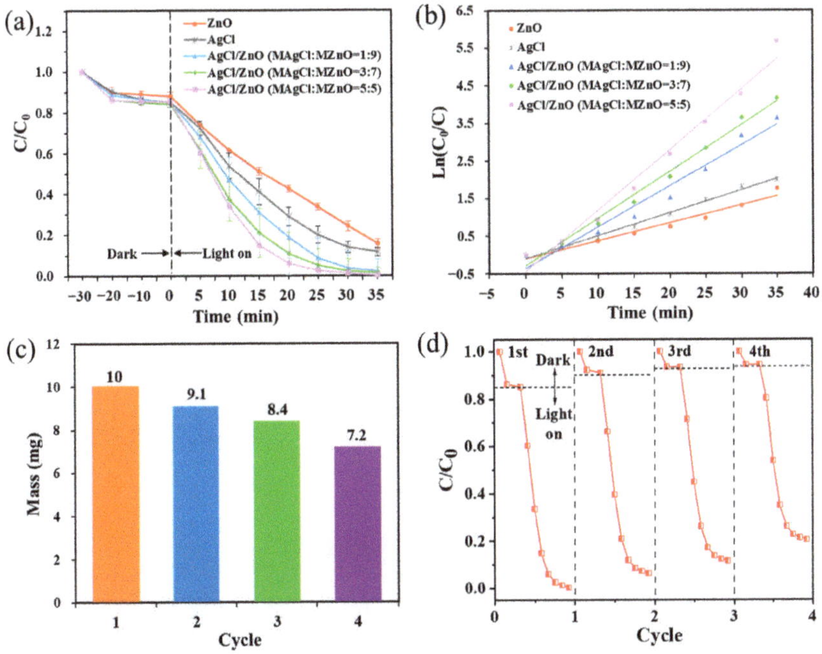

Figure 6. (**a**) Photodegradation activity and (**b**) corresponding first-order kinetic fitting curves of different photocatalytic NPs toward MB under simulated solar light irradiation. (**c**) Mass and (**d**) photocatalytic performance of AgCl/ZnO (M_{AgCl}:M_{ZnO} = 5:5) after each cycle.

3.2. Loading Photocatalytic NPs Onto Electrospun Nanofibers

In order to improve the recycling efficiency of the photocatalysts, the aforementioned photocatalytic particles were loaded onto nanofiber membranes via multifluid electrospinning. With PCL as the core layer in the coaxial electrospinning process, fibers with improved mechanical properties were obtained to compensate for the weakened mechanical performance of the porous fibers.

As shown in Figure 7a–e, SEM observations revealed that there were numerous irregular pores and a minor quantity of photocatalytic particles on the surfaces of electrospun nanofibers F1–F5. The presence of these irregular pores can be attributed to the rapid evaporation of HFIP in the sheath solution during electrospinning, resulting in a decrease in the surface temperature of the fibers. Consequently, the surrounding water molecules undergo condensation and form water droplets attached to the surfaces of nanofibers. Additionally, the core solvent, CF, is volatile and diffuses to the surfaces of nanofibers as HFIP evaporates. CF is sparingly soluble in water, resulting in the formation of irregular structures when CF comes into contact with water. Once water droplets and CF completely evaporate, irregular pores are left behind on the surfaces of nanofibers. The successful loading of the photocatalytic nanoparticles and the core–sheath structure of the fibers could be further verified through TEM analysis, as shown in Figure 7f.

The nitrogen adsorption–desorption isotherm of F5, which corresponds to a type II isotherm [31] that is suitable for mesoporous materials, is shown in Figure 7g. The specific surface area of F5 was calculated to be 293.54 m^2/g. The pore size distribution curve shown in Figure 7h revealed that the pores on F5 mainly consisted of micropores, mesopores, and a small number of macropores, and the average pore size was 6.22 nm.

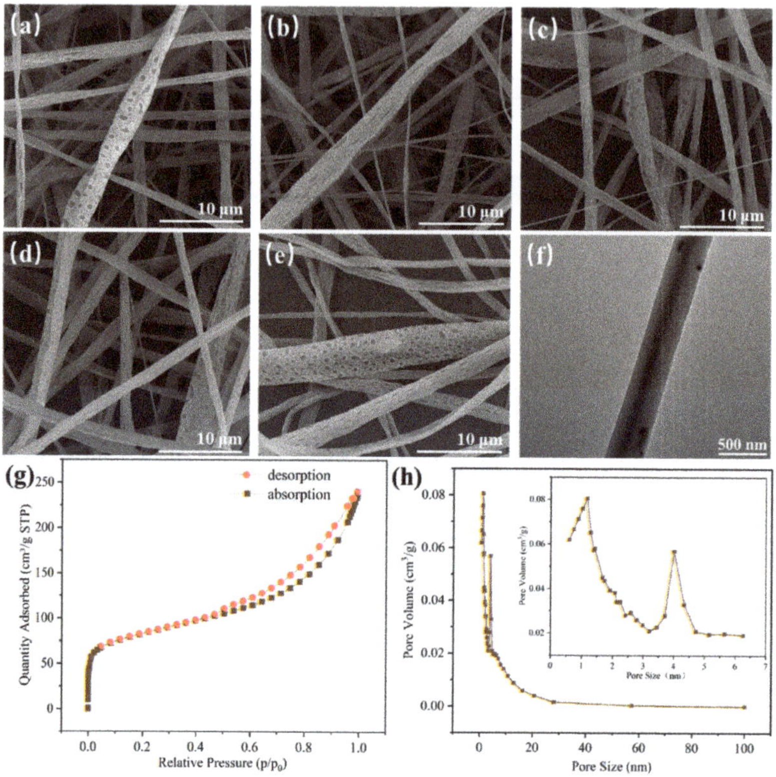

Figure 7. (**a**–**e**) SEM images of F1–F5, (**f**) TEM image, (**g**) nitrogen adsorption–desorption isotherm, and (**h**) pore size distribution curve of F5.

Figure 8a shows the XRD patterns of PCL, AgCl NPs, ZnO NPs, and coaxial electrospun nanofiber membranes F1–F5. PCL had two diffraction peaks at around 21.7° and 24.0°, indicating its semicrystalline nature. PCL is composed of two structural forms, namely, ordered crystalline and disordered amorphous regions. The polymerization of PCL involves the ring-opening polymerization of ε-caprolactone, resulting in crystalline regions in the middle of the polymer chains and amorphous regions surrounding them. Therefore, PCL is considered a semicrystalline polymer. The peaks at $2\theta = 31.8°, 34.4°, 36.3°, 47.5°$, and 56.6° observed in the XRD patterns of the F1 and F3–F5 nanofiber membranes correspond to the (100), (002), (101), (102), and (110) crystal planes of ZnO, respectively, indicating the successful loading of ZnO NPs onto the surfaces of the nanofibers. The peaks at $2\theta = 27.8°, 32.2°, 46.2°, 54.8°$, and 57.5° observed in the XRD patterns of the F2 and F3–F5 nanofiber membranes correspond to the (111), (200), (220), (311), and (222) crystal planes of AgCl, respectively, indicating the successful loading of AgCl NPs onto the surfaces of the nanofibers. With increasing AgCl contents, the diffraction peaks of AgCl in F3–F5 gradually increased, whereas the diffraction peaks of ZnO gradually decreased.

PL is used to study the transfer and separation of photo-induced charge carriers in electrospun nanofiber membranes [32]. Typically, the lower the peak intensity of PL, the lower the recombination rate of photo-induced charge carriers and the higher the photocatalytic activity. As shown in Figure 8b, F1 and F2 exhibited higher peak PL intensities than F3–F5, and the peak PL intensities of F3–F5 decreased because the number of AgCl/ZnO heterojunctions increased. This result was consistent with the photocatalytic MB degradation efficiency of unloaded photocatalytic particles.

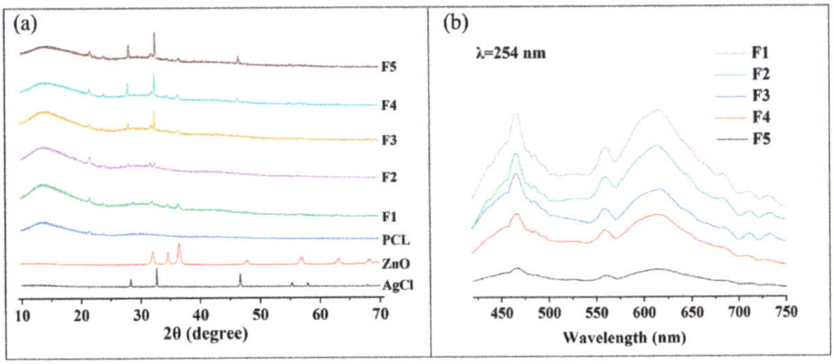

Figure 8. (**a**) XRD patterns of PCL, AgCl NPs, ZnO NPs, and nanofiber membranes F1–F5 and (**b**) photoluminescence spectra of nanofiber membranes F1–F5.

Analyses and characterizations of the functional groups in the polymer substrates and coaxial electrospun nanofiber membranes were performed, as shown in Figure 9a. In the infrared spectrum of CA, the peak at 1752 cm^{-1} corresponds to the C=O stretching vibration of the carbonyl group, whereas the peaks at 1241 cm^{-1} and 1430 cm^{-1} correspond to C-O tensile vibration and C-H tensile vibration, respectively [33]. In the infrared spectrum of PCL, the peaks of C-H, C=O, and C=O stretching vibrations appear at 2954 cm^{-1}, 1733 cm^{-1}, and 1173 cm^{-1}, respectively [34]. After loading the photocatalytic particles onto the nanofibers, a noticeable shift in the C=O vibration peak toward lower wave numbers was observed, which suggested the formation of hydrogen bonds between CA and the photocatalytic particles.

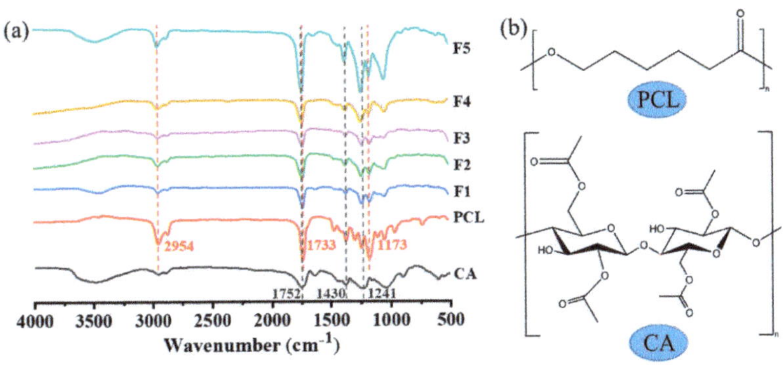

Figure 9. (**a**) FTIR spectra of the raw materials PCL and CA, and nanofiber membranes F1–F5. (**b**) Molecular structures of PCL and CA.

3.3. Photocatalytic Performance of the Nanofiber Membranes

Photocatalyst-loaded porous nanofiber membranes F1–F5 were added to a 100 mL 10 mg/L MB solution. As shown in Figure 10a, after 30 min of dark treatment, the adsorption and desorption of MB on the porous nanofiber membranes reached equilibrium, and the absorption of MB was greater than when using photocatalytic nanoparticles. Subsequently, after simulated solar light irradiation for 70 min, the adsorption degradation rates of MB on F1–F5 were 85.67%, 90.07%, 95.18%, 97.50%, and 98.47%, respectively. The correlation coefficients of the kinetic fitting were 0.96428, 0.98785, 0.98032, 0.97579, and 0.97996, as shown in Figure 10b, indicating compliance with the first-order kinetic model. In addition, a five-cycle repeated experiment was conducted on F5. The cyclic degradation efficiency and the remaining mass of the membrane were recorded after each cycle. As presented on the left y-axis in Figure 10c, even after five cycles, F5 maintained a photocatalytic efficiency of over 95% with only minor variations. This result suggests that the photocatalyst was firmly fixed on the porous nanofiber membranes and formed a stable structure that maintained good dispersion and catalytic activity even after multiple cycles. The nanofiber membrane possessed a porous structure and a large surface area, allowing the photocatalyst to be exposed on the surface and increasing the area of contact with MB. Additionally, the pore structure could scatter light effectively within the fiber membrane, thereby enhancing the photocatalytic reaction. As shown in Figure S2, the structure of the nanofiber membrane remained after the photocatalytic reaction.

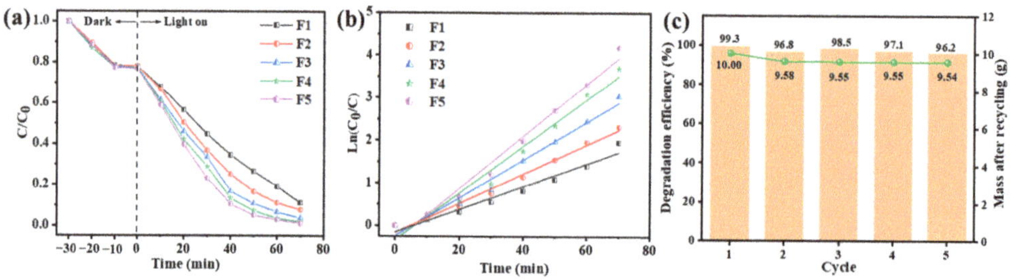

Figure 10. (**a**) Photodegradation activity, (**b**) corresponding first-order kinetic fitting curves of MB under simulated solar light irradiation of F1–F5, and (**c**) cyclic experiment on F5.

3.4. Photocatalytic Mechanism

To gain a deeper insight into the photocatalytic mechanism and band structure of AgCl/ZnO heterojunctions, the photocatalytic nanoparticles were first characterized using UV–Vis absorption spectra, as shown in Figure 11a. The absorption spectrum of pure ZnO exhibited strong absorption below 400 nm, indicating the efficient absorption of UV light due to the charge transfer from the VB (O 2p orbitals) to the CB (Zn 4s orbitals). In comparison with ZnO, the absorption spectra of AgCl NPs and AgCl/ZnO NPs showed strong absorption below 400 nm and between 400 and 800 nm. The bandgap widths of ZnO and AgCl could be calculated using the Kubelka–Munk equation $(\alpha h\nu) = A(h\nu - E_g)^{n/2}$, where the integer n values of the direct semiconductor (ZnO) and the indirect semiconductor (AgCl) were 1 and 4, respectively [35]. As shown in Figure 11b, the bandgap (E_g) values of AgCl and ZnO were 3.25 and 3.38 eV, respectively. The Mott–Schottky [36] curves of AgCl and ZnO, shown in Figure 11c,d, suggest that AgCl and ZnO are n-type semiconductors. The flat-band potentials (E_{FB}) of AgCl and ZnO relative to Ag/AgCl at pH 7 were −0.03 and −0.40 V, respectively. After conversion to the normal hydrogen electrode (NHE), the E_{CB} values were calculated to be −0.033 and −0.403 V ($E_{CB}(NHE) = E_{CB}(Ag/AgCl) + 0.197$). Therefore, the corresponding valence ($E_{VB} = E_g + E_{CB}$) values of AgCl and ZnO were 3.217 and 2.977 eV, respectively.

The I-t testing [37] of F1–F5 was performed in a 0.2 M Na_2SO_4 electrolyte solution to further demonstrate the effective separation of photogenerated charge carriers in the nanofiber membranes. As shown in Figure 12a, when the xenon lamp was turned on to illuminate the sample, the electrons on the sample surface were excited. The concentration of charge carriers on the sample surface increased, resulting in a linear increase in the current density, and a stable state was reached at a certain amplitude. When the xenon lamp was turned off, the number of excited electrons on the sample surface decreased, and the concentration of charge carriers decreased accordingly, leading to a decrease in the current density and a continued stable state. With 100 s as one switching cycle, the peak intensity of the current density when the xenon lamp was turned on and off remained relatively constant in each cycle. The photocurrent densities of F3–F5 were much higher than those of F1 and F2, revealing the strong light absorption capability of the AgCl/ZnO nanoparticles, which could promote the separation of photogenerated charge carriers. F5 exhibited the highest photoresponse activity, which was consistent with the PL results.

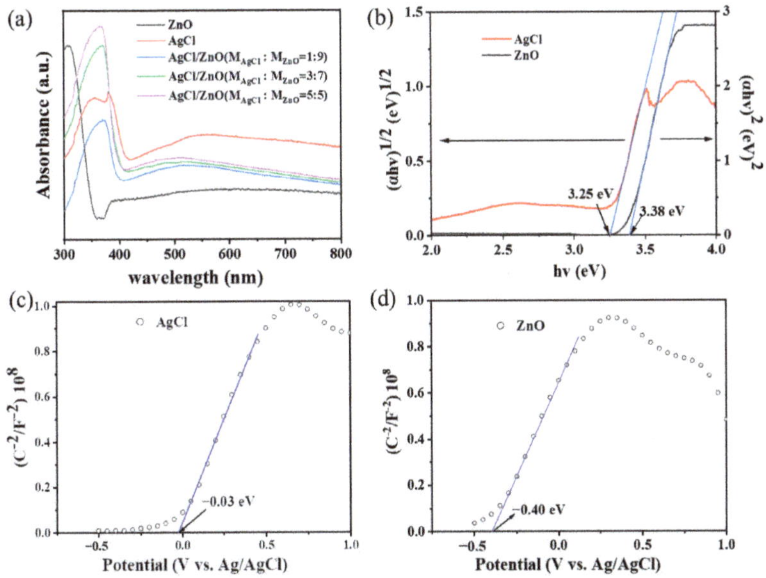

Figure 11. (**a**) UV–visible absorption spectra of AgCl NPs, ZnO NPs, and AgCl/ZnO NPs. (**b**) Tauc curves of ZnO NPs and AgCl NPs. Mott–Schottky curves of (**c**) ZnO NPs and (**d**) AgCl NPs.

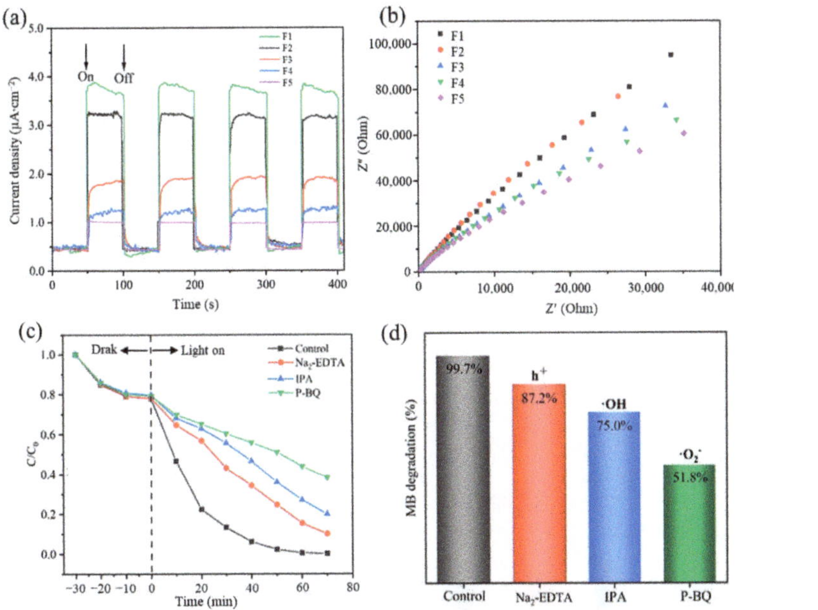

Figure 12. (**a**) Transient photocurrent responses of F1–F5 under simulated solar light irradiation. (**b**) Electrochemical impedance spectra of F1–F5. (**c**) Free radical trapping experiment on F5 under simulated solar light irradiation. (**d**) Degradation efficiency of MB after the addition of free radicals.

The electrochemical impedance spectra were measured in a 0.2 M Na_2SO_4 electrolyte solution to further demonstrate the effective separation of photogenerated charge carriers in F1–F5 [38]. The size of the electrochemical reaction region on an electrode surface, which is often referred to as the electrode polarization layer radius or effective diffusion layer radius, reflects the size of the electrochemical reaction region. Generally, a small electrochemical reaction region on an electrode surface indicates a fast diffusion of reactants, which helps improve the efficiency of electron–hole separation. Figure 12b shows that the radii of F3–F5 were smaller than those of F1 and F2, and as the molar ratio of AgCl to ZnO increased, the radii gradually decreased. F5 exhibited high efficiency in the separation of electrons and holes, indicating that the formed heterojunctions suppressed the recombination of photogenerated carriers to a great extent. This result further confirmed the presence of a larger number of electrons and holes participating in the photocatalytic reaction in F5.

To investigate the influence of free radicals on MB degradation during the photocatalytic process, this study introduced 0.1 mmol Na_2-EDTA, P-BQ, and IPA as free radical scavengers into the F5 system for MB degradation; these scavengers can capture h^+, $\cdot O_2^-$, and ·OH free radicals, respectively. As shown in Figure 12c,d, the addition of P-BQ considerably reduced the degradation efficiency of MB. This finding indicates that $\cdot O_2^-$ plays a dominant role as the key free radical in the degradation process of MB. P-BQ reacts with and neutralizes $\cdot O_2^-$ free radicals, thereby reducing the amount of $\cdot O_2^-$ in the photocatalytic reaction and consequently lowering the rate and efficiency of the photocatalytic reaction [39]. Moreover, P-BQ can be competitively adsorbed onto the photocatalysts, obstructing the adsorption of organic molecules and impeding the progress of the photocatalytic reaction, which also leads to a decrease in photocatalytic efficiency. The addition of Na_2-EDTA and IPA also resulted in a certain degree of photocatalytic efficiency reduction, indicating the participation of h^+ and ·OH. Therefore, h^+, ·OH, and $\cdot O_2^-$ all play significant roles in the photocatalytic decomposition process.

The photocatalyst was in an excited state under simulated solar light irradiation. The electrons (e^-) in AgCl and ZnO underwent a transition from the VB to the CB, leaving corresponding holes in the VB. At this point, the CB potential of AgCl was more positive than that of ZnO, so electrons from the CB of ZnO transitioned to the CB of AgCl and gradually accumulated. Based on the above analysis, a photocatalytic mechanism was proposed in Figure 13. Firstly, under UV light irradiation, photogenerated electron–hole pairs (e^-–h^+) were formed in ZnO NPs. AgCl decomposes under light to produce Ag, which absorbed UV light, and generated Ag^+ and e^-. Secondly, some e^- from the CB reacted with Ag^+ to form Ag and the other e^- were expected to be trapped by O_2 in solution to generate $\cdot O_2^-$ and other oxygen species like HOO· and H_2O_2. The generated radicals can react with dyes [40]. Thirdly, the h^+ could react with AgCl to form Ag^+ and Cl^0, and a part of h^+ generated by ZnO reacted with H_2O to form ·OH. The Cl^0 and ·OH could also react with the dyes. Through the synergistic action of $\cdot O_2^-$, ·OH, h^+ and Cl^0, the MB were degraded into CO_2 and H_2O.

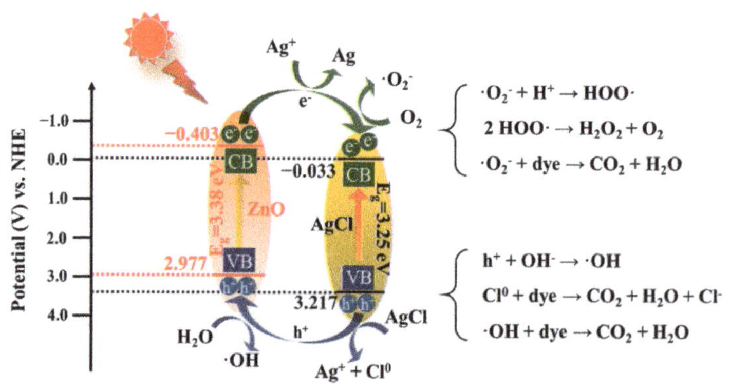

Figure 13. Possible photocatalytic mechanism of the degradation.

4. Conclusions

Water pollution caused by organic dyes necessitates more efficient and sustainable treatment strategies. This study demonstrated a one-step method for fabricating porous nanofiber membranes loaded with functional particles that had large surface areas and high photocatalytic efficiency for MB. With a stable core layer in the coaxial electrospinning process, nanofibers with improved mechanical properties were obtained to compensate for the weakened mechanical performance of the porous fibers. A porous nanofiber membrane loaded with AgCl/ZnO (M_{AgCl}:M_{ZnO} = 5:5) showed the highest photodegradation efficiency for MB, and even after five cycles of repeated experiments, the mass and photocatalytic efficiency of the nanofiber membrane remained above 95%, which successfully addressed the issues of photocatalytic efficiency loss during the photodegradation with AgCl/ZnO NPs. However, the porous structure only existed on the surfaces of the nanofibers, leaving room for further improvement in the structural design. Additionally, when loading functional particles onto the surfaces of porous nanofibers, attention should be paid to potential issues, such as particle detachment, which could affect photocatalytic efficiency. Although this study focused on the photocatalytic degradation of organic dyes, the porous nanofiber membrane could also be applied to address environmental issues such as heavy metal ion adsorption.

Supplementary Materials: The following supporting information can be downloaded at: https://www.mdpi.com/article/10.3390/polym16060754/s1, Figure S1: XRD patterns of ZnO NPs, AgCl NPs, and AgCl/ZnO NPs with different molar ratios after photocatalytic reaction. Figure S2: SEM images of F1, F2, F3, F4, and F5 before (a, b, c, d, e) and after (a1, b1, c1, d1, e1) photocatalytic reaction. Scale bar: 500 nm.

Author Contributions: Conceptualization, Y.Y., S.Z. and X.C.; funding acquisition, Y.Y.; project administration, Y.Y.; writing—original draft, S.Z., X.C., H.L., Z.L., R.Z. and F.Y.; writing—review and editing, Y.Y. and D.Y. All authors have read and agreed to the published version of the manuscript.

Funding: This research was funded by the National Natural Science Foundation of China (Grant No. 51803121).

Data Availability Statement: The raw data supporting the conclusions of this article will be made available by the authors on request.

Conflicts of Interest: The authors declare no conflicts of interest.

References

1. Bobde, P.; Patel, R.K.; Panchal, D.; Sharma, A.; Sharma, A.K.; Dhodapkar, R.S.; Pal, S. Utilization of layered double hydroxides (LDHs) and their derivatives as photocatalysts for degradation of organic pollutants. *Environ. Sci. Pollut. Res.* **2021**, *28*, 59551–59569. [CrossRef]
2. Madhavan, J.; Theerthagiri, J.; Balaji, D.; Sunitha, S.; Choi, M.Y.; Ashokkumar, M. Hybrid advanced oxidation processes involving ultrasound: An overview. *Molecules* **2019**, *24*, 3341. [CrossRef]
3. Liu, T.; Yang, G.; Wang, W.; Wang, C.; Wang, M.; Sun, X.; Xu, P.; Zhang, J. Preparation of C_3N_5 nanosheets with enhanced performance in photocatalytic methylene blue (MB) degradation and H_2-evolution from water splitting. *Environ. Res.* **2020**, *188*, 109741. [CrossRef] [PubMed]
4. Zhu, S.; Wang, D. Photocatalysis: Basic principles, diverse forms of implementations and emerging scientific opportunities. *Adv. Energy Mater.* **2017**, *7*, 1700841. [CrossRef]
5. Li, Y.; Cao, T.; Mei, Z.; Li, X.; Sun, D. Separating type I heterojunction of $NaBi(MoO_4)_2/Bi_2MoO_6$ by TiO_2 nanofibers for enhanced visible-photocatalysis. *Chem. Phys.* **2020**, *533*, 110696. [CrossRef]
6. Liu, Y.; Jiang, Z.; Jia, J.; Robertson, J.; Guo, Y. 2D $WSe_2/MoSi_2N_4$ type-II heterojunction with improved carrier separation and recombination for photocatalytic water splitting. *Appl. Surf. Sci.* **2023**, *611*, 155674. [CrossRef]
7. Li, J.; Feng, J.; Guo, X.; Fang, H.; Chen, J.; Ma, C.; Li, R.; Wang, Y.; Rui, Z. Defect-band bridge photothermally activates type III heterojunction for CO_2 reduction and typical vocs oxidation. *Appl. Catal. B* **2022**, *309*, 121248. [CrossRef]
8. Yang, Q.; Tan, G.; Zhang, B.; Feng, S.; Bi, Y.; Wang, Z.; Xia, A.; Ren, H.; Liu, W. $Cs_{0.33}WO_3$/(t-m)-$BiVO_4$ double Z-type heterojunction photothermal synergistic enhanced full-spectrum degradation of antibiotics. *Chem. Eng. J.* **2023**, *458*, 141378. [CrossRef]
9. Li, D.; Xu, K.; Zhang, C. Improvement of photocatalytic performance by building multiple heterojunction structures of anatase-rutile/BiOI composite fibers. *Nanomaterials* **2022**, *12*, 3906. [CrossRef]

10. Yang, P.; Wu, X.; Zhang, J.; Ma, Q.; Cheng, J.; Wang, H.; He, C.; Li, Z. Synthesis of C^{3+} alcohols through ethanol condensation and carbon-carbon coupling of ethanol with CO_2. *Appl. Catal. A* **2023**, *653*, 119036. [CrossRef]
11. McCormick, W.J.; McCrudden, D.; Skillen, N.; Robertson, P.K.J. Electrochemical monitoring of the photocatalytic degradation of the insecticide emamectin benzoate using TiO_2 and ZnO materials. *Appl. Catal. A* **2023**, *660*, 119201. [CrossRef]
12. Huang, J.; Liu, S.; Long, W.; Wang, Q.; Yu, X.; Li, S. Highly enhanced photodegradation of emerging pollutants by Ag/AgCl/Ta_2O_{5-x} mesocrystals. *Sep. Purif. Technol.* **2021**, *279*, 119733. [CrossRef]
13. Mohsen, P.; Shahnaz, G.; Sima, H.; Baker, R.; Zahra, G.N.S.K.; Zahra, S.; Wang, C. Multifunctional Ag/AgCl/$ZnTiO_3$ structures as highly efficient photocatalysts for the removal of nitrophenols, CO_2 photoreduction, biomedical waste treatment, and bacteria inactivation. *Appl. Catal. A* **2022**, *643*, 118794. [CrossRef]
14. Zhao, M.; Zhou, W.; Lu, M.; Guo, Z.; Li, C.; Wang, W. Novel AgCl nanotubes/BiOCl nanosheets composite with improved adsorption capacity and photocatalytic performance. *J. Alloys Compd.* **2019**, *773*, 1146–1153. [CrossRef]
15. Cai, A.; Guo, A.; Du, L.; Qi, Y.; Wang, X. Leaf-templated synthesis of hierarchical AgCl-Ag-ZnO composites with enhanced visible-light photocatalytic activity. *Mater. Res. Bull.* **2018**, *103*, 225–233. [CrossRef]
16. Lv, H.; Zhang, M.; Wang, P.; Xu, X.; Liu, Y.; Yu, D. Ingenious construction of $Ni(DMG)_2/TiO_2$-decorated porous nanofibers for the highly efficient photodegradation of pollutants in water. *Colloids Surf. A Physicochem. Eng. Asp.* **2022**, *650*, 129561. [CrossRef]
17. Qi, X.; Zhu, Y.; Song, L.; Peng, G.; Qu, W.; Xiong, J. Photocatalytic degradation of PET coupled to green hydrogen generation using flexible $Ni_2P/TiO_2/C$ nanofiber film catalysts. *Appl. Catal. A* **2023**, *656*, 119130. [CrossRef]
18. Yu, D.; Li, J.; Zhang, M.; Williams, G.R. High-quality Janus nanofibers prepared using three-fluid electrospinning. *Chem. Comm.* **2017**, *53*, 4542–4545. [CrossRef] [PubMed]
19. Lee, H.; Kharaghani, D.; Nishino, M.; Song, K.H.; Lee, J.S.; Kim, I.S. Electrospun tri-layered zein/PVP-GO/zein nanofiber mats for providing biphasic drug release profiles. *Int. J. Pharm.* **2017**, *531*, 101–107. [CrossRef] [PubMed]
20. Asare, K.; Hasan, M.F.; Shahbazi, A.; Zhang, L. A comparative study of porous and hollow carbon nanofibrous structures from electrospinning for supercapacitor electrode material development. *Surf. Interfaces* **2021**, *26*, 101386. [CrossRef]
21. Sun, R.; Yan, G.; Zhang, X.; Li, Z.; Chen, J.; Wang, L.; Wu, Y.; Wang, Y.; Li, H. Fe-ZIF-derived hollow porous carbon nanofibers for electromagnetic wave absorption. *Chem. Eng. J.* **2023**, *455*, 140608. [CrossRef]
22. Kossyvaki, D.; Barbetta, A.; Contardi, M.; Bustreo, M.; Dziza, K.; Lauciello, S.; Athanassiou, A.; Fragouli, D. Highly porous curcumin-loaded polymer mats for rapid detection of volatile amines. *ACS Appl. Polym. Mater.* **2022**, *4*, 4464–4475. [CrossRef]
23. El-Samak, A.A.; Ponnamma, D.; Hassan, M.K.; Al-Maadeed, M.A. A stable porous vessel for photocatalytic degradation of Azocarmine G dye. *Microporous Mesoporous Mater.* **2022**, *341*, 111994. [CrossRef]
24. Tu, H.; Li, D.; Yi, Y.; Liu, R.; Wu, Y.; Dong, X.; Shi, X.; Deng, H. Incorporation of rectorite into porous polycaprolactone/TiO_2 nanofibrous mats for enhancing photocatalysis properties towards organic dye pollution. *Compos. Commun.* **2019**, *15*, 58–63. [CrossRef]
25. Rong, F.; Lu, Q.; Mai, H.; Chen, D.; Caruso, R.A. Hierarchically porous $WO_3/CdWO_4$ fiber-in-tube nanostructures featuring readily accessible active sites and enhanced photocatalytic effectiveness for antibiotic degradation in water. *ACS Appl. Mater. Interfaces* **2021**, *13*, 21138–21148. [CrossRef] [PubMed]
26. Cao, X.; Chen, W.; Zhao, P.; Yang, Y.; Yu, D. Electrospun porous nanofibers: Pore-forming mechanisms and applications for photocatalytic degradation of organic pollutants in wastewater. *Polymers* **2022**, *14*, 3990. [CrossRef]
27. Tian, X.; Wu, H.; Hu, X.; Wang, Z.; Ren, C.; Cheng, Z.; Dou, L.; Lin, Y. Enhanced photocatalytic performance of ZnO/AgCl composites prepared by high-energy mechanical ball milling. *New J. Chem.* **2022**, *46*, 9155–9171. [CrossRef]
28. Hou, W.; Guo, H.; Wu, M.; Wang, L. Amide covalent bonding engineering in heterojunction for efficient solar-driven CO_2 reduction. *ACS Nano* **2023**, *17*, 20560–20569. [CrossRef]
29. Poudel, M.B.; Kim, H.J. Confinement of Zn-Mg-Al-layered double hydroxide and α-Fe_2O_3 nanorods on hollow porous carbon nanofibers: A free-standing electrode for solid-state symmetric supercapacitors. *Chem. Eng. J.* **2022**, *429*, 132345. [CrossRef]
30. Feng, Z.; Yu, J.; Sun, D.; Wang, T. Visible-light-driven photocatalysts Ag/AgCl dispersed on mesoporous Al_2O_3 with enhanced photocatalytic performance. *J. Colloid Interface Sci.* **2016**, *480*, 184–190. [CrossRef]
31. Jamnongkan, T.; Intaramongkol, N.; Kanjanaphong, N.; Ponjaroen, K.; Sriwiset, W.; Mongkholrattanasit, R.; Wongwachirakorn, P.; Lin, K.Y.A.; Huang, C.F. Study of the enhancements of porous structures of activated carbons produced from durian husk wastes. *Sustainability* **2022**, *14*, 5896. [CrossRef]
32. Dong, T.; Wang, P.; Yang, P. Synthesis of magnetic $Ag_3PO_4/Ag/NiFe_2O_4$ composites towards super photocatalysis and magnetic separation. *Int. J. Hydrogen Energy* **2018**, *43*, 20607–20615. [CrossRef]
33. Brites, M.D.; Ceron, A.A.; Costa, S.M.; Oliveira, R.C.; Ferraz, H.G.; Catalani, L.H.; Costa, S.A. Bromelain immobilization in cellulose triacetate nanofiber membranes from sugarcane bagasse by electrospinning technique. *Enzyme Microb. Technol.* **2020**, *132*, 109384. [CrossRef]
34. Erdal, N.B.; Hakkarainen, M. Construction of bioactive and reinforced bioresorbable nanocomposites by reduced nano-graphene oxide carbon dots. *Biomacromolecules* **2018**, *19*, 1074–1081. [CrossRef]
35. Zhang, J.; Ding, E.; Xu, S.; Li, Z.; Fakhri, A.; Gupta, V.K. Production of metal oxides nanoparticles based on poly-alanine/chitosan/reduced graphene oxide for photocatalysis degradation, anti-pathogenic bacterial and antioxidant studies. *Int. J. Biol. Macromol.* **2020**, *164*, 1584–1591. [CrossRef]

36. Mou, H.; Song, C.; Zhou, Y.; Zhang, B.; Wang, D. Design and synthesis of porous Ag/ZnO nanosheets assemblies as super photocatalysts for enhanced visible-light degradation of 4-nitrophenol and hydrogen evolution. *Appl. Catal. B* **2018**, *221*, 565–573. [CrossRef]
37. Shi, W.; Ren, H.; Huang, X.; Li, M.; Tang, Y.; Guo, F. Low cost red mud modified graphitic carbon nitride for the removal of organic pollutants in wastewater by the synergistic effect of adsorption and photocatalysis. *Sep. Purif. Technol.* **2020**, *237*, 116477. [CrossRef]
38. Sang, Y.; Cao, X.; Dai, G.; Wang, L.; Peng, Y.; Geng, B. Facile one-pot synthesis of novel hierarchical Bi_2O_3/Bi_2S_3 nanoflower photocatalyst with intrinsic p-n junction for efficient photocatalytic removals of RhB and Cr (VI). *J. Hazard. Mater.* **2020**, *381*, 120942. [CrossRef] [PubMed]
39. Xiao, L.; Li, G.; Chu, X.; Yan, G. Visible-light-driven photocatalytic removal of Cr (VI) and rhodamine B by monoclinic $BiVO_4$-diatomite composite. *Mater. Res. Express.* **2019**, *6*, 115904. [CrossRef]
40. Xu, Y.; Xu, H.; Li, H.; Xia, J.; Liu, C.; Liu, L. Enhanced photocatalytic activity of new photocatalyst Ag/AgCl/ZnO. *J. Alloys Compd.* **2011**, *509*, 3286–3292. [CrossRef]

polymers

Article

Preparation of Cationic Polyacrylamide Suspension and Its Application in Oilfield Wastewater Treatment

Zhongjin Wei, Wenjun Long, Shaohua Li, Yu Zhao, Siting Yu and Fengshan Zhou *

Engineering Research Center of Ministry of Education for Geological Carbon Storage and Low Carbon Utilization of Resources, Beijing Key Laboratory of Materials Utilization of Nonmetallic Minerals and Solid Wastes, National Laboratory of Mineral Materials, School of Material Sciences and Technology, China University of Geosciences, Beijing 100083, China; 3003190032@cugb.edu.cn (Z.W.); longwj@cugb.edu.cn (W.L.); 2103210080@email.cugb.edu.cn (S.L.); 2103210079@email.cugb.edu.cn (Y.Z.); 2103210078@email.cugb.edu.cn (S.Y.)

* Correspondence: zhoufs@cugb.edu.cn

Abstract: Cationic polyacrylamide (CPAM) solid particle is one of the most commonly used organic polymer flocculants in oilfield wastewater treatment, but it poses some problems, such as a slow dissolution rate and an easy formation into a "fish-eye" in the process of diluting into aqueous solution. However, the current liquid CPAM products also have some problems, such as low effective content, poor storage stability, degradation in a short time, and high preparation costs. In this paper, a CPAM suspension was successfully prepared with 50.00% CPAM fine powder, 46.87% oil phase solvent, 0.63% separating agent, 1.56% emulsifying and dispersing agent, and 0.94% rheology modifier. This suspension has an effective content of 50.00%. It also showed no separation in 7 days of storage at room temperature, no separation in 30 min of centrifugation at a speed of 2000 rpm, and diluted to a 0.40% solution in just 16.00 min. For 1000 NTU of diatomite-simulated wastewater, the optimal turbidity removal rate of the suspension was 99.50%, which was higher than the optimal turbidity removal rate of 98.40% for the inorganic flocculant polymeric aluminum chloride (PAC). For oilfield wastewater, the optimal turbidity removal rate of the CPAM suspension was 35.60%, which was higher than the optimal turbidity removal rate of 28.40% for solid particle CPAM. In a scale-up test, the CPAM suspension achieved a good application effect.

Keywords: cationic polyacrylamide; flocculant; suspension; oilfield wastewater

1. Introduction

Oilfield wastewater mainly comes from oilfield extraction water, drilling fluid waste fluid, refinery wastewater, etc. After mixing, its composition is very complex, and its main characteristic is that it contains a large number of oil droplets of different sizes and suspended matter [1–4]. The treatment methods of oilfield wastewater mainly include the physical separation method, biological treatment method, and chemical flocculation method [5]. The physical separation method is mainly used to remove oil and suspended solids from oily wastewater, etc. The main treatment processes are the gravity separation process, membrane separation process, air flotation process, and so on. The physical separation method has the advantages of simple operation and low cost, but there are obvious shortcomings including that the separation is not thorough enough [6]. The biological treatment method employs microorganisms in some wastewater organic matter decomposition into some relatively simple substances, mainly used in the treatment of oilfield wastewater with high water quality requirements; the disadvantage of the biological method is that it needs to be based on the characteristics of the sewage for the cultivation of special biological strains of bacteria, and sewage water quality changes need to be cultivated to produce different strains of biological bacteria [7,8]. The chemical flocculation method can make the oil droplets and suspended matter in wastewater quickly settle and flocculate, which is one of the main treatment methods for oilfield wastewater [9,10].

The organic polymer flocculant cationic polyacrylamide (CPAM) has the characteristics of a low additive amount, good turbidity removal and water purification effect, and high COD removal efficiency, and it has become the most commonly used polymer flocculant in the oilfield wastewater treatment process [11–15].

Polyacrylamide solid particles need to be prepared into a diluent of a certain concentration for use, which will undergo three stages of wetting, swelling, and dissolution during the dilution process [16]: wetting is the gradual solventization (hydration) of the surface of the polymer particles; swelling is the hydration of the polymer particles' surface by the osmotic pressure of the solvent, during which small water molecules continue to penetrate into the polymer particles, resulting in the gradual swelling of the particles; and when the distance between the molecules within the swelling polymer particles reaches a certain value, the linear polymer molecular clusters begin to disperse, forming a solution, which is the process of dissolution. The dilution process of all water-soluble polymer solid particles is the same. The external fine particles will first quickly get wet, swell, and dissolve, forming a locally highly viscous solution, which will make a large number of solvated water molecules adhere to the outside of the polymer, preventing the internal small particles from further contact with the solvent water molecules, resulting in the external particles of the polymer being dissolved, while the internal particles will not even come into contact with the water molecules. These internal particles are "encapsulated" to make it difficult to reach the water molecules, forming a "fish-eye" that is difficult to dissolve. Once the "fish-eye" is formed, it will cause the waste of the flocculant, or it will block the pipeline and pump head, which will affect the production of the system. In addition, due to the slow dissolution rate of solid particle flocculants, a specialized device is needed to dissolve these flocculants, and the device operates continuously for 24 h, increasing equipment maintenance costs and energy consumption.

The solubility problem of CPAM solid particles has brought great trouble to the oilfield wastewater treatment process, restricting the improvement and innovation of the process. In recent years, researchers have begun to study the preparation technology of liquid cationic polyacrylamide products [17–19], but there are still many technical difficulties to be solved in these technologies, such as a low effective content, poor storage stability, poor water purification effect, complex preparation process, high production costs, etc. Liquid CPAM products represented by "water-in-oil" emulsions and "water-in-water" emulsions have received extensive attention from many researchers due to their fast dissolution rate [20–24], and the advantages and disadvantages of different products are shown in Table 1.

Table 1. Comparison of advantages and disadvantages of different CPAM products.

CPAM Product	Advantages	Disadvantages
Solid particles	Convenient transportation, stable storage	Slow dissolving speed, easy to form fish-eye
Low-concentration solution	Fast dissolving speed, convenient preparation	Inconvenient transportation, low effective content ($\leq$10%), easy to deteriorate, odor
Water-in-water emulsion	Fast dissolving speed	Easy to separate layers, easy to degrade
Water-in-oil emulsion	Fast dissolving speed, good stability	Low effective content ($\leq$30%), high price, complex preparation processes
Suspension (this work)	No fish-eye, high effective content, simple preparation process	Dissolving speed is relatively slow, loss of solution viscosity

Franklin [25] prepared a polyacrylamide water-in-oil microemulsion, which had good stability and was not easy to separate. However, the viscosity of the product was relatively high. When the polymer content was 28.10%, the apparent viscosity of the microemulsion reached more than 321 mPa·s, and its mobility was too poor to be pumped in practical applications. Liu [26] prepared a water-in-water cationic polyacrylamide emulsion with a viscosity of 150~500 mPa·s, but its effective content was only about 15%. The preparation of this "water-in-water" emulsion required a strict monomer concentration. In the polymerization process, once the monomer concentration was too high, it was easy to form a gel rather than a stable emulsion. The "water-in-oil" emulsion prepared by Wang had a stable performance [27], no separation after 3 months at room temperature, and a dissolution time of about 10 min. However, the preparation process of a "water-in-oil" emulsion is very complicated, and it needs to be synthesized under the conditions of no metal ions and no oxygen, so the production cost is high and the cost performance of the product is relatively low. In short, the preparation technology of these liquid products is still immature, and there are obvious defects, such as a low effective content of the solid phase, with the effective content generally being no more than 30%, which brings great inconvenience to its transportation and production. Another disadvantage includes a poor storage stability, with long-term storage resulting in serious degradation. Finally, these liquid products have a low molecular weight, compared with solid particles and not only require greatly increasing the dosage of the agent, but at the same time, their flocculation and purification performance on wastewater is also poor and it is difficult to meet most of application requirements.

Due to the many defects of CPAM products existing in the current market, with for example, solid-type CPAM (CPAM particles) having the problems of a slow dissolution speed and easy formation into a "fish-eye", while liquid-type CPAM products (water-in-water emulsions, water-in-oil emulsions, low-concentration aqueous solutions) have the problems of a low effective content, poor storage stability, expensiveness, and so on, the development of a novel type of CPAM flocculant with a fast dissolution speed, high effective content, and good storage stability is imperative. Therefore, the objective of this paper is to develop a novel type of CPAM, which, in comparison with the current work, has the following features: (1) good solubility, in that it can be quickly dissolved, not form a "fish-eye", and solve the problems of CPAM waste, poor application effect, and pipeline blockage caused by the poor solubility of current solid CPAM products; (2) good stability, solving problems such as spoilage in summer, freezing in winter, and easy degradation of current liquid products; and (3) high effective content, reducing storage, transportation, packaging, and other costs, with the current liquid-type CPAM effective content being too low and emulsion-type CPAM being limited by its stability in the polymerization process, usually requiring a low concentration of monomers. Once a monomer concentration is too high, burst polymerization will occur a stable emulsion cannot be obtained. Meanwhile, solution-type CPAM is limited by its fluidity, and when a solution with a concentration of more than 10% is prepared, the solution will be too viscous and difficult to make flow.

In this paper, a novel type of liquid CPAM product, a CPAM suspension, was prepared, the optimal preparation process of the suspension was first studied, and then the dissolution performance, storage stability, flocculation, and water purification performance of the suspension were evaluated. Finally, a scale-up test was conducted to apply the produced CPAM suspension to oilfield wastewater treatment to verify the performance of the suspension for practical applications.

2. Materials and Methods

2.1. Materials and Instruments

Cationic polyacrylamide 4190 (CPAM 4190) was purchased from SNF China Co., Ltd., Taixing, China. #5 White Oil (#5WO) was purchased from Shenzhen Zhongruntong Chemical Co., Ltd., Shenzhen, China. Fumed silica (F-Silica 5000 mesh) was purchased from Yangshan Yuanfeng Powder Material Co., Ltd., Qingyuan, China. Organic bentonite (O-Bent) was purchased from CNPC Bohai Drilling Engineering Co., Ltd., Tianjin, China. Polyaluminum chloride (PAC) was purchased from Nanjing Xiangling Environmental Protection Technology Co., Ltd., Nanjing, China. Sodium dodecyl sulfate (SDS), sodium dodecyl benzene sulfonate (SDBS), polysorbate 80 (Tween 80), sorbitan tristearate 60 (Span 60), sorbitan tristearate 80 (Span 80), hexadecyl trimethyl ammonium bromide (CTAB), and dodecyl phenol polyoxyethylene ether 10 (OP-10) were purchased from Beijing Yili Fine Chemicals Co., Ltd., Beijing, China. Except for CPAM 4190, #5WO, F-Silica, and O-Bent, which were industrial-grade, the other reagents were analytical-grade.

The six-speed rotary viscometer, ZNN-D6, and digital display high-speed mixer, GJ-2S, used were manufactured by Qingdao Tongda Special Instrument Co., Ltd., Qingdao, China. The multi-functional high-speed blender, SUS-304, employed was manufactured by Wuyi Hainan Electric Co., Ltd., Haikou, China. The standard sample sieve used was manufactured by Beijing Chemical Glass Station Bioanalytical Technology Co., Ltd., Beijing, China. The desktop high-speed centrifuge was manufactured by Changsha Dongwang Experimental Instrument Co., Ltd., Changsha, China. The high-speed adjustable homogeneous emulsifier, FSV-2, was manufactured by Changzhou Jintan Jingda Instrument Manufacturing Co., Ltd., Changzhou, China. The digital display laser turbidity meter SGZ-2 was manufactured by Shanghai Yuefeng Instrument Co., Ltd., Shanghai, China. The optical microscope, CX-43, was manufactured by Olympus (China) Co., Ltd., Beijing, China.

2.2. Preparation Method of CPAM Suspension

Firstly, the industrial-grade CPAM solid particles were crushed by a pulverizer for 2 min and then poured out, the crushed CPAM solid particles were then sieved through a 100-mesh standard sample sieve, and a certain amount of separating agent was added and mixed well with a stirrer to obtain CPAM fine powder for use.

Then, a certain amount of rheology modifier was added into the oil phase solvent, and the solution was emulsified and stirred for 2 min with a high-speed homogenizer to obtain the suspension solvent.

Finally, the CPAM fine powder was added with the emulsifying and dispersing agent together into the suspension solvent and emulsified and dispersed at a high speed for 10 min with a high-speed homogenizer to obtain a stable CPAM suspension. The preparation schematic is shown in Figure 1.

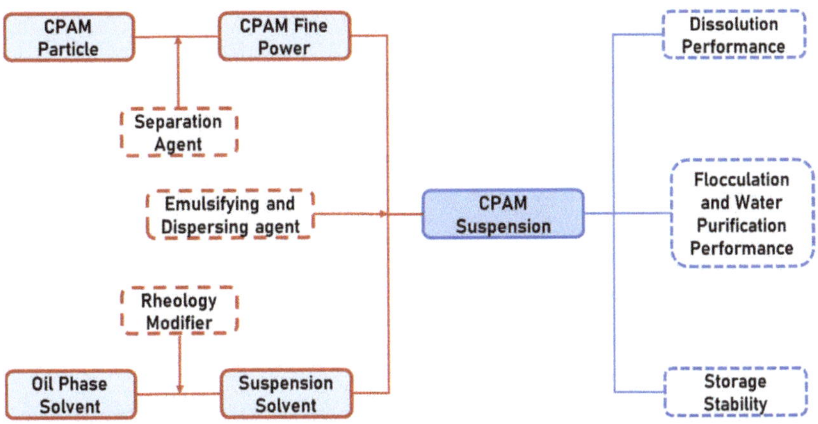

Figure 1. Schematic diagram of the CPAM suspension preparation.

2.3. Performance Evaluation Methods for Suspension

2.3.1. Evaluation Method of Solubility

Non-Newtonian fluid rheological performance test—A certain volume of non-Newtonian fluid was taken, a special measuring cup was filled by it for viscosity testing, the readings of the six-speed rotational viscometer at different speeds were recorded, and the following formulas were used to calculate the rheological parameters:

$$AV = R_{600}/2 \tag{1}$$

$$PV = R_{600} - R_{300} \tag{2}$$

$$YP = 0.5(R_{300} - PV) \tag{3}$$

where

R_{600} = the reading of the viscometer at 600 r/min (dia);
R_{300} = the reading of the viscometer at 300 r/min (dia);
AV = apparent viscosity (mPa·s);
PV = plastic viscosity (mPa·s);
YP = yield point (Pa).

Dissolution time test—Under the speed of 200 rpm, a certain mass of suspension was slowly added dropwise to the beaker containing deionized water, and the apparent viscosity value of the solution was recorded at intervals until the apparent viscosity value no longer increased, which was the dissolution time for the sample to fully form an aqueous solution.

2.3.2. Evaluation Method of Stability

A certain volume of CPAM suspension was placed in a test tube and the height from the bottom of the test tube to the liquid surface was recorded as L_1. After storing the test tube at room temperature for 1 day, 3 days, and 7 days, the height of the separated supernatant was recorded as L_2, and the following formula was used to calculate the storage separation rate of the suspension:

$$S = L_2/L_1 \times 100\% \tag{4}$$

where

S = storage separation rate (%);
L_1 = height of the suspension (cm);
L_2 = height of separated supernatant (cm).

A certain mass of CPAM suspension was weighed and recorded as M_1, poured into a centrifuge tube and sealed, placed into a high-speed centrifuge, and then centrifuged at 2000 rpm for 30 min. The mass of the centrifuged supernatant was weighed and recorded as M_2, and the following formula was used to calculate the centrifugal separation rate of the suspension:

$$C = M_2/M_1 \times 100\% \tag{5}$$

where

C = centrifugal separation rate (%);
M_1 = height of the suspension (g);
M_2 = height of separated supernatant (g).

2.3.3. Evaluation Method of Flocculation and Purification Performance

The simulated wastewater preparation—A certain mass of diatomite was weighed, a certain volume of tap water was added, and a high-speed mixer was used to mix the simulated wastewater well with a turbidity of about 1000 NTU. The oilfield wastewater was taken from the combined wastewater treatment stations of the oilfield of PetroChina Daqing, including La-I-1, Xing-1, and Xing-20.

Wastewater turbidity removal rate test—A turbidimeter was first calibrated with 400 NTU standard solution. After the reading of the turbidimeter found it to be stable, a certain volume of wastewater was taken and put into the test bottle of the turbidimeter. The original turbidity of the wastewater was recorded as U_1. A certain amount of flocculant was added to the wastewater, stirred thoroughly, and it was allowed to stand for 2 min. The upper clear liquid was then taken and its turbidity was tested as U_2. The following formula was used to calculate the turbidity removal rate of the flocculant on the wastewater:

$$T = U_2/U_1 \times 100\% \tag{6}$$

where

T = turbidity removal rate (%);
U_1 = turbidity of original wastewater (NTU);
U_2 = turbidity after turbidity removal (NTU).

2.3.4. Determination of Oil Content

In this study, the oil content in the wastewater was determined using the spectrophotometric method [28], whereby the oily wastewater (from Daqing Bluestar Environmental Protection Engineering Co., Ltd. (Daqing, China) wastewater treatment station) was placed in a dispensing funnel, a certain amount of gasoline and hydrochloric acid was added to extract the oil in the wastewater, the extract was dehydrated by anhydrous calcium chloride and then filtered, the filtrate was evaporated to remove the gasoline in the water bath at 80 °C, and the standard oil samples were obtained to be used.

The standard oil sample was used to prepare the standard oil solution with an oil concentration of 5.00 mg/mL, and then 0.00 mL, 0.50 mL, 1.00 mL, 1.50 mL, 2.00 mL, 5.00 mL, and 10.00 mL of the standard oil solution was put into a 50 mL colorimeter tube. Then, the absorbance was tested by a spectrophotometer (λ = 430 nm), and the relationship between the oil content and absorbance was obtained, as shown in Table 2. A standard curve was calculated by the least squares method: $y = -0.27 + 42.11x$.

Table 2. Relationship between oil content and absorbance.

Oil Content y (mg)	Absorbance x
0	0
2.50	0.062
5.00	0.123
7.50	0.187
10.00	0.248
25.00	0.613
50.00	1.186

Hydrochloric acid and gasoline were added to a certain volume of wastewater for extraction, the extracted oil phase was put in a colorimeter tube, gasoline was added to dilute it to the scale, and a spectrophotometer was used to test its absorbance. To find out the oil content on a standard curve, the oil content calculation formula used was the following:

$$O = M_o/V_o \times 10^3 \tag{7}$$

where

O = oil content (mg/L);
M_o = oil content found on the standard curve (mg);
V_o = volume extracted from wastewater (mL).

2.3.5. Optical Microscope

Slides and coverslips were first cleaned using ethanol and then dried in an oven at 85 °C. Samples of the CPAM suspension were taken with a dropper, a drop was placed on a slide and covered by a coverslip, and the suspension sample was observed using the optical microscope Olympus CX-43 at different magnifications.

3. Results and Discussion

3.1. Influencing Factors of the Suspension

3.1.1. Effect of the Separating Agent

The finer polymer particles are, the easier it is for them to agglomerate, so to prevent the fine CPAM particles in the water from forming a "fish-eye", it was necessary to add a separating agent to disperse the particles. F-Silica [29,30] has the characteristics of a small particle size, large specific surface area, and strong adsorption, which can be effectively adsorbed on the surface of CPAM particles, making the suspension more stable [31]. The adsorption mechanism of F-Silica is shown in Figure 2.

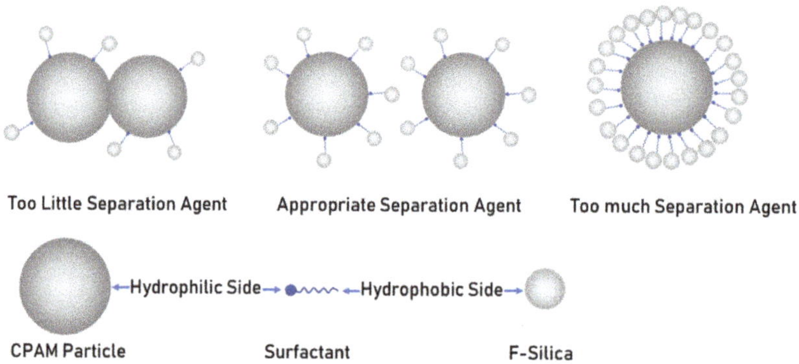

Figure 2. The adsorption mechanism of F-Silica.

F-Silica is prepared using a fumed phase method, and the surface of its particles contain silicon–oxygen bonds, which are hydrophobic, while the surface of CPAM particles contain hydrophilic groups such as amine groups, which are hydrophilic. The suspension contains surfactants with an amphoteric molecular structure; with hydrophilic groups at one side and hydrophobic groups at the other side, the hydrophilic side can be attached to the surface of CPAM particles, while the hydrophobic side can be in contact with the surface of the F-Silica. Therefore, the surfactant can be used as a connecting thread to link the two types of particles, F-Silica and CPAM, and this is the mechanism by which F-Silica adsorbs on the surface of CPAM particles. If the amount of F-Silica is too little, it cannot play a role in separation, and the CPAM particles will be in contact with each other; if the amount of F-Silica is too much, it will cover all the CPAM particles, which is not conducive to dissolution, so the amount of F-Silica should be controlled in an appropriate range.

In this study, different masses of F-Silica were added to the suspension to test the apparent viscosity of the suspension and the dissolution time of different concentrations of diluent, and the results are shown in Figure 3.

As can be seen from Figure 3a, the apparent viscosity of the suspension showed a trend of decreasing and then increasing with the increase in the mass of F-Silica, and the apparent viscosity of the suspension reached a minimum of 136.00 mPa·s at its addition of 0.40 g/60.00 g. From Figure 3b, it can be seen that by adding 0.40 g/30.00 g of F-Silica, the dissolution time of 0.20% diluent was shortened from 22.00 min to 18.00 min. However, the dissolution time was not further shortened by the addition of F-Silica more than 0.40 g/30.00 g. This is because with the addition of F-Silica, the overall structure of the CPAM particles in the suspension was fluffier and easier to flow, but with the excessive addition of F-Silica, the internal friction between particles increases, which makes it difficult for the particles to slide into the suspension, thus increasing the apparent viscosity of the suspension. The presence of F-Silica creates a barrier between the polymer's fine particles, which renders it more difficult to agglomerate in water and more conducive to dissolution. F-Silica adsorbed on the surface of the polymer particles only plays a segregation role, and adding too much is not conducive to the diffusion of particles in water, so the amount of F-Silica additives selected as 0.40 g/60.00 g was appropriate.

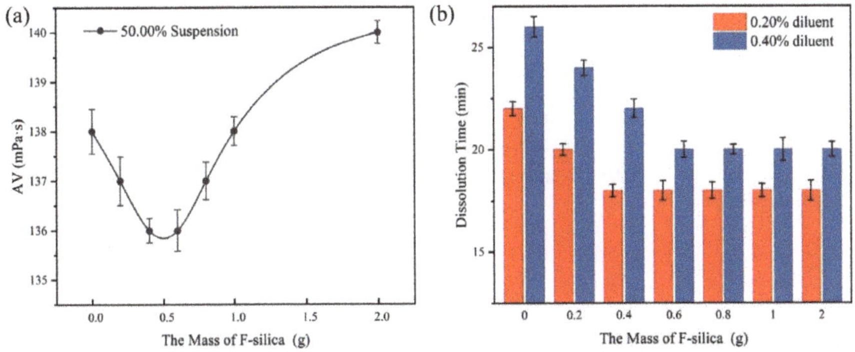

Figure 3. Effects of F-Silica: (**a**) trend of the apparent viscosity of the suspension (at 50% effective content of the suspension); (**b**) trend of dissolution time (dilution concentrations of 0.20% and 0.40%).

3.1.2. Effects of the Emulsifying and Dispersing Agent

In this study, WO was used as the solvent of the suspension, but WO could not be dissolved in water, so it was necessary to add surfactants with hydrophilic and non-hydrophilic groups to the suspension so that WO could be quickly dispersed and dissolved in water, and at the same time, to promote the rapid diffusion of the polymer's fine particles in water.

Both the type and amount of surfactant added affect the emulsification and dispersion of WO. In this study, 1.00 g/60.00 g of different types of surfactants was added to the suspension and then the complete dissolution time of the suspension in water was tested; the results are shown in Figure 4.

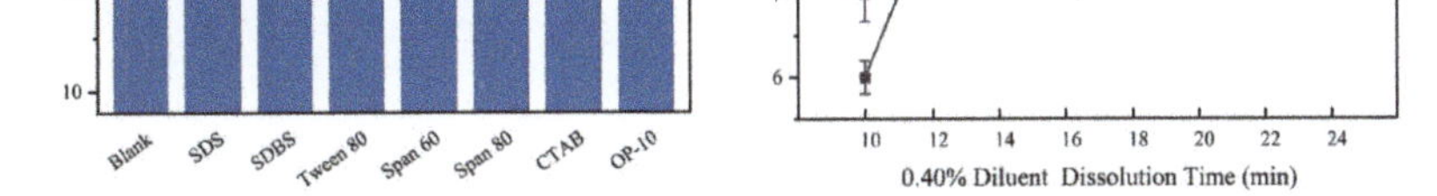

Figure 4. Effects of emulsifying and dispersing agent: (**a**) dissolution time (dilution concentration of 0.40%); (**b**) synergistic effect of CTAB and Span 80 (dilution concentration of 0.40%).

As shown in Figure 4a, it can be seen that the emulsification effects of SDS and SDBS on WO were not good, and they did not emulsify or disperse WO in water, resulting in no improvement in the dissolution time of the suspension. Meanwhile, Tween 80 and Span 60 had some emulsification effect, which could accelerate the dissolution of the suspension to a certain extent. The dissolution time of the suspension in water was obviously accelerated, and the shortest dissolution time was 18.00 min. Since the use of OP emulsifiers is restricted by relevant regulations due to their harmful effects on the environment [32], CTAB and

Span 80 were used as the main and auxiliary emulsifiers for the suspension from the perspective of environmental friendliness.

After Span 80/CTAB was added to the suspension at different mass ratios, the dissolution time of the 0.4% diluent was tested and the results are shown in Figure 4b. It can be seen that when the mass ratio of Span 80:CTAB was 1:1, the apparent viscosity of the diluent reached a maximum value of 11.00 mPa·s in the shortest time, which took only 16.00 min. This is because under the synergistic effect of Span 80 and CTAB, after the suspension was added to water, WO molecules wrapped around the surfaces of the fine CPAM particles; they rapidly emulsified and dispersed in water because the surfaces of the polymer particles came into rapid contact with water molecules to achieve rapid dissolution.

3.1.3. Effects of the Rheology Modifier

The density of the WO used in this study was about 0.85 g/cm^3, while the density of the CPAM particles was about 1.32 g/cm^3, and the polymer particles added to the WO directly settled, so it was necessary to improve the shearing force of the WO. In the field of oilfield chemical drilling, bentonite is modified into organic bentonite (O-Bent), which can be dispersed and swelled in WO to form a spatial network structure and increase the viscosity and shearing force of oil-based drilling fluids (Figure 5d), which is used to enhance the cutting transportation of drilling fluids [23–35]. Therefore, in this study, different masses of O-Bent were first added to the suspension as a rheology modifier, and then the rheology and separation rate of the suspension were tested. The experimental results are shown in Table 3 and Figure 5.

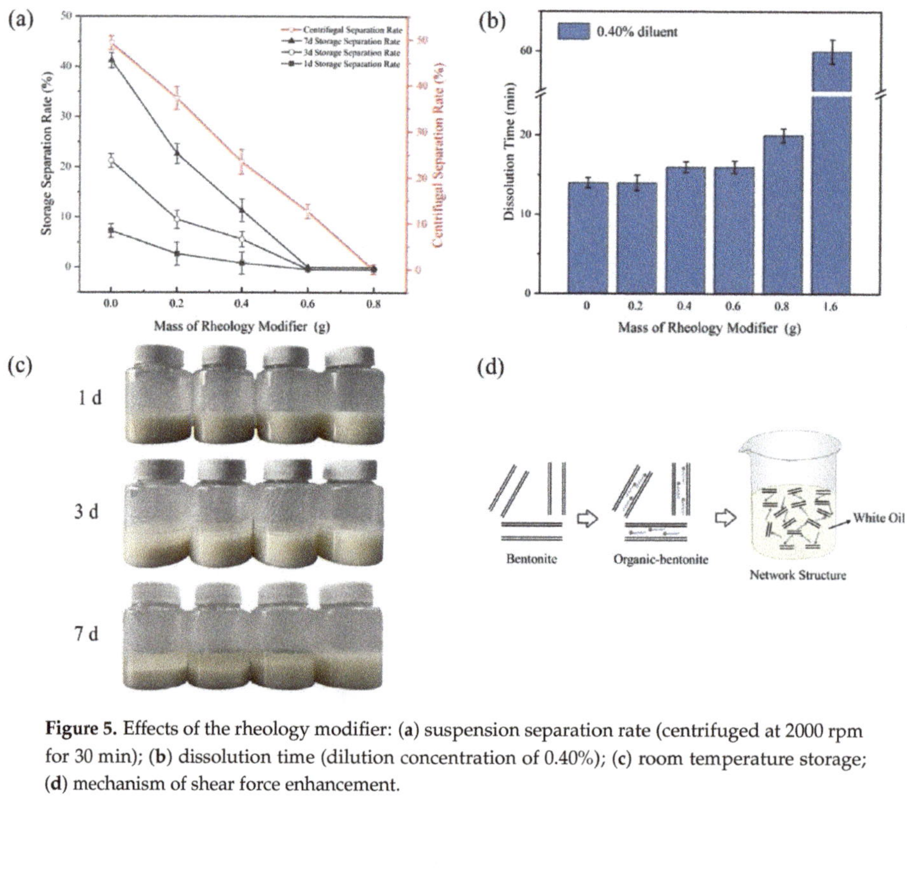

Figure 5. Effects of the rheology modifier: (**a**) suspension separation rate (centrifuged at 2000 rpm for 30 min); (**b**) dissolution time (dilution concentration of 0.40%); (**c**) room temperature storage; (**d**) mechanism of shear force enhancement.

Table 3. Effects of the rheology modifier on the suspension.

Mass of Rheology Modifier (g)	R_{600} (dia)	R_{300} (dia)	AV (mPa·s)	PV (mPa·s)	YP (Pa)
0.00	203.00	104.00	101.50	99.00	2.50
0.20	230.00	120.00	115.00	110.00	5.00
0.40	251.00	132.00	125.50	119.00	6.50
0.60	274.00	147.00	137.00	127.00	10.00
0.80	296.00	162.00	148.00	134.00	14.00
1.60	>300.00	177.00	>150.00	>123.00	>27.00

As can be seen from Table 3, with the increase in the addition of O-Bent, the viscosity and dynamic shear of the suspension were significantly improved, but when the mass of the rheology modifier was 1.60 g/60.00 g, the suspension became gelatinous and difficult to make flow. It was difficult for it to disperse after adding it to the water, which indicates that the addition of a rheology modifier should be sparing. As can be seen from Figure 5a, when the mass of the rheology modifier was 0.60 g/60.00 g, the suspension was able to be stored for 7 days without separation, which indicates that when stored at room temperature, O-Bent can provide a solvent with enough shear force so that the solvent can suspend CPAM particles. When the mass of the rheology modifier was 0.80 g/60.00 g, the suspension did not separate when stored at room temperature for 7 days, nor did it separate when centrifuged at 2000 rpm for 30 min, but the apparent viscosity of the suspension was higher. The diffusion of the suspension was slower after the drop was added to the water, resulting in a longer time for the complete dissolution of the suspension in the water (as shown in Figure 5b), and given the comprehensive consideration of its stability and dissolvability, the mass of the rheology modifier being 0.60 g/60.00 g was more appropriate.

As can be seen from Figure 6, the size of most polymer particles in the suspension was <100 μm and the polymer particles were uniformly distributed in the suspension, with obvious boundaries between the particles. Furthermore, there was no aggregation of the particles to precipitate, which indicates that the rheology modifier played a role in improving the shear force, so that the suspension had enough shear force to suspend the polymer particles. In addition, it is obvious that there were many fine particles attached to the surface of the polymer particles, which indicates that the separating agent F-Silica was indeed adsorbed at the surface of the polymer particles, making the polymer particles dispersed from each other.

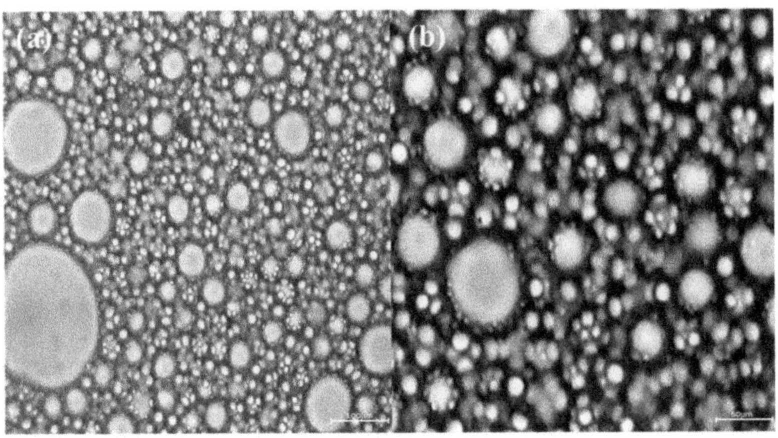

Figure 6. Image of the suspension under an optical microscope: (**a**) 200× magnification; (**b**) 400× magnification.

Considering all these influencing factors, the proportions of each component of the suspension were obtained as shown in Table 4, and the suspension prepared according to these proportions can have good suspension stability and be completely dissolved in a short time.

Table 4. Percentages of components in the suspension.

Function	Component	Mass (g)	Percentage (%)
Flocculant	CPAM fine powder	32.00	50.00
Solvent	#5WO	30.00	46.87
Separating agent	F-Silica	0.40	0.63
Emulsifying and	Span 80	0.50	0.78
dispersing agent	CTAB	0.50	0.78
Rheology modifier	O-Bent	0.60	0.94

3.2. Comparison of the Products' Performance

3.2.1. Comparison of the Dissolution Performance

In this study, original particles of CPAM, fine particles of the CPAM crushed to a 100 mesh, and a 50% content of CPAM suspension were prepared at different concentrations of diluent. The rheology and complete dissolution time of the diluents were tested, and the results are shown in Table 5.

Table 5. Comparison of dissolution performances of different samples.

CPAM Sample	Concentration (%)	AV (mPa·s)	PV (mPa·s)	YP (Pa)	Dissolution Time (min)
Original particle	0.10	6.50	6.00	0.50	64.00
	0.20	12.00	12.00	0.50	96.00
	0.40	30.00	28.00	2.00	Fish-eye, 120.00+
Fine particle	0.10	5.00	6.00	0.00	Fish-eye, 86.00
	0.20	10.00	9.00	1.00	Fish-eye, 120.00+
	0.40	27.50	26.00	1.50	Fish-eye, 120.00+
Suspension (50%)	0.20	5.50	5.00	0.50	12.00
	0.40	10.50	10.00	0.50	16.00
	0.80	29.00	27.00	2.00	22.00

As can be seen from Table 5, CPAM fine particles easily form a "fish-eye" and are difficult to dissolve in water no matter how much concentrated diluent is prepared; at 0.40% diluent, the AV value began to decline, but the formed "fish-eye" still failed to dissolve completely. The original CPAM particles started to resemble a "fish-eye" when the concentration reached 0.40%, making the dissolution time more than 120.00 min; benefiting from the liquid form of the product, the CPAM suspension can be quickly dispersed in water, and the CPAM particles have a smaller particle size, so the dissolution time is greatly shortened (0.20% diluent dissolved in 12.00 min). Since the CPAM suspension needs to be mechanically refined during the preparation process, this process will break part of the molecular chain of the linear polyacrylamide, which will lead to a certain degree of viscosity decrease in the diluent. In the preparation of 0.20% CPAM diluent, compared with the original particles, although the apparent viscosity of the suspension decreased by 12.50%, the dissolution time could be shortened by 83.33%, and did not produce a "fish-eye"; the decrease in apparent viscosity was worthwhile.

3.2.2. Comparison of Flocculation and Water Purification Performance

In this study, we prepared original CPAM particles into a diluent with a concentration of 0.10%; the CPAM suspension, into a diluent with a concentration of 0.20%; and the inorganic flocculant PAC, into a diluent with a concentration of 10.00%. Then, the prepared diluent was added to the simulated wastewater and oilfield wastewater according to different dosing amounts, the turbidity before and after dosing was tested, and then the turbidity removal rate was calculated, as shown in Figure 7.

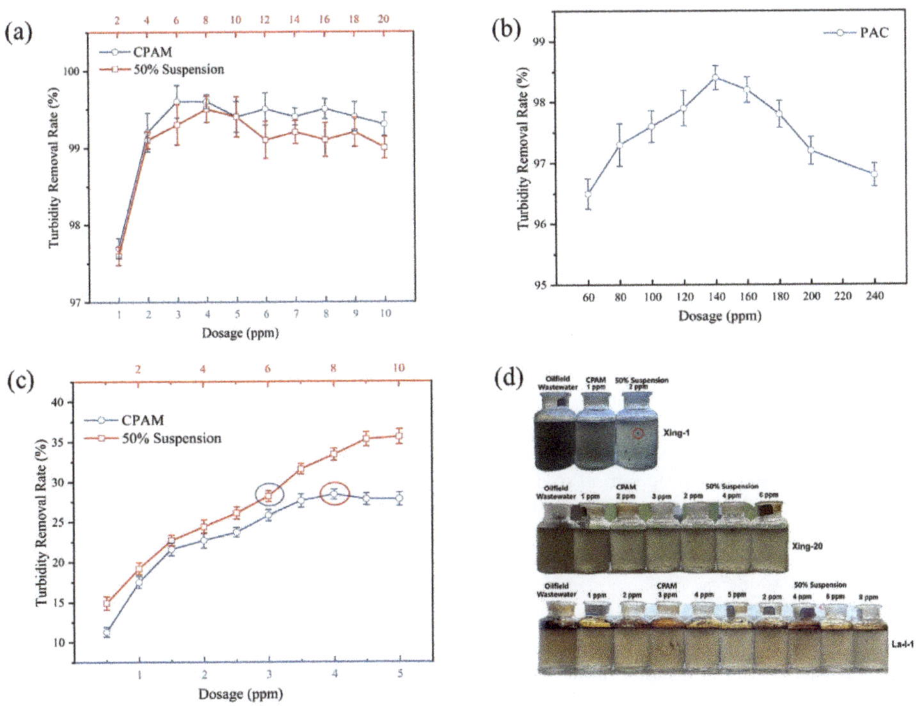

Figure 7. Comparison of flocculation and water purification performance of different products: (**a**) simulated wastewater (organic polymer flocculant, dilution concentrations of 0.20%); (**b**) simulated wastewater (inorganic flocculant, dilution concentrations of 10.00%); (**c**) oilfield wastewater (dilution concentrations of 0.20%); (**d**) water purification effect.

As can be seen from Figure 7a, for the diatomite-simulated wastewater, the best turbidity removal effect was achieved at a polymer dosing of 4.00 ppm, and the turbidity removal effects of the original CPAM original particles and the CPAM suspension were basically the same, with optimal turbidity removal rates of 99.60% and 99.50%, respectively. Conversely, the inorganic flocculant PAC at a 140.00 ppm dosage showed the best turbidity removal effect for simulated wastewater and the turbidity removal rate was 98.40% (Figure 7b), which was lower than the CPAM suspension. For the oilfield wastewater (Xing-1 combined wastewater treatment station), the turbidity removal rate of the CPAM suspension was higher than that of the original CPAM particles under the same polymer dosage, with the optimal turbidity removal rate of the CPAM suspension being 35.60%, and that of the original CPAM particles, being 28.40% (Figure 7c). When the dosage of the CPAM suspension was 1.5 times the optimal dosage of the original CPAM particle, the turbidity removal rates were basically the same, with 28.40% and 28.20%, respectively. This is due to the addition of cationic surfactants in the suspension, as cationic surfactants can assist the polyacrylamide adsorption of suspended particles in wastewater through electrostatic action, whereby the more negatively charged suspended particles are in wastewater, the more pronounced is this auxiliary flocculation effect [36]. As shown in Figure 7d, for the oilfield wastewater from the Xing-1 combined wastewater treatment station, the wastewater with the addition of the CPAM suspension was obviously clearer and brighter than that with the addition of the original CPAM particles, and the flocs aggregated out of the wastewater were obviously larger.

3.2.3. Comparison with Other Study

In this study, we compared the performances of a CPAM suspension with the current liquid CPAM study of the same type, and the results are shown in Table 6.

Table 6. Comparative data from the CPAM suspension and other studies.

CPAM Sample	Effective Content (%)	Molecular Mass (million)	Dissolution Time (min)	Apparent Viscosity (mPa·s)	Market Price (CNY/t)
Particle	100.00	12.32	96.00	–	15,000.00
Suspension	50.00	11.57	16.00	136.00	11,500.00
Water-in-oil emulsion [25]	28.10	4.50~6.50	10.00	321.00	–
Water-in-water emulsion (a Chinese brand)	30.00	0.35	6.00	108.00	30,000.00

As can be seen from Table 6, the effective content of the suspension can reach 50.00%, which is higher than the emulsion-type CPAM (28.10% and 30.00%, respectively); although the emulsion-type CPAM has the advantage of a better dissolution speed, due to its preparation process being more complex, the high requirements for its instrumentation, and its subsequent preparation cost being high, the market price of the emulsion with a content of 30% is about CNY 3000.00/t, while the market price of the 50% content suspension is about CNY 11,500.00/t, so the CPAM suspension is more cost-effective.

3.3. *Application*

3.3.1. Background of the Wastewater Station

We conducted a scale-up test and produced 108.00 kg of CPAM suspension, which was applied to the wastewater treatment station of Daqing Bluestar Environmental Protection Engineering Co., and the wastewater treatment process at the wastewater station is shown in Figure 8.

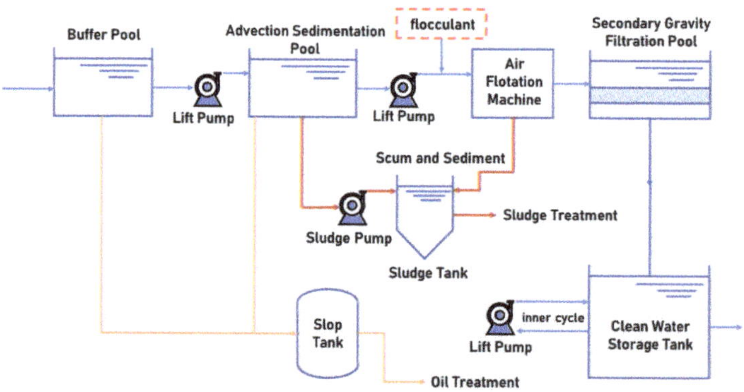

Figure 8. The wastewater treatment process for the wastewater station.

As shown in Figure 8, there is a buffer pool in the wastewater station for buffering incoming water, with a design volume of 150 m^3, steel mixed structure, internal corrosion, and seepage prevention, with a mud hopper at the bottom and oil scraping equipment on the surface of the pool. There are two advection sedimentation pools, the sedimentation part of the effective water depth is 3.0 m, the sedimentation part of the effective volume is 270 m^3, the pool length is 20.0 m, and the pool width is 4.5 m. The sludge tank volume is

35 m^3, the internal setup consists of a scum scraper and sediment scraper, and the sludge discharge mode for the use of the bottom of the sludge pump involves a sludge hopper that emits intermittent sludge discharge. There are two air flotation machines in the station, and the displacement of the air flotation machine lifting pump is 50 m^3/h. The CPAM suspension flocculant diluent is added at the inlet of the air flotation machine. After the suspended matter in the wastewater is flocculated, the flocculant will float and be scraped away by the scum scraper at the top of the air flotation machine. There are eight secondary gravity filtration pools, the effective volume of the sedimentation part is 216 m^3, and the length, width, and height of the tanks are all 3.0 m.

3.3.2. Application Effects

The CPAM suspension with a concentration of 1.20% added to a 500 L dosing tank for dilution, with stirring for 20 min to fully dissolve it, and then use of a dosing pump with a displacement of 500 L/h of dosing. The specific dosing parameters are shown in Table 7 and the field dosing equipment is shown in Figure 9.

Table 7. Dosing parameters.

CPAM Sample	Wastewater Treatment Capacity (m^3/d)	Dosing Concentration (mg/L)	Daily Dosage (kg)	Dosing Days (d)	Total Dosage (kg)
Particle Suspension	300.00	60.00	18.00	1	18
		120.00	36.00	3	108

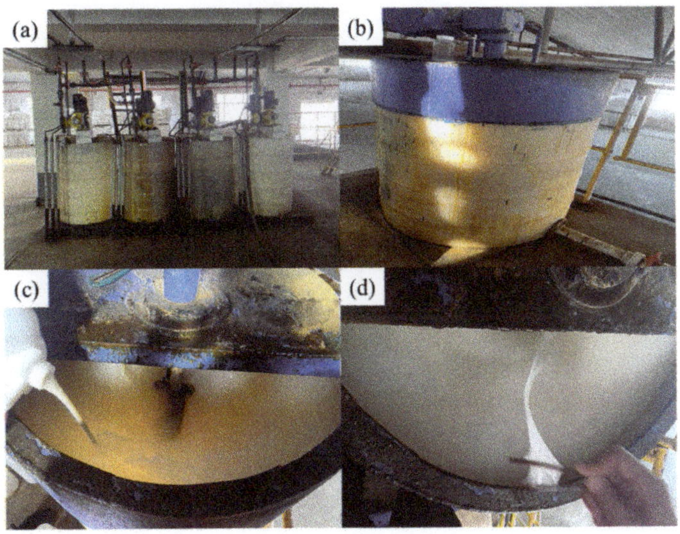

Figure 9. Field dosing situation: (**a**) CPAM particle dosing tank; (**b**) CPAM suspension dosing tank; (**c**) CPAM suspension for dilution; (**d**) after full dissolution.

As can be seen from Figure 8, the CPAM suspension dissolved very well in the dosing tank, at a concentration of 1.20%, the suspension could be uniformly dispersed in the water, and it could be completely dissolved within 20 min, without a "fish-eye" situation. We observed the flocculation of the wastewater in the air flotation machine before and after dosing, as shown in Figure 9. After the completion of dosing, wastewater samples were taken at the outlet to test their turbidity and oil content, and the experimental results are shown in Table 7.

As can be seen from Figure 10, in the air flotation machine, no obvious flocs were floating in the wastewater before dosing, while after dosing, the wastewater had obvious, large, and dense floating flocs; in the air flotation machine outlet, the wastewater in the bottle after dosing was significantly clearer, and there was an obvious layer of flocs settling at the bottom of the bottle. From Table 8, it can also be seen that after treatment with the CPAM suspension flocculant, the turbidity and oil content of the wastewater were significantly reduced, the turbidity was reduced from 1126.00 NTU to 554.62 NTU (turbidity removal rate of 50.74%), and the oil content was reduced from 68.55 mg/L to 20.36 mg/L (oil removal rate of 70.30%). Meanwhile, after treatment with CPAM particles, the turbidity of the wastewater was reduced to 756.23 NTU (turbidity removal rate of 32.84%), and the oil content was reduced to 27.24 mg/L (oil removal rate of 60.26%), which indicates that the flocculation and oil removal effects of the CPAM suspension on wastewater were better than those of the CPAM particles.

Figure 10. Wastewater before and after field dosing: (**a**) wastewater in the air flotation machine before dosing; (**b**) wastewater in the air flotation machine after dosing; (**c**) wastewater at the outlet before dosing; (**d**) wastewater at the outlet after dosing.

Table 8. Wastewater quality indicator testing.

Wastewater Sample		Turbidity (NTU)	Oil Content (mg/L)
Original		1126.00	68.55
After adding CPAM particles		756.23	27.24
After adding CPAM suspension	First day	536.21	18.36
	Second day	569.32	22.68
	Third day	558.34	20.04
	Average	554.62	20.36

4. Conclusions

A CPAM suspension prepared from CPAM solid particles can reach an effective content of 50%, apparent viscosity of 136.00 mPa·s and show no separation after 7 days of storage at room temperature, as well as no separation after 30 min of centrifugation at a speed of 2000 rpm. When diluted in a 0.40% aqueous solution, compared with CPAM solid particles, its apparent viscosity decreased by 12.5%, while its dissolution time was shortened by 81.40%, which was only 16.00 min.

The optimal composition ratio of the CPAM suspension was 50.00% CPAM fine powder, 46.87% oil phase solvent, 0.63% separating agent, 1.56% emulsifying and dispersing agent, and 0.94% rheology modifier.

The best turbidity removal rate of the CPAM suspension for diatomite-simulated wastewater was 99.50%, slightly lower than the CPAM solid particles of the best turbidity removal rate of 99.60%, but better than the inorganic flocculant PAC's best turbidity removal rate of 98.40%. The best turbidity removal rate of the CPAM suspension for oilfield wastewater was 35.60%, which is better than the best turbidity removal rate of the CPAM solid particles at 28.40%; when an additional amount of CPAM suspension was 1.5 times the additional amount of CPAM solid particles, the same turbidity removal and purification effect could be achieved.

In the scale-up test, the CPAM suspension, formulated with a 1.20% concentration, could be completely dissolved within 20 min; the CPAM suspension was superior to the CPAM particles in flocculation and the oil removal of wastewater, with the turbidity removal rate of the CPAM suspension being 50.74% and the oil removal rate being 70.30%, while the turbidity removal rate of the CPAM particles was only 32.84% and the oil removal rate was 60.26%.

Author Contributions: All authors contributed to the conception and design of the study. Material preparation, data collection, and analyses were performed by Z.W., W.L. and S.L. Original draft preparation, Z.W. Review and editing, Z.W. and F.Z. Software, Z.W., Y.Z. and S.Y. All authors have read and agreed to the published version of the manuscript.

Funding: This research was funded by the Fundamental Research Funds for the Central Universities, Grant No. 2-9-2019-141 and No. 4-2-2022-002.

Institutional Review Board Statement: Not applicable.

Informed Consent Statement: Not applicable.

Data Availability Statement: Data are available from the authors upon request.

Conflicts of Interest: There were no conflicts of interest in the submission of this manuscript and the manuscript was approved by all authors for publication.

References

1. Liu, X.Y. Study on the Synthesis of NFP Modified Cationic Polyacrylamide and Its Flocculation Performance for Oily Wastewater. Master's Thesis, Shaanxi University of Science and Technology, Xi'an, China, 2020. Available online: https://d.wanfangdata.com.cn/thesis/ChJUaGVzaXNOZXdTMjAyMzAxMTISCFkzNzMyNjI0GghsMjlhd2szbQ== (accessed on 16 June 2023).
2. Yu, L.; Han, M.; He, F. A review of treating oily wastewater. *Arab. J. Chem.* **2017**, *10*, S1913–S1922. [CrossRef]
3. Peng, Y.; Liu, Y.; Dai, J.; Cao, L.; Liu, X. A sustainable strategy for remediation of oily sewage: Clean and safe. *Sep. Purif. Technol.* **2020**, *240*, 116592. [CrossRef]
4. Jamaly, S.; Giwa, A.; Hasan, S.W. Recent improvements in oily wastewater treatment: Progress, challenges, and future opportunities. *J. Environ. Sci.* **2015**, *37*, 15–30. [CrossRef] [PubMed]
5. Medeiros, A.D.M.d.; Silva Junior, C.J.G.d.; Amorim, J.D.P.d.; Durval, I.J.B.; Costa, A.F.d.S.; Sarubbo, L.A. Oily Wastewater Treatment: Methods, Challenges, and Trends. *Processes* **2022**, *10*, 743. [CrossRef]
6. Zhang, W.; Wang, Y.; Wang, X.; Wang, Y.; Zhang, X.; Xiong, Y.; Duan, M.; Fang, S. The effect of molecular structure on the properties of cationic polyacrylate and its flocculation-flotation performance of oilfield oily wastewater treatment. *J. Appl. Polym. Sci.* **2022**, *139*, e52700. [CrossRef]
7. Mahdy, S.A. Biodegradability enhancement of oily wastewater by an SBR treatment methods. *AIP Conf. Proc.* **2023**, *2809*, 040004. [CrossRef]
8. Zhang, X.; Cheng, J.; Liu, X.; Yue, S.; Wang, X.; Shi, Y.; Qiao, N. Computer assisted design and flow field analysis of a multi-tube airlift reactor for biological treatment of oily wastewater. *J. Water Process Eng.* **2023**, *56*, 104411. [CrossRef]
9. Wang, C.; Zhang, X.; Ning, P.; Yuan, W.; Bai, J. Research progress and prospect of oil-bearing wastewater treatment technology. *Chem. Ind. Eng. Prog.* **2021**, *40*, 451–462. [CrossRef]
10. Wang, Q. Application of oil-bearing wastewater treatment technology in oilfield. *China Pet. Chem. Stand. Qual.* **2023**, *43*, 193–195. [CrossRef]

11. Swerin, A.; Ödherg, L. Flocculation of cellulosic fibre suspensions by a microparticulate retention aid system consisting of cationic polyacrylamide and anionic montmorillonite. *Eff. Contact Time Shear. Level Electrolyte Conc.* **1996**, *11*, 22–29. [CrossRef]
12. Lee, K.E.; Morad, N.; Poh, B.T.; Teng, T.T. Comparative study on the effectiveness of hydrophobically modified cationic polyacrylamide groups in the flocculation of kaolin. *Desalination* **2011**, *270*, 206–213. [CrossRef]
13. Chi, N.; Liu, J.; Lei, M.; Feng, L. Preparation of amphiphilic cationic polyacrylamide (CPAM) with cationic microblock structure to enhance printing and dyeing sludge dewatering and condition performance. *Environ. Sci. Pollut. Res.* **2023**, *30*, 13079–13093. [CrossRef] [PubMed]
14. Wang, Y.; He, Y.; Lin, A. Synthesis and optimization of cationic polyacrylamide flocculant PAD. *Chin. J. Environ. Eng.* **2020**, *14*, 1453–1464. [CrossRef]
15. Ji, G.; Xu, W. Treatment of oily wastewater by new cationic flocculant. *Appl. Chem. Ind.* **2017**, *46*, 2091–2093,2098. [CrossRef]
16. Yan, R.X. *Water-Soluble Polymer*; Chemical Industry Press: Beijing, China, 1998; pp. 12–25.
17. Aveyard, R.; Binks, B.P.; Clint, J.H. Emulsions stabilised solely by colloidal particles. *Adv. Colloid Interface Sci.* **2003**, *100–102*, 503–546. [CrossRef]
18. Tseng, C.M.; Lu, Y.Y.; El-Aasser, M.S.; Vanderhoff, J.W. Uniform polymer particles by dispersion polymerization in alcohol. *J. Polym. Sci. Part A Polym. Chem.* **1986**, *24*, 2995–3007. [CrossRef]
19. Ahmed, S.F.; Poehlein, G.W. Kinetics of Dispersion Polymerization of Styrene in Ethanol. 1. Model Development. *Ind. Eng. Chem. Res.* **1997**, *36*, 2597–2604. [CrossRef]
20. Esquena, J. Water-in-water (W/W) emulsions. *Curr. Opin. Colloid Interface Sci.* **2016**, *25*, 109–119. [CrossRef]
21. Fu, Z.B.; Peng, X.H. Preparation conditions for cationic polyacrylamide water-in-water emulsion by dispersion polymerization. *Petrochem. Technol.* **2010**, *39*, 1140–1145. Available online: https://d.wanfangdata.com.cn/periodical/ChlQZXJpb2RpY2FsQ0hJTmV3UzIwMjMwNDI2Eg1zeWhnMjAxMDEwMDEzGghkNHNpaGh4cw== (accessed on 27 March 2023).
22. Jahandideh, H.; Ganjeh-Anzabi, P.; Bryant, S.L.; Trifkovic, M. The Significance of Graphene Oxide-Polyacrylamide Interactions on the Stability and Microstructure of Oil-in-Water Emulsions. *Langmuir* **2018**, *34*, 12870–12881. [CrossRef]
23. Chen, Z.; Zhou, Y.; Chen, C.; Wang, R. Polymerization of cationic polyacrylamide as flocculant with inverse emulsion and its characterization. *J. Lanzhou Univ. Technol.* **2019**, *45*, 16–21. [CrossRef]
24. Yang, B.; Sun, B. Synthesis of cationic polyacrylamide water-in-water emulsion. *Contemp. Chem. Ind.* **2018**, *47*, 491–493,497. [CrossRef]
25. Franklin, E.S.; Ryan, S.T.; Todd, J.O. Water-in-Oil Polyacrylamide-Based Microemulsions and Related Methods. U.S. Patent US20190169418A1, 6 June 2019. Available online: https://www.freepatentsonline.com/9428630.html (accessed on 16 June 2023).
26. Liu, J.; Huang, Z.; Shu, X.; Ren, Q. A preparation Method of Cationic Polyacrylamide Water-in-Water Emulsion. China Patent CN201210536403.9, 3 April 2013. Available online: https://d.wanfangdata.com.cn/patent/ChJQYXRlbnROZXdTMjAyMjEyMDcSEENOMjAxMjEwNTM2NDAzLjkaCHlzcmV2N3Fi (accessed on 16 June 2023).
27. Wang, Q.; Qin, X.; Hu, H.; Zhou, T. A Fast Dissolving Water-in-Oil Cationic Polyacrylamide Emulsion Preparation Method. China Patent CN201010133804.0, 5 October 2011. Available online: https://d.wanfangdata.com.cn/patent/ChJQYXRlbnROZXdTMjAyMjEyMDcSEENOMjAxMDEwMTMzODA0LjAaCG44NmpmbnE5 (accessed on 16 June 2023).
28. *SY/T 5329-2012*; CNPC Enterprise Standard. Water Quality Standard and Practice for Analysis of Oilfield Injecting Waters in Clastic Reservoirs. China National Petroleum Corporation: Beijing, China, 2012. (In Chinese)
29. Gun'ko, V.M.; Mironyuk, I.F.; Zarko, V.I.; Voronin, E.F.; Turov, V.V.; Pakhlov, E.M.; Goncharuk, E.V.; Nychiporuk, Y.M.; Vlasova, N.N.; Gorbik, P.P.; et al. Morphology and surface properties of fumed silicas. *J. Colloid Interface Sci.* **2005**, *289*, 427–445. [CrossRef] [PubMed]
30. Xiu, H.; Duan, X.; Zhang, T.; Zhao, Y.; Guo, Z.; Wu, C.; Wang, Y. Determination of physical adsorption water of fumed silica and its relationship with surface silanol concentration. *Polym. Mater. Sci. Eng.* **2021**, *37*, 102–108,149. [CrossRef]
31. Huang, Y.; Kobayashi, M. Direct Observation of Relaxation of Aqueous Shake-Gel Consisting of Silica Nanoparticles and Polyethylene Oxide. *Polymers* **2020**, *12*, 1141. [CrossRef]
32. *SY/T 6787-2010*; CNPC Enterprise Standard. Technical Requirements of Environmental Protection for Water-Soluble Oilfield Chemicals. China National Petroleum Corporation: Beijing, China, 2010. (In Chinese)
33. Jordan, J.W.; Hook, B.J.; Finlayson, C.M. The Organophilic Bentonites. II. Organic Liquid Gels. *J. Phys. Colloid Chem.* **1950**, *54*, 1196–1208. [CrossRef]
34. Li, Z.B. Emulsion and Organic Clay Dispersion Stability in Oil-Based Drilling Fluid. Master's Thesis, Shandong University, Jinan, China, 2017. [CrossRef]
35. Pan, Y. Research on the Modification Mechanism of Organic Clay Used in Oil-Based Drilling Fluids. Master's Thesis, China University of Petroleum, Qingdao, China, 2014.
36. Chang, M.; Ma, X.; Dong, X.; Fan, Y.; Chen, R. The Synergistic Effects of Al^{3+} and Chitosan on the Solid–Liquid Separation of Coal Wastewater and Their Mechanism of Action. *Polymers* **2022**, *14*, 3970. [CrossRef]

polymers

MDPI

Article

Myrica esculenta Leaf Extract—Assisted Green Synthesis of Porous Magnetic Chitosan Composites for Fast Removal of Cd (II) from Water: Kinetics and Thermodynamics of Adsorption

Anjali Yadav [1], Sapna Raghav [2], Nirmala Kumari Jangid [1,*], Anamika Srivastava [1], Sapana Jadoun [3], Manish Srivastava [4,*] and Jaya Dwivedi [1]

[1] Department of Chemistry, Banasthali Vidyapith, Banasthali 304022, India; raoanjli001@gmail.com (A.Y.); jayadwivedi@banasthali.in (J.D.)
[2] Department of Chemistry, Nirankari Baba Gurubachan Singh Memorial College, Sohna 122103, India
[3] Departamento de Química, Facultad de Ciencias, Universidad de Tarapacá, Avda. General, Velásquez, Arica 1775, Chile; sjadoun@academicos.uta.cl
[4] Department of Chemistry, University of Allahabad, Prayagraj 211002, India
* Correspondence: nirmalajangid.111@gmail.com (N.K.J.); sagermanish1@gmail.com (M.S.)

Abstract: Heavy metal contamination in water resources is a major issue worldwide. Metals released into the environment endanger human health, owing to their persistence and absorption into the food chain. Cadmium is a highly toxic heavy metal, which causes severe health hazards in human beings as well as in animals. To overcome the issue, current research focused on cadmium ion removal from the polluted water by using porous magnetic chitosan composite produced from Kaphal (*Myrica esculenta*) leaves. The synthesized composite was characterized by BET, XRD, FT-IR, FE-SEM with EDX, and VSM to understand the structural, textural, surface functional, morphological-compositional, and magnetic properties, respectively, that contributed to the adsorption of Cd. The maximum Cd adsorption capacities observed for the Fe_3O_4 nanoparticles (MNPs) and porous magnetic chitosan (MCS) composite were 290 mg/g and 426 mg/g, respectively. Both the adsorption processes followed second-order kinetics. Batch adsorption studies were carried out to understand the optimum conditions for the fast adsorption process. Both the adsorbents could be regenerated for up to seven cycles without appreciable loss in adsorption capacity. The porous magnetic chitosan composite showed improved adsorption compared to MNPs. The mechanism for cadmium ion adsorption by MNPs and MCS has been postulated. Magnetic-modified chitosan-based composites that exhibit high adsorption efficiency, regeneration, and easy separation from a solution have broad development prospects in various industrial sewage and wastewater treatment fields.

Keywords: adsorption; cadmium removal; chitosan; MNPs; wastewater treatment

1. Introduction

Water is the main component for the survival of all living organisms. However, it is gravely endangered due to the massive amount of pollution caused by domestic, industrial, and agricultural actions. Water scarcity and quality have emerged as major issues for long-term development [1,2]. Many water contaminants, including organic and inorganic constituents, have been mentioned in the literature [3]. Heavy metal-contaminated water bodies pose serious problems owing to their toxic behavior and bioaccumulation [4–7].

Cadmium is one of the non-essential heavy metals that is regarded as highly noxious and carcinogenic due to its non-biodegradability and ability to bioaccumulate in the environment. It is classified as a human carcinogen by the Environmental Protection Agency, (EPA), US, with a maximum permissible limit of 0.005 mg/L in drinking water [8].

Various methods for removing Cd (II) from the aqueous environment that is often used include chemical precipitation, membrane treatment, ion exchange, electro-dialysis, membrane flotation, electrochemical methods, ultra-filtration, etc. [9]. However, these processes, have several limitations, including lower effectiveness, high energy requirements, expensive disposal, and incomplete removal. Adsorption is extensively used for wastewater treatment due to its, quick response, low cost, simple operation, absence of sludge production, and reusability [10–12].

Various types of adsorbents have been used to remove Cd from wastewater. Among them, natural polysaccharide polymers, particularly chitosan, and its derivatives, have gained the attention of researchers owing to their renewable nature, sustainability, and adsorption efficiency [13–15]. Chitosan (CS), which is formed by the deacetylation of chitin, is present in the shells of crustacean crabs and shrimps, insect carapaces, and fungal and plant cell walls. CS, as one of nature's most abundant biopolymers, has sparked widespread scientific interest due to its low cost, non-toxicity, inadequate hydrophilicity, biocompatibility, and biodegradability. It has many functional groups, including -NH_2 and -OH, which facilitate good sorption of heavy metal ions. Raw chitosan, like many other widely used materials, cannot be efficiently separated from aqueous medium using traditional separation methods. As a result, attempts have been made to recycle CS by combining it with magnetic nanoparticles [16,17]. Because magnetic chitosan composites benefit from both chitosan (excellent adsorption performance) and magnetic material (easy magnetic separation), combining CS with a magnetic component is an effective way to address the above-mentioned inadequacies [18,19].

Due to their high chelating capacity and ease of magnetic separation, porous magnetic chitosan (MCS) material has been considered an efficient adsorbent for Cd removal [20]. A schematic representation of the application of MCS for the adsorption of Cd (II) and the reuse of the adsorbent is provided in Figure 1. The combination of MNPs and CS successfully excludes chitosan's inherent drawbacks for application as an adsorbent, and its separation and reuse. Furthermore, the -OH groups on the surface of MNPs can form a hydrogen bonding network with chitosan by interacting with its -OH and -NH_2 groups. As a result, the stability and integrity of the composite are retained even under acidic or alkaline environments [21]. Various chitosan-based materials were synthesized with magnetic NPs because of their lower internal diffusion resistance and higher specific surface area. MNPs, on the other side, get easily oxidized in air and are chemically active resulting in aggregation and loss of magnetism. Because of their small size, these NPs cause secondary contamination. Magnetic cores with larger sizes, on the other hand, can aid in the magnetic separation process [22]. Porous Magnetic chitosan composites have been used extensively and successfully to eliminate pollutants such as dyes, heavy metal ions, and other organic contaminants.

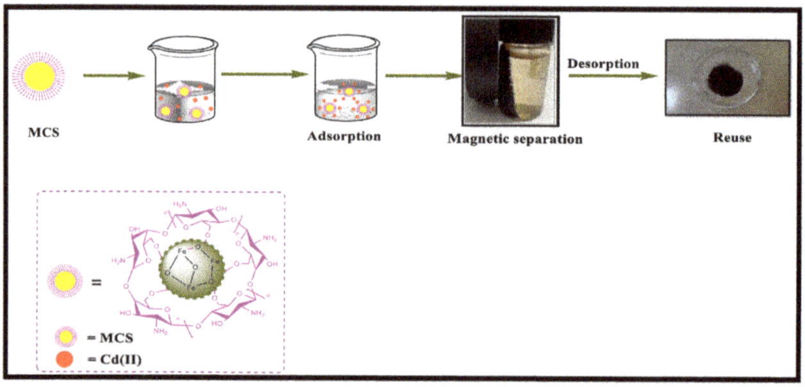

Figure 1. Systematic illustration of magnetic chitosan composite for Cd(II) removal.

During the modification process, functional groups such as hydroxyl and carboxyl groups increased. Various studies are performed for the removal of Cd (II) by magnetically modified material, and the results showed that after the material was magnetically modified, the pH value, specific surface area, and polar oxygen-containing functional groups all increased, resulting in a saturated adsorption capacity. The mechanism of MCS to remove Cd is surface complexation and electrostatic adsorption. The strong affinity of iron oxide for Cd can enhance the complexation between them.

Plant extracts have been proposed as an easy and convenient alternative to chemical and physical methods for the preparation of metallic nanoparticles and their composites in recent years owing to the concerns of sustainability and environmental deterioration using conventional non-renewable sources. MNPs are one of these nanoparticles that have caught the interest of many researchers. *Myrica esculenta* leaf extract contains a variety of bioactive phytoconstituents, including phenolic compounds, glycosides, alkaloids, triterpenoids, and volatile oils [23].

To the best of our knowledge, there has not been much research done to assess the effectiveness of magnetic chitosan employing leaf extract of *Myrica esculenta* (Kaphal) for removing cadmium from aqueous solutions. Therefore, our aim is to develop a green synthetic process to produce chitosan—Fe_3O_4 composite using *Myrica esculenta* leaf extract as a reducing agent. This has resulted in the development of a novel porous magnetic chitosan composite in the present work that turned out to be an efficient material for Cd (II) ion removal. The MCS composite as well as the Fe_3O_4 nanoparticles were characterized using XRD, BET sorptometry, FT-IR, FE-SEM, EDX, and VSM for evaluating their physico-chemical properties and also for determining the effect of such properties on the adsorption process parameters, namely, contact time, pH, adsorbent dose, and initial concentration of the analyte (Cd^{2+}). The potential of MNPs and magnetic chitosan composite as potential adsorbents for Cd (II) was further probed using kinetic models and adsorption isotherms. Moreover, the reusability of the composite was investigated up to seven cycles.

2. Materials and Method

2.1. Materials

$FeCl_2{\cdot}4H_2O$, Chitosan (degree of deacetylation $\geq$ 95%; viscosity: 100–200 mPa.s), $FeCl_3{\cdot}6H_2O$, and acetic acid (CH_3COOH), used during the preparation for adsorbents were of analytical-grade and were procured from Sigma Aldrich (St. Louis, MO, USA). Cd $(NO_3)_2{\cdot}4H_2O$ was purchased from E. Merck, Mumbai, India. All aqueous solutions for the adsorption studies were prepared using Millipore Milli-Q® ultrapure water (Jaipur, Rajasthan). As Cd (II) has been chosen as the model contaminant in this investigation, a solution of Cd (II) of desired concentrations was prepared and used as an adsorbate solution. Dried Kaphal leaves were sourced from the farmlands of Uttarakhand.

2.2. Preparation of Aqueous Leaf Extract of *Myrica esculenta* (Kaphal)

To remove any pollutants, the Kaphal leaves were thoroughly washed with distilled water after being collected from the highlands of Uttarakhand, India. The leaves were manually cut into little pieces, air-dried for a week at room temperature, and then manually ground in a home kitchen grinder; 4 g of the resulting fine powder was then added into double-distilled water and stirred at 70 °C for 45 min. It was then filtered, and the filtrate was stored at 4 °C (Figure 2). The synthesis process for Fe_3O_4 nanoparticles and Fe_3O_4/Chitosan Composite is shown in the Supplementary file.

2.3. Characterization of the Adsorbents

The XRD of the MNPs and the CS composite were recorded on (Bruker D8 Discover X-ray Diffractometer, Karlsruhe, Germany) using Cu Kα radiation. The XRD pattern was recorded in the 2θ range of 10° and 70°.

By using a surface area analyzer N_2 adsorption desorption studies were carried out on Quantachrome Autosorb iQ Surface Analyzer, CIQTEK, Hefei, China). For the analysis of the surface area, the samples were first degassed for 3.0 h to remove volatile gases and were then placed in a surface area analyzer for N_2 adsorption-desorption.

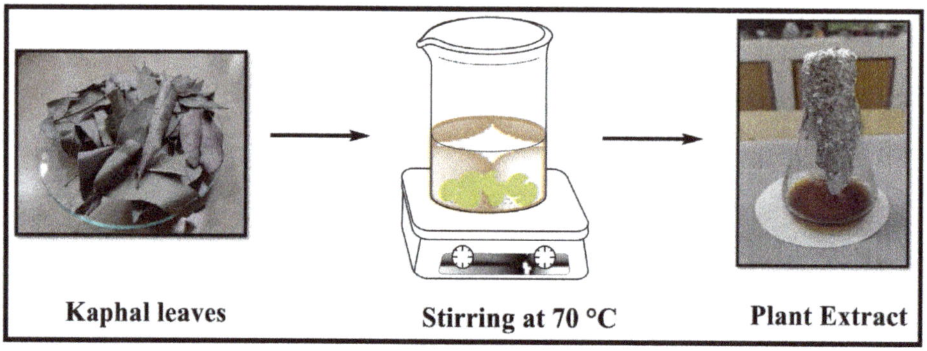

Figure 2. Preparation of plant extract of *Myrica esculenta* (Kaphal).

A Perkin-Elmer FT–IR spectrometer was used to record the FTIR spectra. A spectrum of the adsorbents in the range of 400–4000 cm^{-1} to characterize the nature of chemical bonding and the type of surface functional group.

FE-SEM was used to investigate the morphology of MNPs and MCS composites as well as their elemental composition. FE-SEM equipped with EDAX (FEI QUANTA FEG250, Oregon, OH, USA) and an INCA Energy X-MAX-50, Oregon, OH, USA) was employed to characterize the morphology and chemical composition of the adsorbents used in the study.

A LakeShore 7404 (Lakeshore cryotronics, Westerville, OH, USA) vibrating sample magnetometer was used to determine the magnetization properties of the adsorbents. The magnetic sample is positioned on the sample holder and placed between the electromagnetic poles, normally horizontally, for VSM measurement.

2.4. Batch Adsorption Studies for the Adsorption of Cd (II) Using MNPs and Chitosan/Fe_3O_4 Composite as Adsorbents

To study the optimal adsorption performance and investigate the adsorption mechanism, batch adsorption experiments were performed. Different parameters were examined in fixed ranges, including, adsorbent dose (0.01–0.1 g), pH (2–11), temperature (303, 313, and 323 K), contact time (5–50 min), initial metal concentration (10–100 mgL^{-1}), and adsorption/desorption studies, to better understand the potential of MNPs and MCS as adsorbents for Cd (II). The C_0 (initial concentration) values were used to conduct the adsorption isotherms. After the adsorption process, atomic absorption spectroscopy was utilized to examine the heavy metal ion concentration remaining in the filtrate. The removal effectiveness (%) and equilibrium adsorption capacity (q_e) of the adsorbents, namely, MNPs and MCS composite. For Cd (II) was given by Equations (1) and (2).

$$\% \text{ Adsorption} = \frac{(C_0 - C_e) \times 100}{C_0} \tag{1}$$

$$\text{Adsorption Capacity}(q_e) = \frac{(C_0 - C_e) \times V}{m} \tag{2}$$

where, V is the volume of metal ion solution, m is the mass of adsorbent dose, C_0 is the initial metal ion concentrations, C_e is the residual Cd (II) ion concentration, and q_e is the adsorption capacity at C_e, respectively.

3. Result and Discussion

3.1. Adsorbent Characterization

3.1.1. X-ray Diffraction

To understand the chemical and physical structure of the magnetic particles encapsulated in the CS matrix, X-ray diffraction (XRD) is a highly valuable technique. Figure 3a,b represent the XRD patterns of MNPs and porous MCS composite. The crystalline phase of magnetic NPs and porous magnetic chitosan composite were analyzed by XRD.

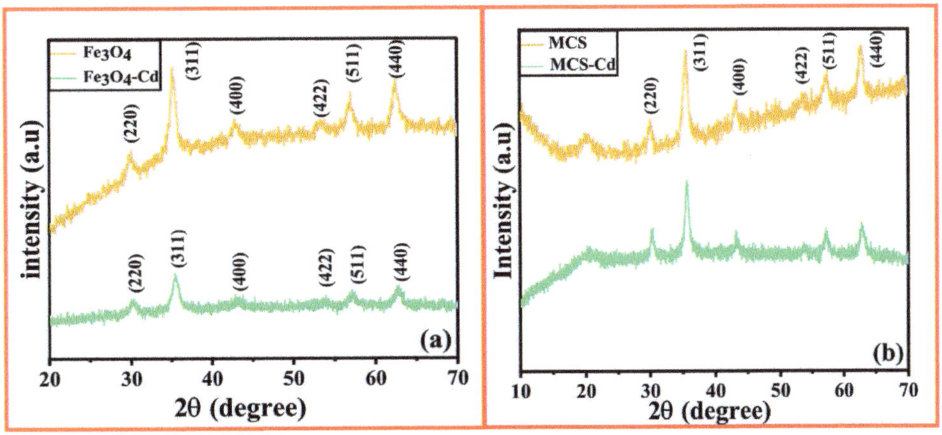

Figure 3. XRD spectra for (**a**) MNPs [before and after Cd (II) adsorption], (**b**) Porous MCS adsorbent [before and after Cd (II) adsorption].

The diffraction peaks at 2θ values of 30.2°, 35.6°, 43.2°, 53.6°, 57.2°, and 62.9° correspond to the (220), (311), (400), (422), (511), and (440) planes of MNPs (JCPDS file number 01-075-0033) [24]. These peaks closely resemble the standard XRD pattern of Fe_3O_4. Thus, the XRD pattern demonstrated the formation of magnetic particles. Diffraction peaks at 2θ values in the range of 21° to 28° were attributed to amorphous chitosan (Figure 3b). Moreover, the peaks typical of Fe_3O_4 were also present in the composite as expected [25].

3.1.2. BET Sorptometry for Evaluating the Textural Properties of Adsorbents

N_2 gas adsorption-desorption isotherms were observed in the relative pressures (p/p_0) values ranging from 0 to 1 to estimate the surface area and porosity values of the synthesized adsorbents. BET sorptometry was performed to investigate the average pore radius, surface area, and pore volume of the material. About 25.0 mg of material was degassed at 300 °C for analysis. According to IUPAC standards, the N_2 adsorption-desorption isotherms are Type IV, which agrees with the mesoporous nature of the composite [26]. The surface area and total pore volume of MNPs and MCS were 105 m^2/g, 173 m^2/g, and 0.3410 cc/g, and 0.4305 cc/g, respectively (Table 1). The Langmuir surface area of the MNPs and MCS was 201.084 m^2/g and 536.934 m^2/g, respectively (Figure 4). Thus, the higher surface area is responsible for the higher adsorption capacity of porous MCS compared to MNPs.

Table 1. Surface parameters of MNPs and MCS adsorbents before cadmium adsorption.

Parameters	MNPs	MCS
BET specific surface area (m^2/g)	105	173
Langmuir surface area (m^2/g)	201	537
Average pore size (Å)	64.36	49.77
Total pore volume (cc/g)	0.3410	0.4305

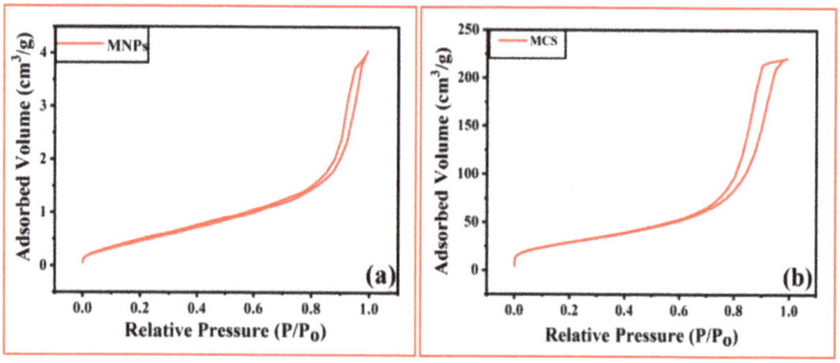

Figure 4. N_2 sorption isotherm of (**a**) MNPs, (**b**) MCS composite.

The Langmuir curve and BET multipoint of MNPs and porous MCS are expressed in detail in the Supporting Information (Figure S3).

3.1.3. Fourier Transform Infrared (FT-IR) Spectroscopy

For understanding the nature of chemical bonding and the kind of surface functionalities the adsorbents used in this study were examined using FT-IR spectroscopy. The FT-IR spectra of chitosan and MNPs are shown in Figure 5a. Likewise, the FT-IR spectra of the MCS composite before and after the adsorption process are shown in Figure 5b.

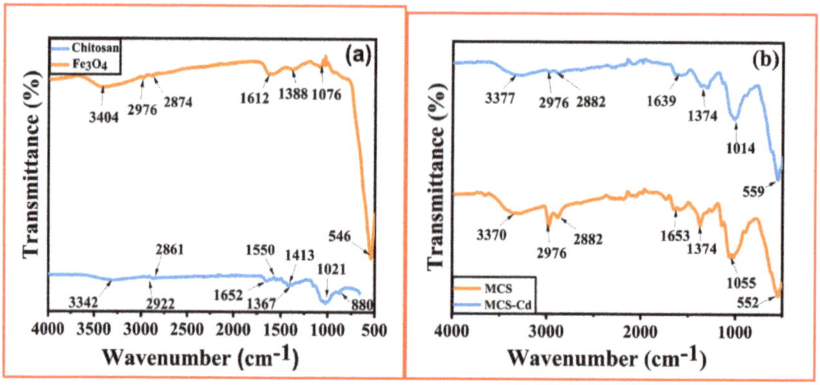

Figure 5. TIR spectra for (**a**) Chitosan and Fe_3O_4 nanoparticles, (**b**) MCS adsorbent [before and after Cd (II) adsorption].

The absorption bands in the range of 3200–3400 cm^{-1} corresponded to N–H and O–H stretching vibrations of CS (Figure 5a). The bands at 2922 cm^{-1} and 2861 cm^{-1} were due to the C-H symmetric and unsymmetric stretching vibrations of the $-CH_2$ groups in CS. The -NH deformation, C-N axial deformation, $-CH_3$ bending vibration, and stretching vibration of C-O-C in the chitosan structure appeared at, around 1413, 1367, and 1021 cm^{-1}. The existence of chitosan in the composite and its structural integrity was confirmed by the appearance of peaks at 1652 cm^{-1} and 1550 cm^{-1}, which corresponds to the N-H bending vibration of primary amine [27].

The absorption bands characteristic of MNPs were observed in the FT-IR spectrum (Figure 5a). Metal oxygen bonds typical of MNPs were observed in the range of 400 and 850 cm^{-1}. The absorption band at 546 cm^{-1} was attributed to the Fe-O stretching vibration of MNPs. This band is very sharp and has strong intensity, indicating the crystallinity of the sample. The broad bands around 3404 cm^{-1} and 1612 cm^{-1} were due to the O–H stretching vibration of surface-adsorbed water molecules. The band located around 1388 cm^{-1}

and 1076 cm^{-1} were attributed to the unsymmetric and symmetric stretching vibrations of COO- [28].

Similar peaks were observed in the FT-IR spectrum of the MCS composite. In addition, some specific bands correspond to the original formation of new bonds between chitosan and Fe_3O_4 and some bands of chitosan have moved slightly from their original positions owing to atomic-level interactions between chitosan and MNPs. In the FT-IR spectrum of the MCS composite (Figure 5b), specific bands were observed at 3370 cm^{-1} which was attributed to N-H stretching vibration. The band at 1652 and 1550 cm^{-1} was attributed to C=O stretching vibration in the CS, for the N-H scissoring from the primary amine due to free amino groups in the CS and was compared with the standard chitosan [29]. The peak at 552 cm^{-1} for the Fe-O group was due to bare magnetic nanoparticles. This spectrum (Figure 5b) unambiguously showed the presence of both CS and MNPs in the composite [30]. The NMR spectrum of *Myrica esculenta* leaf extract has been already discussed by Nguyan et al. [31].

3.1.4. FE-SEM-EDX

FE-SEM was used to find out the morphology of MNPs and MCS composite, and EDAX was utilized to determine the elemental composition. SEM and EDX images of both the adsorbents were displayed in Figure 6 (EDX result shown in Figure S4). Spherical-shaped particles with agglomeration were observed in synthesized magnetic NPs (Figure 6a) as well as MCS (Figure 6b) composite in the FE-SEM images. Fe_3O_4 NPs exhibited a smooth surface, and the particles had an irregular shape, as observed in Figure 6a. In Figure 6b, the FE-SEM image of MCS was shown to exhibit a rough, granular surface after chitosan was bonded to Fe_3O_4. Agglomeration of the particles on the surface of the composite cannot be ruled out [32]. EDX spectra of the Fe_3O_4 sample confirm the presence of iron (Figure S4). The EDX spectra of MNPs and MCS are shown in Supporting Information (Figure S4). The mass percentage of carbon in MCS composite is higher than in MNPs, which proved MCS composite had lesser hydrophilicity properties.

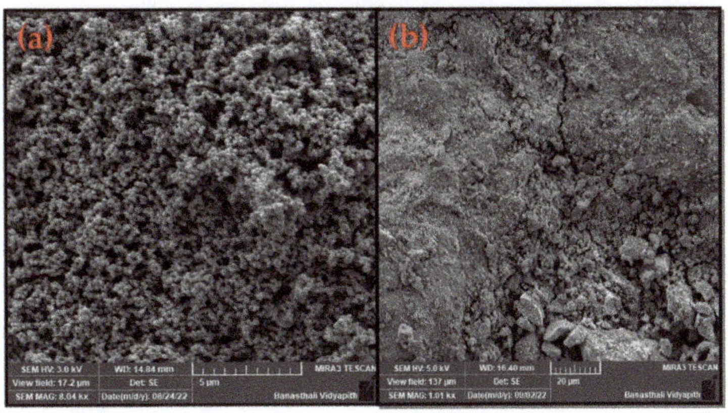

Figure 6. FE-SEM image of (**a**) MNPs, and (**b**) MCS composite.

3.1.5. Vibrating Sample Magnetometer Study for Evaluating the Magnetic Property of the Adsorbents

To evaluate the adsorbents' magnetic properties, VSM (Vibrating-sample magnetometer) tests were utilized [20]. The magnetic properties of investigated magnetic materials were studied by measuring the magnetization curves (Figure 7). The results showed that all magnetization curves pass through the origin, which indicated that there was no residual magnetization occurring in test samples and these materials have superparamagnetism [33]. By changing H between O_e +10,000 and −10,000 O_e, magnetization hysteresis was produced. MNPs had a saturation magnetization of 55.070 emu/g. The saturation magnetization of MCS decreased to 24.186 emu/g after the formation of composite between MNPs and chitosan, indicating that the adsorbent was still super-paramagnetic and that it would be easier to separate the adsorbent from solution by gravity and magnetism within a short period of time [34]. The lower M_s value for MCS could be explained by the quenching of surface magnetic moment in the material owing to the presence of non-magnetic species, such as CS as expected [35]. Even though MCS' saturation magnetization value was obviously lower than that of MNPs, the application of a magnet allowed the adsorbent to immediately aggregate and to be separated in the 20 s [32].

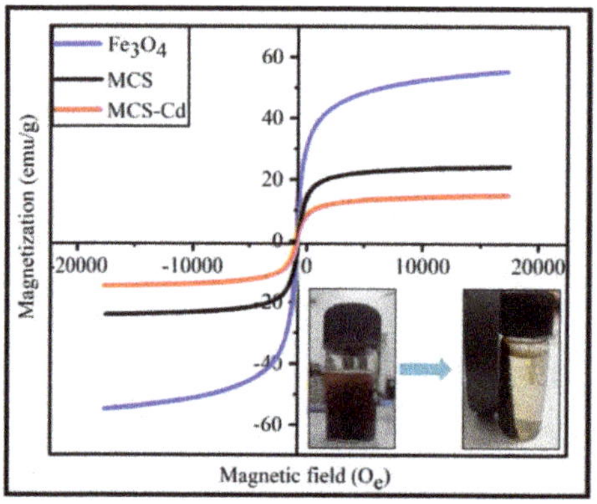

Figure 7. Magnetization curves of the Fe_3O_4, MCS composite, and MCS composite after Cd (II) adsorption.

3.2. Adsorption Experiments

3.2.1. Study of Kinetics of Adsorption of Cd (II) onto the Adsorbents, MNPs, and MCS Composite

To find out the appropriate rate expression and adsorption mechanism, the adsorption kinetics was examined [36]. Several models were examined for the adsorption of contaminants in water onto the surface of adsorbents and also to identify the main mechanism of such adsorption process and the kinetics of adsorption [37]. Adsorption kinetics was investigated in this work by fitting Lagergren's pseudo-first-order (PFO) and Ho's second-order models. The adsorption process was determined using the Elovich and intra-particle diffusion (IPD) models [38].

The PFO kinetic model represents weak interaction between sorbate and sorbent predominantly proceeding via physisorption. The PFO kinetic model is represented below [39]

$$\ln (q_e - q_t) = \ln q_e - k_1 \times t \quad (3)$$

PSO kinetic model is based on chemisorption. Pseudo–second order kinetics is represented as [40].

$$t/q_t = 1/(k_2 \times q_e^2) + t/q_e \quad (4)$$

The IPD model is given by Equation (5).

$$q_t = k_{id} \times \sqrt{t} + C \quad (5)$$

The Elovich model is specified by Equation(6).

$$q_t = (1/\beta) \ln (\alpha \times \beta \times t) + (1/\beta) \ln t \quad (6)$$

where, k_1 (min^{-1}) and k_2 (g/mg/min) are the rate constants of PFO and PSO, k_{id}—intraparticle diffusion rate constant (mg/(gmin$^{-0.5}$)), constant C is the y-intercept, α is the initial rate of adsorption, β is the desorption constant [41].

Four kinetic models were studied, and the kinetic parameters deduced from various models are summarized in Table S1 [42,43].

From the R^2 analysis of kinetic models, the best-fitted model for Cd (II) adsorption on MNPs at 50 mg/L was Elovich, and at 100 mg/L was IPD. Also, from the R^2 analysis of kinetic models, the PSO model was the best fit for MCS. The pseudo-second-order model's correlation coefficient (R^2) was observed to be significantly greater than other models employed for MCS adsorbent, which means that the mechanism of adsorption was governed by this model (Figure 8).

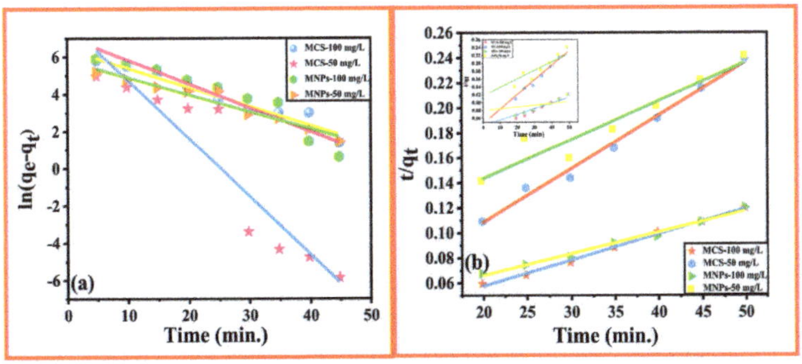

Figure 8. (**a**) PFO and (**b**) PSO for MNPs and MCS composite.

3.2.2. Adsorption Isotherm Models

Adsorption isotherms describe how adsorbents and adsorbates interact in aqueous media at the attained saturation point. The most popular isotherm models, including Langmuir, Freundlich, Dubinin–Radushkevich, and Temkin isotherm models, were utilized.

The Langmuir isotherm is characterized by single-layer sorption onto a surface with countless open sites without interaction between the adsorbate molecules under the assumption that the material is regular and homogeneous. The model's primary problem is its assumption that the energies of the adsorbent sites at each location are uniform. Equation (7) describes the Langmuir isotherm's linear for [44]

$$\frac{1}{q_e} = \frac{1}{K_L q_{max} C_e} + \frac{1}{q_{max}} \quad (7)$$

where q_{max} represents the material's maximum adsorption capacity, and K_L is the Langmuir adsorption constant (L/mg) representing the attraction of binding sites [45].

R_L, the dimensionless constant separation factor is employed to express the important characteristic of the Langmuir isotherm:

$$R_L = \frac{1}{1 + K_L C_0}$$

The value of R_L shows whether the adsorption isotherm is favorable, linear, or unfavorable. The value of R_L was found in the range of 0.0717–0.898 for the MCS composite. This shows the efficient interaction between the MCS composite and cadmium ions [46]. Figure 9a shows the Langmuir plot for MNPs and MCS composite. The values of K_L and q_{max} are calculated using the slope and intercept of the linear regression plot of $1/C_e$ vs. $1/q_e$, which also provides the R^2 value, which indicates how well the experimental results correspond with the mathematical isotherm model.

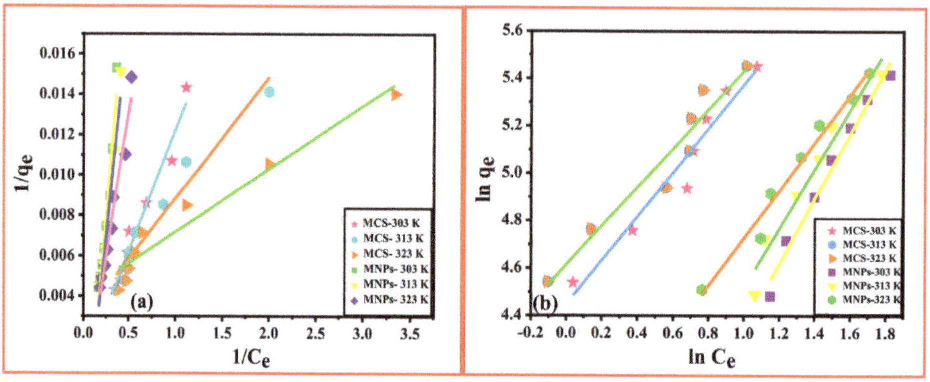

Figure 9. (**a**) Langmuir, (**b**) Freundlich curves for Fe_3O_4 and MCS composite as adsorbents for Cd (II) at different temperatures.

If adsorption happens on heterogeneous surfaces, the Freundlich adsorption isotherm is an empirical equation that is utilized to describe multilayer (physisorption) adsorption as well as monolayer (chemisorption) [42].The linearized Freundlich isotherm equation is explained by Equation (8).

$$\text{Logq}_e = \log K_F + \frac{1}{n}\log C_e \tag{8}$$

where K_F is the Freundlich constant of the adsorbent, and n is the adsorption intensity of the adsorbent. 1/n is the adsorption intensity which signifies the heterogeneity of the adsorbent sites as well as the relative distribution of energy. K_F and n are dimensionless constants. The value of n > 1 and 1/n < 1 indicates the favorable condition for adsorption. In this study, the value of n ranged between 1.075 and 1.222 which is greater than 1 for MCS composite, indicating favorable adsorption. The graph of $\ln C_e$ Vs $\ln q_e$ gives a straight line with intercept K_F and slope 1/n as shown in Figure 9b. The R^2 values are 0.97, 0.95, and 0.97 for MNPs at 303, 313, and 323 K and 0.94, 0.92, 0.92 for MCS at 303 K, 313 K and 323 K (Table S2) [45].

The Temkin Isotherm Model was used to characterize the adsorption considering the interaction between adsorbate and adsorbent, which resulted in a linear reduction in isotherm when the heat of adsorption of all molecules in the layer was included Equation (9) provides the isotherm:

$$q_e = \beta \ln K_T + \beta \ln C_e \tag{9}$$

where, $\beta = RT/b$.

β represents the heat of adsorption (J/mol) and K_T is the Temkin isotherm constant (L/g). The plot of q_e Vs $\ln C_e$ produces a straight line with slope β and intercept $\beta \ln K_T$. *b* is the Temkin constant which is associated with the sorption heat (J/ mg) (Figure S6).

The adsorption is characterized by a uniform distribution of binding energies. The binding energies were 199.28, 185.86, and 146.10 J/mol for MNPs and 132.85, 94.68, and 69.43 J/mol for MCS at 308, 303, and 298 K, respectively [46–48].

The D-R isotherm can be used to find out the adsorption mechanism.

Dubinin-Radushkevich (D-R) isotherm (Dubinin, 1960) is:

$$\ln q_e = \ln q_m - \beta \varepsilon^2 \tag{10}$$

ε^2 and b are constants. The D-R constants q_m and b were calculated from the slope and intercept of $\ln q_e$ Vs ε^2 [37]. The q_m calculated by the D-R model was 640.699 mg/g and 598.7594 mg/g for MNPs and MCS composite at 303 K, respectively (Figure S4).

Four isotherm models were studied, and the significant parameters are shown in Table S2.

3.3. Thermodynamic Studies

To investigate the spontaneity, feasibility, and endo/exothermic nature, thermodynamic data plays an important role. The effect of solution temperature (20–50 °C) on Cd (II) ion adsorption was investigated under optimal conditions. Standard entropy (S), enthalpy (H), and Gibbs free energy (G) were calculated as thermodynamic parameters. The Van't Hoff equation was used to calculate these thermodynamic parameters.

$$K_d = \frac{q_e(w/v)}{C_e} \tag{11}$$

$$\ln K_d = \Delta S^0/R - \Delta H^0/RT \tag{12}$$

$$\Delta G^0 = -RT \ln K_d \tag{13}$$

where ΔG^0 is the Gibbs free energy, ΔS^0 is the entropy, and ΔH^0 is the enthalpy, K_d is the distribution coefficient for the adsorption process.

Equation (11) can be used to calculate the value of K_d after the values of q_e and C_e have been determined experimentally. Using the value of K_d in Equation (12), the values of enthalpy and entropy can be calculated. Using Equation (12), the slope and intercept of the plot ln (K_d) vs. 1/T in Figure 10 would give the values ΔH^0 and ΔS^0, respectively.

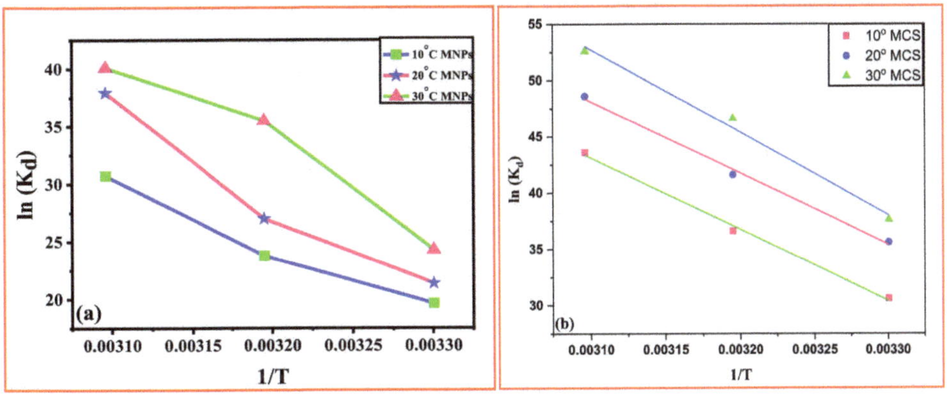

Figure 10. Linear dependence of ln (Kd) on 1/T based on adsorption thermodynamics for (**a**) MNPs and (**b**) MCS at various temperatures.

From Table S3, the positive value of ΔH shows that the adsorption process is endothermic, which confirms the chemisorption nature of adsorption. The negative value of ΔG at all three temperatures for MNPs and at 303 K for MCS composite indicated that the spontaneity of the process was favored at these temperatures. The positive value of ΔS showed a rise in randomness during the adsorption of Cd (II).

3.4. Reusability of Adsorbent

Metal ion desorption from the sorbent and adsorbent regeneration are critical challenges in terms of adsorbent reusability. The major goals of the regeneration process are to recover useful components from the adsorbed phase and to restore the sorption capability of the exhausted material [49]. Figure 11 displays the results of seven adsorption-regeneration cycles (Figure 11a,b). Using various agents including HCl (0.05 M, 0.1 M), and NaOH (0.05 M, 0.1 M), the batch technique was used to evaluate the desorption of the sorbed cadmium from Fe_3O_4 and MCS composite. It was discovered that 0.1 M HCl produced the greatest amount of Cd (II) desorption (98%) while using 0.1 M NaOH, the desorption efficiency was observed to be 84%. Adsorption, as well as desorption cycles, were repeated seven times utilizing the same adsorbents to assess the adsorbent's reusability. After each cycle of adsorption, the solid portion of the adsorbent was centrifuged and further mixed in DI water. This mixture was then agitated for about 1/2 h. The remaining suspension was once again used for a different batch experiment. Seven rounds of a similar process were performed. The removal capacity of the regenerated sorbent gradually reduced in comparison to the original adsorbent. As a result, MCS was an effective reusable adsorbent that could be used to recover Cd (II) ions from aqueous medium. The percentage removal of Cd (II) decreased from the first to seventh cycle (i) 99.99% to 80.44% for MCS, (ii) 95.1% to 42% for MNPs. This demonstrates that the MCS may be reused at least seven times while retaining good sorption efficiency. As a result, it was possible to conclude that magnetic chitosan composite can be employed as an efficient material in the removal process.

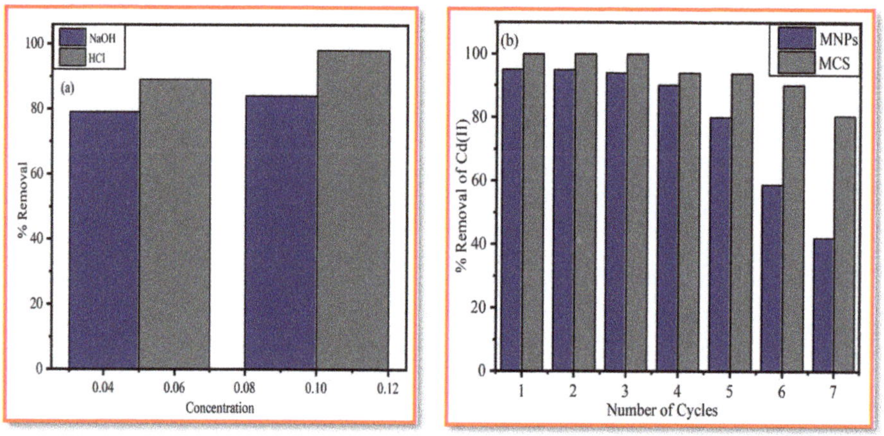

Figure 11. Regeneration studies using (**a**) 0.1 M HCl and 0.1 M NaOH, (**b**) Regeneration studies of MNPs and MCS (0.1 M HCl).

3.5. Comparative Studies

$MnFe_2O_4$/CS microspheres were prepared by coating chitosan on $MnFe_2O_4$. Maximum sorption capacity was observed to be 60.6 mg/g for Cd (II) removal. The experiments showed that the composite could maintain sorption capacities after three cycles of adsorption–regeneration [50]. A new nanobiosorbent based on methionine-glutaraldehyde Schiff's base-modified cross-linked chitosan magnetic beads was prepared by Salehi et al. (2020) for the elimination of Cd (II). The utmost sorption capacity was observed to be 163.9 mg/g [51]. Chitosan and sodium tripolyphosphate cross-linked chitosan beads were synthesized for cadmium removal from an aqueous medium by Babakhani and Sartaj (2020). The maximum sorption capacity was observed to be 99.87 mg/g [52]. Attapulgite/$CoFe_2O_4$@SiO_2-chitosan/EDTA was synthesized by solvothermal and sol-gel techniques. The maximum sorption capacity was observed to be 127.79 mg/g [53]. Fan et al. (2017) studied the sorption of cadmium by using magnetic chitosan nanoparticles. The maximum sorption capacity was observed to be 36.42 mg/g for Cd (II). Relatively, the maximum sorption capacity for pure Fe_3O_4 nanoparticles was observed to be 13.04 mg/g [54]. Li et al. (2017) prepared chitosan/polyethylenimine grafted magnetic gelatin to eliminate cadmium from wastewater. According to the Cd (II) sorption results, the process can be explained by a monolayer forming on the surface of the material with 321 mg/g from the Langmuir isotherm [55]. The relative adsorption capacity of various adsorbents for Cd (II) and the relative adsorption performance of MCS-based adsorbents for heavy metal ions removal are shown in Tables 2 and 3, respectively.

Table 2. The relative adsorption capacity of various adsorbents for Cd (II).

S. No.	Materials	Q_e(mg/g)	Reference
1	Egg-albumen-formaldehyde-based magnetic polymeric resin	149.3	[56]
2	Mesoporous magnetic nanocomposite	158.68	[57]
3	Amino-decorated magnetic metal-organic framework	693.0	[58]
4	Carboxymethyl chitosan/sodium alginate/graphene oxide@ Fe_3O_4 beads	86.28	[59]
6	Poly(γ-glutamic acid) modified magnetic Fe_3O_4-GO-(o-MWCNTs) hybrid nanocomposite	625.00	[60]
7	Citric acid- and Fe_3O_4-modified sugarcane bagasse	33.2	[61]
8	Fe_3O_4@Biuret-formaldehyde pre polymeric resin	92.6	[62]
9	Fe_3O_4/$FeMoS_4$/MgAl-LDH nanocomposite	140.50	[63]
10	Fe_3O_4@PDA microspheres	296.4	[64]
11	Fe_3O_4/SiO_2/PP	30.1	[65]
12	Fe_3O_4 nanoparticles	290	This work
13	Fe_3O_4/Chitosan composite	426	This work

Table 3. The relative adsorption performance of MCS-based adsorbents for heavy metal ions removal.

S. No.	Materials	Heavy Metals	Q_e(mg/g)	Reference
1.	Magnetic chitosan composite	Ni(II) Cu(II) Pb(II)	108.9 216.8 220.9	[35]
2.	Magnetic chitosan nanocomposites modified with graphene oxide and polyethyleneimine	As(V) Hg(II)	220.26 124.84	[66]
3.	Chitosan magnetic beads modified with cysteine glutaraldehyde Schiff's base	Cu(II) Cr(VI)	156.49 138.53	[67]
4.	PEI-grafted magnetic gelatin	Pb(II) Cd (II)	341 321	[68]
5.	Magnetic chitosan/polyethyleneimine embedded hydrophobic sodium alginate composite	Cr(VI) Cu(II)	87.53 351.03	[69]
6.	Magnetic graphene oxide/chitosan composite beads	Ni(II)	80.48	[70]
7.	Magnetic Fe_3O_4/Chitosan nanoparticles	Pb(II) Cd (II)	79.24 36.42	[54]
8.	Magnetic chitosan/graphene oxide (MCGO) materials	Pb(II)	76.94	[71]
9.	Xanthate-modified cross-linked magnetic chitosan/poly(vinyl alcohol) particles	Pb(II) Cu(II)	59.855 139.797	[72]
10.	Magnetic anaerobic granule sludge/chitosan composite	Pb(II) Cu(II)	97.97 83.65	[73]

3.6. Adsorption Studies

3.6.1. Variation of Adsorbent Dose

The amount of the adsorbent used is a crucial factor in calculating sorption capacity. Studies were carried out by changing the sorbent dosage from 0.01 to 0.1 g/L, while all other factors such as contact duration, temperature, and rpm were held constant to optimize the sorbent dose for the elimination of Cd (II) ions from the aqueous medium. Adsorption capacity experienced a substantial reduction with increasing adsorbent doses. This may be explained by the fact that a greater sorbent dose makes it easier for Cd (II) ions to access active sites on the pores of magnetic chitosan composite, which enhances removal rates. More easily accessible functional groups and adsorption sites for metal ions were responsible for the increment. However, higher adsorbent dosage most certainly enhances particle interactions, like aggregation, which is due to high sorbent concentration. As a result, the adsorbent's active surface area is significantly reduced, which lowers its capacity for adsorption. The adsorption process achieves the equilibrium point due to the overcrowding of adsorbent particles brought on by the overlapping of adsorption sites after a gradual decline in the sorption capacity of Cd (II) ions from the wastewater. The decline in the ratio of Cd (II) per mass unit of sorbent might potentially account for this [20,22]. The adsorption capacity for MCS fell from 613.75 mg/g to about 249.99 mg/g for $C_o(100)$ and from 488.7 mg/g to 124.99 mg/g for $C_o(50)$ as the sorbent dosage rose from 0.01 g/L to 0.1 g/L. Also, the adsorption capacity for MNPs fell from 363.75 mg/g to 248.75 mg/g for $C_o(100)$ and from 238.75 mg/g to 123.875 mg/g for $C_o(50)$ as the sorbent dosage rose from 0.01 g/L to 0.1 g/L, as shown in Figure 12a.

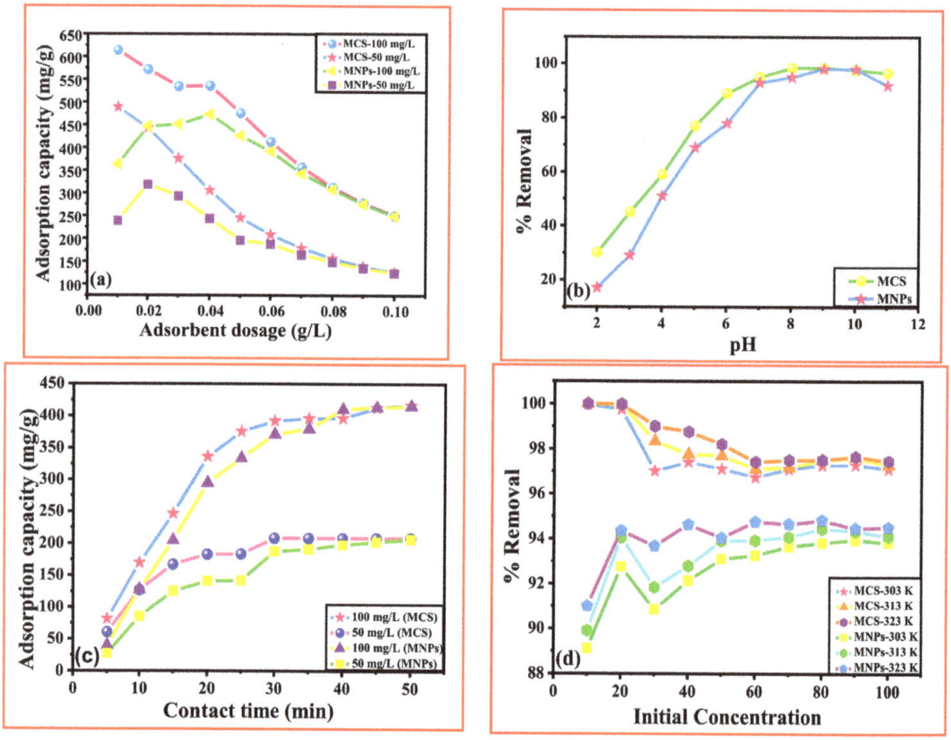

Figure 12. Effect of (**a**) adsorbent dose, (**b**) pH, (**c**) contact time, (**d**) initial concentration for adsorption of Cd (II) on Fe_3O_4 and MCS composite, respectively.

3.6.2. Variation in pH

The solution's pH is a very important factor that strongly affects the sorption process because it influences the charge of the surface by protonation and deprotonation of the material and degree of ionization. By conducting equilibrium adsorption experiments at various pH levels, the impact of the suspending medium's pH on Cd (II) removal was investigated. In the current study, the influence of solution pH on the sorption of Cd (II) by MNPs and MCS was examined throughout a pH range of 2–11 (Figure 12b). Temperature, dose, rpm, dosage, and other factors were all held constant. Low adsorption efficiency was found at low pH which was caused by competition between the Cd (II) ion and the H^+ ion for placements on material active sites. Additionally, the concentration of adsorbent positive charges is large in acidic conditions, which results in electrostatic repulsion between the magnetic composite and the Cd (II), decreasing the efficiency of Cd (II) elimination. The removal of cadmium reduces when the pH rises from 2 to 6 because the positive charges on the adsorbent surface are reduced, repulsive forces are stronger, and the positive charges are less attractive [74]. The electrostatic attraction was strengthened as the pH increased, which progressively raised the degree of deprotonation of functional groups and improved the sorption capacity. The utmost removal was seen between pH 6.0 and 8.0 [75].

3.6.3. Variation of Contact Time

It has a pronounced effect on the elimination of adsorbate species from aqueous medium. To assess the Cd (II) adsorption behavior by MNPs and magnetic chitosan composite, the influence of time of contact between the adsorbent and adsorbate was optimized in this work by altering the contact duration from 5 to 50 min. By separating the supernatant at different times, the effect of contact time on sorption capacity was measured. The adsorption rate was quick in the beginning, as seen in Figure 12c. The initial strong absorption of Cd (II) ion was owing to high adsorbate and adsorbent interactions with low solute-solute interaction and more adsorption site accessibility. As the time increased, equilibrium was attained in the adsorption as more and more active sites were occupied. However, after equilibrium was reached, mass transfer diminished and the repulsion of the adsorbate molecule on the surface and inside the solution increased.

It took around 50 min to achieve equilibrium for uptake of Cd (II) [76]. The increment in time can offer plenty of chances for Cd(II) to adsorb to the composite surface. Figure 12c shows that on raising the contact time from 5 to 50 min, the sorption capacity increased from 81.458 to 416.65, 60.625 to 208.33, 39.79 to 414.79, and 27.29 to 206.458 mg/g for MCS (100 mg/L), MCS (50 mg/L), MNPs (100 mg/L), and MNPs (50 mg/L).

The results demonstrated that the adsorbents showed fast removal of Cd (II) within 50 min which is faster when we compare with the reported literature of other adsorbents employed for the Cd (II) removal [77].

3.6.4. Variation of Concentration

The elimination of Cd (II) by adsorption on MNPs and magnetic composite (MSC) was examined in relation to the effect of initial concentration. It appears that the metal ion's initial concentration is significant and influences the sorption capacity. Because the initial concentration of Cd (II) may offer the driving force required to transfer a mass of Cd (II) between the water phase and the adsorbent surface, the concentration of Cd (II) can also impact the performance of the sorption process.

With the rise in the concentration of Cd (II), there was a decrease in % removal. The experiments were performed at different initial concentrations varying from 10 to 100 mg/L. The graph (Figure 12d) illustrates that with an increase in the concentration of Cd (II) ions, there is a decrease in Cd (II) removal % and a rise in sorption capacity. The presence of active surfaces and sites in the MNPs and MCS composite structure for the assignment of Cd (II) can be related to improving efficiency at low initial concentrations of Cd (II). Additionally, when the concentration of cadmium rises, the %-age of Cd (II) removal decreases, which is explained by the saturation of the MSC's active sites above a specific level of Cd (II), at which point the system enters equilibrium and no further Cd (II) adsorption takes place [22].

The % removal of Cd (II) by adsorption onto the adsorbent decreases from 99.96 to 97.09, 99.99 to 97.25, and 99.999 to 97.45 for MCS (303 K, 313 K, and 323 K) and from 89.1 to 93.81, 89.9 to 94.1, and 91.0 to 94.5 for MNPs (303 K, 313 K, and 323 K), respectively (Figure 12d).

4. Adsorption Mechanism

The Cd (II) ion has an empty d orbit, and both nitrogen and oxygen atoms have lone electron pairs that can attach metal ions to create the complex via electron pair sharing. Fe_3O_4 enclosed in CS NPs and MNPs not only provides rich functional groups (-OH, -COOH, and -CO-NH-) as a binding site for heavy metal ions but also has a greater specific surface area to maximize functional group utilization. Functional groups on Fe_3O_4-loaded CS NPs and MNPs are freely available for metal ion coordination bonding.

Since negatively charged adsorbents might be quickly absorbed by positively charged groups, such as $–OH^{2+}$ and $–NH^{+}_3$ in acidic conditions, the electrostatic attraction was a potential adsorption mechanism. The negatively charged functional groups like COO- and AOA on the surface of Fe_3O_4 -loaded CS NPs and the Cd (II) ion may establish an electrostatic interaction. Therefore, under the influence of both coordination bonds with the oxygen atom in Fe_3O_4 and electrostatic attraction, the adsorbents demonstrated a significant adsorption capacity for the Cd (II) ion.

The amines and secondary alcohol functional groups are the major sites for complexing, according to the FTIR analysis of the magnetic chitosan composite before and after Cd (II) adsorption (Figure 13). There is a slight shift in the position of various bands in the MCS after the adsorption of the metal ion, Cd (II). From the XRD data, there was a slight shift in the diffraction peaks after the adsorption of Cd (II). Also, from the magnetization results, the decrease in magnetization indirectly asserted the formation of a composite between CS and MNPs.

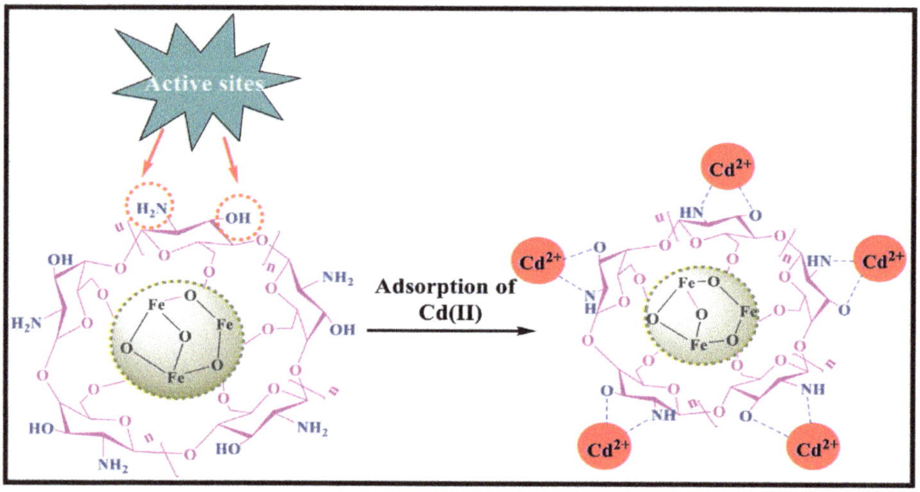

Figure 13. Proposed mechanism for Cd (II) capture by MCS composite [78].

The oxygen of the alcohol and the nitrogen of the amino group has a pair of electrons that can add themselves to a proton by coordinated covalent bonds. While nitrogen has a larger capacity to donate its pair of electrons to a cadmium ion to form a complex through a coordinated and covalent bond, oxygen has a stronger attraction of the electron pair by the atom nucleus. This leads us to suggest that the mechanism shown in Figure 13 governs the formation of the complexes between Cd (II) and adsorbent. In this mechanism, the cadmium ion acts as a Lewis acid that may take in electron pairs due to its vacant orbitals. The amine and hydroxyl groups, on the other hand, which have non-shared electron pairs, serve as Lewis bases by giving their electron pair.

5. Conclusions

In the present study, magnetic chitosan composition was synthesized via a green route and thoroughly characterized by a variety of physical-chemical techniques. Afterward, the composite materials were successfully utilized for the adsorption of Cd (II). Elaborate studies on the kinetics and thermodynamic aspects of adsorption were carried out.For comparison, MNPs were also used as adsorbents for Cd (II). MCS showed excellent and high adsorption capacity compared to native MNPs. The adsorption capacity values of

426 mg/g for MCS and 290 mg/g for NPs were observed. The BET surface area values of MCS and NPs were found to be 178 and 105 m^2/g, respectively. Both adsorbents showed second-order kinetics. The optimum process parameters include an adsorbent dose of 0.05 g/L, a contact time of 50 min, and a pH value of 6.0 with an initial Cd (II) concentration of 100 and 50 mg/L. The adsorption processes on both the adsorbents were feasible and spontaneous as evident from the thermodynamic parameters. The composite material showed good regeneration capacity with an 80.44% removal tendency up to the seventh cycle while MNPs showed low removal capacity of 42% by the seventh and the last cycle. This research provides some insights into the aspects that influence the design of adsorbents with superior performance and easy recovery for Cd (II) ion absorption. Magnetic composites have promising applications in water treatment. Its outstanding selective adsorption considerably improves the material's adsorption efficiency, and it may be specially treated for wastewater enriched with various compounds. Furthermore, the magnetic chitosan adsorption material is easy and simple to recycle, with a high recycling rate. These benefits can lower wastewater treatment costs and increase economic efficiency. So, from the present study, it is concluded that the green synthesized MCS composite with MNPs enhanced the stability of hybrid material and also improved the removal capacity of Cd (II) with good regeneration ability and effective field study results.

Supplementary Materials: The following supporting information can be downloaded at: https://www.mdpi.com/article/10.3390/polym15214339/s1, Figure S1: Preparation of Fe_3O_4 nanoparticles using kaphal leaf extract, Figure S2: Preparation of chitosan/Fe_3O_4 composite, Figure S3: (a) Langmuir curve, (b) BET multipoint of MNPs and MCS, Figure S4: EDX images of (a) MNPs and, (b) MCS composite, Figure S5: title (a) IPD, (b) Elovich kinetic curves for MNPs and MCS composite, respectively, Figure S6: (a)Temkin, (b) D-Radsorption isotherm curves for MNPs and MCS composite respectively, Figure S7: (a) Langmuir, (b) Freundlich non-linear isotherm curves for MNPs and MCS composite respectively, Table S1: Kinetic parameters of PFO, PSO, IPD, Elovich models for Fe_3O_4 and MCS composite, Table S2: Adsorption parameters of four isotherm models, Table S3: Thermodynamic parameters for MNPs and magnetic chitosan composite under optimum conditions title; Table S4: Effect of coexisting ions on the adsorption of Cd(II) by MCS composite.

Author Contributions: Writing—original draft, A.Y.; review and editing, S.R.; investigation, formal analysis, data curation, N.K.J.; conceptualization, methodology, A.S.; data curation, writing—review and editing, S.J.; Supervision, M.S.; methodology, data curation, J.D. All authors have read and agreed to the published version of the manuscript.

Funding: This research is receiving no funding.

Data Availability Statement: Data are available upon request.

Acknowledgments: We are grateful to the authorities of the Department of Chemistry, Banasthali Vidyapith (India) for the moral support and for providing the necessary facilities for the overall experimental work. Author Sapana Jadoun is grateful to Universidad de Tarapacá for its support.

Conflicts of Interest: The authors declare no conflict of interest.

References

1. Mahvi, A.H.; Balarak, D.; Bazrafshan, E. Remarkable Reusability of Magnetic Fe_3O_4-Graphene Oxide Composite: A Highly Effective Adsorbent for Cr (VI) Ions. *Int. J. Environ. Anal. Chem.* **2023**, *103*, 3501–3521. [CrossRef]
2. Li, L.; Zou, J.; Han, Y.; Liao, Z.; Lu, P.; Nezamzadeh-Ejhieh, A.; Liu, J.; Peng, Y. Recent advances in Al (iii)/In (iii)-based MOFs for the detection of pollutants. *New J. Chem.* **2022**, *46*, 19577–19592. [CrossRef]
3. Adelabu, I.O.; Saleh, T.A.; Garrison, T.F.; Al Hamouz, O.C.S. Synthesis of Polyamine-CNT Composites for the Removal of Toxic Cadmium Metal Ions from Wastewater. *J. Mol. Liq.* **2020**, *297*, 111827. [CrossRef]
4. Begum, S.; Yuhana, N.Y.; Saleh, N.M.; Kamarudin, N.H.N.; Sulong, A.B. Review of Chitosan Composite as a Heavy Metal Adsorbent: Material Preparation and Properties. *Carbohydr. Polym.* **2021**, *259*, 117613. [CrossRef]
5. Gupta, K.; Joshi, P.; Gusain, R.; Khatri, O.P. Recent Advances in Adsorptive Removal of Heavy Metal and Metalloid Ions by Metal Oxide-Based Nanomaterials. *Coord. Chem. Rev.* **2021**, *445*, 214100. [CrossRef]
6. Mahmoud, M.E.; Mohamed, A.K. Novel Derived Pectin Hydrogel from Mandarin Peel Based Metal-Organic Frameworks Composite for Enhanced Cr (VI) and Pb (II) Ions Removal. *Int. J. Biol. Macromol.* **2020**, *164*, 920–931. [CrossRef] [PubMed]

7. Yanrong, C.A.I.; Weili, J.; Chun, C. Research Progress on Removal of Metal Ions from Water by Marine Waste Biomass Based Adsorbent. *J. Light Ind.* **2022**, *37*, 100–110.
8. Zhang, H.; Dang, Q.; Liu, C.; Yu, D.; Wang, Y.; Pu, X.; Liu, Y.; Liang, Y.; Cha, D. Fabrication of Methyl Acrylate and Tetraethylenepentamine Grafted Magnetic Chitosan Microparticles for Capture of Cd (II) from Aqueous Solutions. *J. Hazard. Mater.* **2019**, *366*, 346–357. [CrossRef]
9. Imran, M.; Khan, Z.U.H.; Iqbal, J.; Shah, N.S.; Muzammil, S.; Ali, S.; Muhammad, N.; Aziz, A.; Murtaza, B.; Naeem, M.A. Potential of Siltstone and Its Composites with Biochar and Magnetite Nanoparticles for the Removal of Cadmium from Contaminated Aqueous Solutions: Batch and Column Scale Studies. *Environ. Pollut.* **2020**, *259*, 113938. [CrossRef]
10. Karimi-Maleh, H.; Ayati, A.; Davoodi, R.; Tanhaei, B.; Karimi, F.; Malekmohammadi, S.; Orooji, Y.; Fu, L.; Sillanpää, M. Recent Advances in Using of Chitosan-Based Adsorbents for Removal of Pharmaceutical Contaminants: A Review. *J. Clean. Prod.* **2021**, *291*, 125880. [CrossRef]
11. Ahamad, T.; Naushad, M.; Alshehri, S.M. Fabrication of Magnetic Polymeric Resin for the Removal of Toxic Metals from Aqueous Medium: Kinetics and Adsorption Mechanisms. *J. Water Process Eng.* **2020**, *36*, 101284. [CrossRef]
12. Zheng, M.; Chen, J.; Zhang, L.; Cheng, Y.; Lu, C.; Liu, Y.; Singh, A.; Trivedi, M.; Kumar, A.; Liu, J. Metal organic frameworks as efficient adsorbents for drugs from wastewater. *Mater. Today Commun.* **2022**, *31*, 103514. [CrossRef]
13. Doshi, B.; Ayati, A.; Tanhaei, B.; Repo, E.; Sillanpää, M. Partially Carboxymethylated and Partially Cross-Linked Surface of Chitosan versus the Adsorptive Removal of Dyes and Divalent Metal Ions. *Carbohydr. Polym.* **2018**, *197*, 586–597. [CrossRef]
14. Li, D.; Tian, X.; Wang, Z.; Guan, Z.; Li, X.; Qiao, H.; Ke, H.; Luo, L.; Wei, Q. Multifunctional Adsorbent Based on Metal-Organic Framework Modified Bacterial Cellulose/Chitosan Composite Aerogel for High Efficient Removal of Heavy Metal Ion and Organic Pollutant. *Chem. Eng. J.* **2020**, *383*, 123127. [CrossRef]
15. Karimi, F.; Ayati, A.; Tanhaei, B.; Sanati, A.L.; Afshar, S.; Kardan, A.; Dabirifar, Z.; Karaman, C. Removal of Metal Ions Using a New Magnetic Chitosan Nano-Bio-Adsorbent; A Powerful Approach in Water Treatment. *Environ. Res.* **2022**, *203*, 111753. [CrossRef]
16. Chen, B.; Long, F.; Chen, S.; Cao, Y.; Pan, X. Magnetic Chitosan Biopolymer as a Versatile Adsorbent for Simultaneous and Synergistic Removal of Different Sorts of Dyestuffs from Simulated Wastewater. *Chem. Eng. J.* **2020**, *385*, 123926. [CrossRef]
17. Liu, J.; Chen, Y.; Han, T.; Cheng, M.; Zhang, W.; Long, J.; Fu, X. A Biomimetic SiO_2@Chitosan Composite as Highly-Efficient Adsorbent for Removing Heavy Metal Ions in Drinking Water. *Chemosphere* **2019**, *214*, 738–742. [CrossRef]
18. Chen, B.; Zhao, H.; Chen, S.; Long, F.; Huang, B.; Yang, B.; Pan, X. A Magnetically Recyclable Chitosan Composite Adsorbent Functionalized with EDTA for Simultaneous Capture of Anionic Dye and Heavy Metals in Complex Wastewater. *Chem. Eng. J.* **2019**, *356*, 69–80. [CrossRef]
19. Huang, Y.; Hu, C.; An, Y.; Xiong, Z.; Hu, X.; Zhang, G.; Zheng, H. Magnetic Phosphorylated Chitosan Composite as a Novel Adsorbent for Highly Effective and Selective Capture of Lead from Aqueous Solution. *J. Hazard. Mater.* **2021**, *405*, 124195. [CrossRef]
20. Shaumbwa, V.R.; Liu, D.; Archer, B.; Li, J.; Su, F. Preparation and Application of Magnetic Chitosan in Environmental Remediation and Other Fields: A Review. *J. Appl. Polym. Sci.* **2021**, *138*, 51241. [CrossRef]
21. Mohseni-Bandpi, A.; Kakavandi, B.; Kalantary, R.R.; Azari, A.; Keramati, A. Development of a Novel Magnetite–Chitosan Composite for the Removal of Fluoride from Drinking Water: Adsorption Modeling and Optimization. *RSC Adv.* **2015**, *5*, 73279–73289. [CrossRef]
22. Wu, K.; Su, D.; Liu, J.; Saha, R.; Wang, J.-P. Magnetic Nanoparticles in Nanomedicine: A Review of Recent Advances. *Nanotechnology* **2019**, *30*, 502003. [CrossRef] [PubMed]
23. Kabra, A.; Martins, N.; Sharma, R.; Kabra, R.; Baghel, U.S. Myrica Esculenta Buch.-Ham. Ex D. Don: A Natural Source for Health Promotion and Disease Prevention. *Plants* **2019**, *8*, 149. [CrossRef] [PubMed]
24. De Jesús Ruíz-Baltazar, Á.; Reyes-López, S.Y.; de Lourdes Mondragón-Sánchez, M.; Robles-Cortés, A.I.; Pérez, R. Eco-Friendly Synthesis of Fe_3O_4 Nanoparticles: Evaluation of Their Catalytic Activity in Methylene Blue Degradation by Kinetic Adsorption Models. *Results Phys.* **2019**, *12*, 989–995. [CrossRef]
25. Appu, M.; Lian, Z.; Zhao, D.; Huang, J. Biosynthesis of Chitosan-Coated Iron Oxide (Fe_3O_4) Hybrid Nanocomposites from Leaf Extracts of *Brassica oleracea* L. and Study on Their Antibacterial Potentials. *3 Biotech* **2021**, *11*, 271. [CrossRef]
26. Rai, P.; Usman, U.L.; Banerjee, S. Sequestration of Cadmium Ions from Aqueous Phase Using Magnetic Chitosan Composite. *Mater. Today Proc.* **2022**, *49*, 3375–3383. [CrossRef]
27. Jaafari, J.; Barzanouni, H.; Mazloomi, S.; Farahani, N.A.A.; Sharafi, K.; Soleimani, P.; Haghighat, G.A. Effective Adsorptive Removal of Reactive Dyes by Magnetic Chitosan Nanoparticles: Kinetic, Isothermal Studies and Response Surface Methodology. *Int. J. Biol. Macromol.* **2020**, *164*, 344–355. [CrossRef]
28. Bouafia, A.; Laouini, S.E.; Khelef, A.; Tedjani, M.L.; Guemari, F. Effect of Ferric Chloride Concentration on the Type of Magnetite (Fe_3O_4) Nanoparticles Biosynthesized by Aqueous Leaves Extract of Artemisia and Assessment of Their Antioxidant Activities. *J. Clust. Sci.* **2021**, *32*, 1033–1041. [CrossRef]
29. Varma, R.; Vasudevan, S. Extraction, Characterization, and Antimicrobial Activity of Chitosan from Horse Mussel Modiolus Modiolus. *ACS Omega* **2020**, *5*, 20224–20230. [CrossRef]

30. Subedi, N.; Lähde, A.; Abu-Danso, E.; Iqbal, J.; Bhatnagar, A. A Comparative Study of Magnetic Chitosan (Chi@Fe_3O_4) and Graphene Oxide Modified Magnetic Chitosan (Chi@Fe_3O_4GO) Nanocomposites for Efficient Removal of Cr (VI) from Water. *Int. J. Biol. Macromol.* **2019**, *137*, 948–959. [CrossRef]
31. Nhiem, N.X.; Van Kiem, P.; Van Minh, C.; Tai, B.H.; Cuong, N.X.; Thu, V.K.; Kim, Y.H. A new monoterpenoid glycoside from *Myrica esculenta* and the inhibition of angiotensin I-converting enzyme. *Chem. Pharm. Bull.* **2010**, *58*, 1408–1410. [CrossRef] [PubMed]
32. Hu, D.; Lian, Z.; Xian, H.; Jiang, R.; Wang, N.; Weng, Y.; Peng, X.; Wang, S.; Ouyang, X. Adsorption of Pb (II) from Aqueous Solution by Polyacrylic Acid Grafted Magnetic Chitosan Nanocomposite. *Int. J. Biol. Macromol.* **2020**, *154*, 1537–1547. [CrossRef]
33. Zhu, A.; Yuan, L.; Liao, T. Suspension of Fe3O4 Nanoparticles Stabilized by Chitosan and O-Carboxymethylchitosan. *Int. J. Pharm.* **2008**, *350*, 361–368. [CrossRef]
34. Jiang, Z.; Li, N.; Li, P.-Y.; Liu, B.; Lai, H.-J.; Jin, T. One-Step Preparation of Chitosan-Based Magnetic Adsorbent and Its Application to the Adsorption of Inorganic Arsenic in Water. *Molecules* **2021**, *26*, 1785. [CrossRef]
35. Peralta, M.E.; Nisticò, R.; Franzoso, F.; Magnacca, G.; Fernandez, L.; Parolo, M.E.; León, E.G.; Carlos, L. Highly Efficient Removal of Heavy Metals from Waters by Magnetic Chitosan-Based Composite. *Adsorption* **2019**, *25*, 1337–1347. [CrossRef]
36. Liu, Y.; Shen, L. From Langmuir Kinetics to First-and Second-Order Rate Equations for Adsorption. *Langmuir* **2008**, *24*, 11625–11630. [CrossRef] [PubMed]
37. Sarojini, G.; Venkateshbabu, S.; Rajasimman, M. Facile Synthesis and Characterization of Polypyrrole—Iron Oxide—Seaweed (PPy-Fe_3O_4-SW) Nanocomposite and Its Exploration for Adsorptive Removal of Pb(II) from Heavy Metal Bearing Water. *Chemosphere* **2021**, *278*, 130400. [CrossRef] [PubMed]
38. Peers, A.M. Elovich Adsorption Kinetics and the Heterogeneous Surface. *J. Catal.* **1965**, *4*, 499–503. [CrossRef]
39. Revellame, E.D.; Fortela, D.L.; Sharp, W.; Hernandez, R.; Zappi, M.E. Adsorption Kinetic Modeling Using Pseudo-First Order and Pseudo-Second Order Rate Laws: A Review. *Clean. Eng. Technol.* **2020**, *1*, 100032. [CrossRef]
40. Sahoo, T.R.; Prelot, B. Chapter 7—Adsorption Processes for the Removal of Contaminants from Wastewater: The Perspective Role of Nanomaterials and Nanotechnology. In *Micro and Nano Technologies*; Bonelli, B., Freyria, F.S., Rossetti, I., Sethi, R.B.T.-N., Eds.; Elsevier: Amsterdam, The Netherlands, 2020; pp. 161–222. ISBN 978-0-12-818489-9.
41. Ayub, A.; Irfan, A.; Raza, Z.A.; Abbas, M.; Muhammad, A.; Ahmad, K.; Munwar, A. Development of Poly (1-Vinylimidazole)-Chitosan Composite Sorbent under Microwave Irradiation for Enhanced Uptake of Cd (II) Ions from Aqueous Media. *Polym. Bull.* **2021**, *79*, 807–827. [CrossRef]
42. Pavithra, S.; Thandapani, G.; Sugashini, S.; Sudha, P.N.; Alkhamis, H.H.; Alrefaei, A.F.; Almutairi, M.H. Batch Adsorption Studies on Surface Tailored Chitosan/Orange Peel Hydrogel Composite for the Removal of Cr (VI) and Cu (II) Ions from Synthetic Wastewater. *Chemosphere* **2021**, *271*, 129415. [CrossRef] [PubMed]
43. Wang, X.; Li, X.; Peng, L.; Han, S.; Hao, C.; Jiang, C.; Wang, H.; Fan, X. Effective Removal of Heavy Metals from Water Using Porous Lignin-Based Adsorbents. *Chemosphere* **2021**, *279*, 130504. [CrossRef] [PubMed]
44. Jadoun, S.; Fuentes, J.P.; Urbano, B.F.; Yáñez, J. Highly Efficient Remediation of Cu (II) from Water by Novel Poly(o-Phenylenediamine)/Zinc Oxide Nanohybrids: Kinetic, Equilibrium, and Thermodynamic Studies. *J. Water Process Eng.* **2023**, *53*, 103663. [CrossRef]
45. Massoudinejad, M.; Rasoulzadeh, H.; Ghaderpoori, M. Magnetic Chitosan Nanocomposite: Fabrication, Properties, and Optimization for Adsorptive Removal of Crystal Violet from Aqueous Solutions. *Carbohydr. Polym.* **2019**, *206*, 844–853. [CrossRef] [PubMed]
46. Sharma, P.; Yadav, V.; Kumari, S.; Ghosh, D.; Rawat, P.; Vij, A.; Srivastava, C.; Saini, S.; Sharma, V.; Hassan, M.I. Deciphering the Potent Application of Nanobentonite and α-Fe_2O_3/Bentonite Nanocomposite in Dye Removal: Revisiting the Insights of Adsorption Mechanism. *Appl. Nanosci.* **2023**, *13*, 883–897. [CrossRef]
47. Inyinbor, A.A.; Adekola, F.A.; Olatunji, G.A. Kinetics and Isothermal Modeling of Liquid Phase Adsorption of Rhodamine B onto Urea Modified Raphia Hookerie Epicarp. *Appl. Water Sci.* **2017**, *7*, 3257–3266. [CrossRef]
48. Togue Kamga, F. Modeling Adsorption Mechanism of Paraquat onto Ayous (*Triplochiton scleroxylon*) Wood Sawdust. *Appl. Water Sci.* **2019**, *9*, 1. [CrossRef]
49. Manzoor, K.; Ahmad, M.; Ahmad, S.; Ikram, S. Removal of Pb (Ii) and Cd (Ii) from Wastewater Using Arginine Cross-Linked Chitosan–Carboxymethyl Cellulose Beads as Green Adsorbent. *RSC Adv.* **2019**, *9*, 7890–7902. [CrossRef]
50. Zhang, Y.; Wang, Y.; Zhang, Z.; Cui, W.; Zhang, X.; Wang, S. Removing Copper and Cadmium from Water and Sediment by Magnetic Microspheres-$MnFe_2O_4$/Chitosan Prepared by Waste Shrimp Shells. *J. Environ. Chem. Eng.* **2021**, *9*, 104647. [CrossRef]
51. Salehi, N.; Moghimi, A.; Shahbazi, H. Preparation of Cross-Linked Magnetic Chitosan with Methionine-Glutaraldehyde for Removal of Heavy Metals from Aqueous Solutions. *Int. J. Environ. Anal. Chem.* **2022**, *102*, 2305–2321. [CrossRef]
52. Babakhani, A.; Sartaj, M. Removal of Cadmium (II) from Aqueous Solution Using Tripolyphosphate Cross-Linked Chitosan. *J. Environ. Chem. Eng.* **2020**, *8*, 103842. [CrossRef]
53. Wang, Y.; Zhou, R.; Wang, C.; Zhou, G.; Hua, C.; Cao, Y.; Song, Z. Novel Environmental-Friendly Nano-Composite Magnetic Attapulgite Functionalized by Chitosan and EDTA for Cadmium (II) Removal. *J. Alloys Compd.* **2020**, *817*, 153286. [CrossRef]

54. Fan, H.-L.; Zhou, S.-F.; Jiao, W.-Z.; Qi, G.-S.; Liu, Y.-Z. Removal of Heavy Metal Ions by Magnetic Chitosan Nanoparticles Prepared Continuously via High-Gravity Reactive Precipitation Method. *Carbohydr. Polym.* **2017**, *174*, 1192–1200. [CrossRef] [PubMed]
55. Li, J.; Tao, Y.; Chen, S.; Li, H.; Chen, P.; Wei, M.; Wang, H.; Li, K.; Mazzeo, M.; Duan, Y. A Flexible Plasma-Treated Silver-Nanowire Electrode for Organic Light-Emitting Devices. *Sci. Rep.* **2017**, *7*, 16468. [CrossRef] [PubMed]
56. Ahamad, T.; Naushad, M.; Eldesoky, G.E.; Alqadami, A.A.; Khan, A. Synthesis and Characterization of Egg-Albumen-Formaldehyde Based Magnetic Polymeric Resin (MPR): Highly Efficient Adsorbent for Cd (II) Ion Removal from Aqueous Medium. *J. Mol. Liq.* **2019**, *286*, 110951. [CrossRef]
57. Alosaimi, E.H.; Alsohaimi, I.H.; Dahan, T.E.; Chen, Q.; Melhi, S. Adsorptive Performance of Tetracarboxylic Acid-Modified Magnetic Silica Nanocomposite for Recoverable Efficient Removal of Toxic Cd (II) from Aqueous Environment: Equilibrium, Isotherm, and Reusability Studies. *J. Mol. Liq.* **2021**, *334*, 116069. [CrossRef]
58. Mahmoud, M.E.; Amira, M.F.; Seleim, S.M.; Mohamed, A.K. Amino-Decorated Magnetic Metal-Organic Framework as a Potential Novel Platform for Selective Removal of Chromium (Vl), Cadmium (II) and Lead (II). *J. Hazard. Mater.* **2020**, *381*, 120979. [CrossRef]
59. Wu, Z.; Deng, W.; Zhou, W.; Luo, J. Novel Magnetic Polysaccharide/Graphene Oxide@Fe_3O_4 Gel Beads for Adsorbing Heavy Metal Ions. *Carbohydr. Polym.* **2019**, *216*, 119–128. [CrossRef]
60. Wang, L.; Hu, D.; Kong, X.; Liu, J.; Li, X.; Zhou, K.; Zhao, H.; Zhou, C. Anionic Polypeptide Poly (γ-Glutamic Acid)-Functionalized Magnetic Fe_3O_4-GO-(o-MWCNTs) Hybrid Nanocomposite for High-Efficiency Removal of Cd (II), Cu (II) and Ni (II) Heavy Metal Ions. *Chem. Eng. J.* **2018**, *346*, 38–49. [CrossRef]
61. Liu, G.; Liao, L.; Dai, Z.; Qi, Q.; Wu, J.; Ma, L.Q.; Tang, C.; Xu, J. Organic Adsorbents Modified with Citric Acid and Fe_3O_4 Enhance the Removal of Cd and Pb in Contaminated Solutions. *Chem. Eng. J.* **2020**, *395*, 125108. [CrossRef]
62. Naushad, M.; Ahamad, T.; Al-Ghanim, K.A.; Ala'a, H.; Eldesoky, G.E.; Khan, A.A. A Highly Porous Nanocomposite (Fe_3O_4@ BFR) for the Removal of Toxic Cd (II) Ions from Aqueous Environment: Adsorption Modelling and Regeneration Study. *Compos. Part B Eng.* **2019**, *172*, 179–185. [CrossRef]
63. Behbahani, E.S.; Dashtian, K.; Ghaedi, M. Fe_3O_4-FeMoS4: Promise Magnetite LDH-Based Adsorbent for Simultaneous Removal of Pb (II), Cd (II), and Cu (II) Heavy Metal Ions. *J. Hazard. Mater.* **2021**, *410*, 124560. [CrossRef] [PubMed]
64. Sun, C.; Xie, Y.; Ren, X.; Song, G.; Alsaedi, A.; Hayat, T.; Chen, C. Efficient Removal of Cd (II) by Core-Shell Fe_3O_4@ Polydopamine Microspheres from Aqueous Solution. *J. Mol. Liq.* **2019**, *295*, 111724. [CrossRef]
65. Liu, C.; Jiang, X.; Wang, X.; Wang, Q.; Li, L.; Zhang, F.; Liang, W. Magnetic Polyphenol Nanocomposite of Fe_3O_4/SiO_2/PP for Cd (II) Adsorption from Aqueous Solution. *Environ. Technol.* **2022**, *43*, 935–948. [CrossRef] [PubMed]
66. Li, Y.; Dong, X.; Zhao, L. Application of Magnetic Chitosan Nanocomposites Modified by Graphene Oxide and Polyethyleneimine for Removal of Toxic Heavy Metals and Dyes from Water. *Int. J. Biol. Macromol.* **2021**, *192*, 118–125. [CrossRef]
67. Abou El-Reash, Y.G. Magnetic Chitosan Modified with Cysteine-Glutaraldehyde as Adsorbent for Removal of Heavy Metals from Water. *J. Environ. Chem. Eng.* **2016**, *4*, 3835–3847. [CrossRef]
68. Li, B.; Zhou, F.; Huang, K.; Wang, Y.; Mei, S.; Zhou, Y.; Jing, T. Environmentally Friendly Chitosan/PEI-Grafted Magnetic Gelatin for the Highly Effective Removal of Heavy Metals from Drinking Water. *Sci. Rep.* **2017**, *7*, 43082. [CrossRef]
69. Zeng, X.; Zhang, G.; Wen, J.; Li, X.; Zhu, J.; Wu, Z. Simultaneous Removal of Aqueous Same Ionic Type Heavy Metals and Dyes by a Magnetic Chitosan/Polyethyleneimine Embedded Hydrophobic Sodium Alginate Composite: Performance, Interaction and Mechanism. *Chemosphere* **2023**, *318*, 137869. [CrossRef]
70. Le, T.T.N.; Le, V.T.; Dao, M.U.; Nguyen, Q.V.; Vu, T.T.; Nguyen, M.H.; Tran, D.L.; Le, H.S. Preparation of Magnetic Graphene Oxide/Chitosan Composite Beads for Effective Removal of Heavy Metals and Dyes from Aqueous Solutions. *Chem. Eng. Commun.* **2019**, *206*, 1337–1352. [CrossRef]
71. Fan, L.; Luo, C.; Sun, M.; Li, X.; Qiu, H. Highly Selective Adsorption of Lead Ions by Water-Dispersible Magnetic Chitosan/Graphene Oxide Composites. *Colloids Surf. B Biointerfaces* **2013**, *103*, 523–529. [CrossRef]
72. Lv, L.; Chen, N.; Feng, C.; Zhang, J.; Li, M. Heavy Metal Ions Removal from Aqueous Solution by Xanthate-Modified Cross-Linked Magnetic Chitosan/Poly (Vinyl Alcohol) Particles. *RSC Adv.* **2017**, *7*, 27992–28000. [CrossRef]
73. Liu, T.; Han, X.; Wang, Y.; Yan, L.; Du, B.; Wei, Q.; Wei, D. Magnetic Chitosan/Anaerobic Granular Sludge Composite: Synthesis, Characterization and Application in Heavy Metal Ions Removal. *J. Colloid Interface Sci.* **2017**, *508*, 405–414. [CrossRef] [PubMed]
74. Peighambardoust, S.J.; Foroutan, R.; Peighambardoust, S.H.; Khatooni, H.; Ramavandi, B. Decoration of Citrus Limon Wood Carbon with Fe_3O_4 to Enhanced Cd^{2+} Removal: A Reclaimable and Magnetic Nanocomposite. *Chemosphere* **2021**, *282*, 131088. [CrossRef] [PubMed]
75. Homayonfard, A.; Miralinaghi, M.; Shirazi, R.H.S.M.; Moniri, E. Efficient Removal of Cadmium (II) Ions from Aqueous Solution by $CoFe_2O_4$/Chitosan and $NiFe_2O_4$/Chitosan Composites as Adsorbents. *Water Sci. Technol.* **2018**, *78*, 2297–2307. [CrossRef] [PubMed]
76. Wu, D.; Wang, Y.; Li, Y.; Wei, Q.; Hu, L.; Yan, T.; Feng, R.; Yan, L.; Du, B. Phosphorylated Chitosan/$CoFe_2O_4$ Composite for the Efficient Removal of Pb (II) and Cd (II) from Aqueous Solution: Adsorption Performance and Mechanism Studies. *J. Mol. Liq.* **2019**, *277*, 181–188. [CrossRef]

77. Song, X.; Li, L.; Zhou, L.; Chen, P. Magnetic Thiolated/Quaternized-Chitosan Composites Design and Application for Various Heavy Metal Ions Removal, Including Cation and Anion. *Chem. Eng. Res. Des.* **2018**, *136*, 581–592. [CrossRef]
78. Reddy, D.H.K.; Lee, S.M. Application of magnetic chitosan composites for the removal of toxic metal and dyes from aqueous solutions. *Adv. Colloid Interface Sci.* **2013**, *201*, 68–93. [CrossRef]

polymers

MDPI

Article

A Comparative Study of Cr(VI) Sorption by *Aureobasidium pullulans* AKW Biomass and Its Extracellular Melanin: Complementary Modeling with Equilibrium Isotherms, Kinetic Studies, and Decision Tree Modeling

Hala Fakhry [1,2,*], Abeer A. Ghoniem [3], Fatimah O. Al-Otibi [4], Yosra A. Helmy [5], Mohammed S. El Hersh [3], Khaled M. Elattar [6], WesamEldin I. A. Saber [3,*] and Ashraf Elsayed [7]

1 National Institute of Oceanography and Fisheries (NIOF), Cairo 11865, Egypt
2 Department of Aquatic Environmental Science, Faculty of Fish Resources, Suez University, Suez 43518, Egypt
3 Microbial Activity Unit, Department of Microbiology, Soils, Water and Environment Research Institute, Agricultural Research Center, Giza 12619, Egypt; abeer.abdelkhalik@yahoo.com (A.A.G.); m.elhersh@yahoo.com (M.S.E.H.)
4 Botany and Microbiology Department, Faculty of Science, King Saud University, Riyadh 11451, Saudi Arabia; falotibi@ksu.edu.sa
5 Department of Veterinary Science, Martin-Gatton College of Agriculture, Food, and Environment, University of Kentucky, Lexington, KY 40546, USA; yosra.helmy@uky.edu
6 Unit of Genetic Engineering and Biotechnology, Faculty of Science, Mansoura University, Mansoura 35516, Egypt; khaledelattar2@yahoo.com
7 Botany Department, Faculty of Science, Mansoura University, Mansoura 35516, Egypt; ashraf-badawy@mans.edu.eg
* Correspondence: halaelhossany@gmail.com (H.F.); wesameldin.saber@arc.sci.eg (W.I.A.S.)

Abstract: Melanin as a natural polymer is found in all living organisms, and plays an important role in protecting the body from harmful UV rays from the sun. The efficiency of fungal biomass (*Aureobasidium pullulans*) and its extracellular melanin as Cr(VI) biosorbents was comparatively considered. The efficiency of Cr(VI) biosorption by the two sorbents used was augmented up to 240 min. The maximum sorption capacities were 485.747 (fungus biomass) and 595.974 (melanin) mg/g. The practical data were merely fitted to both Langmuir and Freundlich isotherms. The kinetics of the biosorption process obeyed the pseudo-first-order. Melanin was superior in Cr(VI) sorption than fungal biomass. Furthermore, four independent variables (contact time, initial concentration of Cr(VI), biosorbent dosage, and pH,) were modeled by the two decision trees (DTs). Conversely, to equilibrium isotherms and kinetic studies, DT of fungal biomass had lower errors compared to DT of melanin. Lately, the DTs improved the efficacy of the Cr(VI) removal process, thus introducing complementary and alternative solutions to equilibrium isotherms and kinetic studies. The Cr(VI) biosorption onto the biosorbents was confirmed and elucidated through FTIR, SEM, and EDX investigations. Conclusively, this is the first report study attaining the biosorption of Cr(VI) by biomass of *A. pullulans* and its extracellular melanin among equilibrium isotherms, kinetic study, and algorithmic decision tree modeling.

Keywords: artificial intelligence; biosorption; fungal biomass; water treatment; heavy metals

1. Introduction

Admission to hygienic and secure water is crucial for healthy living organisms; however, this reality could not be achieved worldwide owing to the increase in human activity e.g., agricultural practices, and industrialization [1]. Heavy metals are a major environmental anxiety owing to their toxicity, ability to bioaccumulate, and persistence in the food series [2]. Chromium is produced in the environment as a byproduct of several activities such as mining, energy production, steel, and iron industries. However, chromium ions at permissible limits could possess a vital function in the biosynthesis of nucleic acids, and the metabolism of fat, proteins, and carbohydrates. The toxicity of Cr ions was found to arise from oxidation status [3]. The high oxidative form (hexavalent Cr^{6+}) is more toxigenic, and carcinogenic in biological systems compared to the trivalent Cr^{3+} ions, which are less toxic, more soluble, and have stable oxidation status [4].

A variety of water bioremediation processes (e.g., ion exchange, and coagulation-flocculation), are utilized to remove heavy metals. However, these procedures have many disadvantages i.e., expensive, energy-intensive requirements, time-consuming, and undergo drawbacks correlated to recyclability. Therefore, the biotechnological-based treatments, e.g., bacteria, fungi, plants, and algae, are, currently, being documented on leaching heavy metals from contaminated effluents [5,6], since the biological methods are cost-effective and eco-friendly [7].

Melanin is a black or dark brown high molecular polymer, which is biosynthesized by oxidation and polymerization of phenolic compounds [8]. It is associated with some melanized fungal strains, situated in the cell wall or remains extracellularly [9]. Melanin also has many other biological properties, including antimicrobial [10], photoprotective [11], and antioxidant effects. Melanin has been shown to inhibit the growth of bacteria, fungi, and viruses. This is thought to be due to their ability to chelate metal ions and disrupt the cell membranes of microorganisms [12]. Melanin absorbs ultraviolet radiation, which can damage DNA leading to skin cancer [13]. They also help to prevent premature aging of the skin [14].

Both melanized fungi and their melanin are new renewable potential biosorbents for removing heavy metals [15]. Fungal melanin contains a diversity of functional groups, including phenolic, carboxylic, amine, and hydroxyl groups that can bind to metal ions [16]. For instance, copper binds mainly to the carboxyl group in catechol melanin at $pH < 5$, but at $pH > 6$, copper binds primarily to the phenolic hydroxyl group [17]. Another study reported that the proteinaceous fraction containing nitrogen groups can bind iron and copper [18].

Additionally, the biosorbent potential of melanin depends upon the physicochemical features of functional groups [19,20]. Moreover, melanin polymers could also interact with metal ions through their free radical species, forming complexation [21,22]. Another, the thermal tolerance of melanin and insolubility in organic solvents and aqueous solutions ensure the constancy of melanin as an adsorbent to industrial effluents containing metals [23].

Numerous reports discussed the ability of filamentous fungi to bind metals, where fungal mycelium involves an interconnected network of millimeter-to-centimeter-long hyphae [24]. Moreover, fungal biomass can produce massive growth on a large scale, where the biosorption of metal ions by filamentous fungi can occur at a minimal cost [25].

Decision trees (DTs) are a kind of artificial intelligence classified as supervised machine learning and data mining algorithms. DTs are used to solve both classification and regression tasks, determine the relative importance of the independent variable on a response target, clustering analysis, and understand and interpret a wide variety of data types. DTs work by recursively partitioning a data set into minor subsets until each subset is homogeneous concerning the target variable. The decision rules at each node of the tree are learned from the training data and can be used to predict the target variable for new data points [26]. To obtain optimum prediction by DT for new conditions, it is crucial to have an applicable assortment of data. Anyhow, the training and optimization process should be followed by authentication, as the certainty tree can be influenced by overfitting and other affairs thus several roles are recommended to avoid such problems [9,27].

No information is available about the combination of DT with equilibrium isotherms and kinetic studies for the determination of relative variable importance and modeling the biosorption process. Combining equilibrium isotherms and kinetic studies with DTs can provide a holistic understanding of the sorption process. Where equilibrium isotherms describe the system relationship between the amount of solute adsorbed onto a sorbent, kinetics focus on how quickly the sorption process occurs and how the concentration of the solute changes over time, DT, on the other side, can predict the outcome or behavior of all sorption system based on various input parameters, as well as determining the relative importance of the independent variables. As a case study, *Aureobasidium pullulans* and its extracellular melanin were used during the current work in a trial to understand the mechanism by which pollutants are adsorbed onto fungal biomass and melanin. In general, the exhaustion of fungal biomass or melanin in the eradication of heavy metal ions from wastewater is an urgent demand from the environmental arguments of view.

Biomass-based sorbents are a sustainable and cost-effective way to remove Cr(VI) from water and wastewater [28]. These sorbents can be made from a variety of biomass materials, such as agricultural waste, wood waste, and food waste. They can be functionalized with various chemical groups, such as carboxylates, amines, and phosphates, to improve their Cr(VI) sorption capacity. The sorption of Cr(VI) onto biomass-based sorbents is a complex process that is influenced by several factors, including the sorbent material, the functionalization chemistry, the Cr(VI) concentration, the pH, and the temperature. In general, Cr(VI) sorption is favored by low pH values and high temperatures [29]. Biomass-based sorbents are effective in removing Cr(VI) from water and wastewater with high sorption capacities and low regeneration costs [30]. They are promising technologies for the removal of this toxic heavy metal from the environment. The benefits of using functionalized biomass-based sorbents for Cr(VI) removal included [31]: (1) They are renewable and biodegradable, making them ecofriendly. (2) They are low-cost and easy to produce. (3) They can be tailored to specific Cr(VI) removal applications. Functionalized biomass-based sorbents are a promising new technology for the removal of Cr(VI) from water and wastewater. They offer many advantages over traditional sorbents, making them a more sustainable and cost-effective option for environmental remediation [32].

Herein, our investigated study focused on Cr(VI) biosorption through (i) comparing the biomass of *A. pullulans* and its extracellular melanin during isotherm models i.e., Langmuir, Freundlich, and Temkin, (ii) investigating the kinetic behavior by applying pseudo-first-order and pseudo-second-order (iii) modeling the overall biosorption process by DT for both fungal biomass and melanin. Additionally, it confirmed the biosorption process of Cr(VI) ions on fungal biomass and melanin particles by SEM, EDX, and FT-IR analyses. The current study provides complementary information on modeling biosorption using different modeling procedures. Moreover, developing efficient and environmentally friendly methods for heavy metals removal from industrial wastewater. Cr(VI) was chosen as a case study. As fungal biomass and melanin are renewable, and ecofriendly resources that are readily available, it is important to compare the sorption capacity of both biomaterials using two complementary models to determine the most efficient biosorption process for a particular application.

2. Materials and Methods

2.1. Materials

All the used chemicals in the current study were purchased from Sigma Co., St. Louis, USA. HCl (0.1 N), and NaOH (0.1 N) solutions were prepared and utilized to adjust the pH of the solution.

2.2. Preparation of Hexavalent Chromium Solution

Potassium dichromate (Cr(VI), ≥99%) solution was prepared at 1000 mg/L in deionized water and used to prepare a concentration range of 5 to 200 mg/L. The working solution pH was adjusted to the respective level by HCl or NaOH using a pH meter (HI98130, Hanna Instruments, Romania). Trials were performed in a 100 mL working solution in a 250 mL conical flask. The remaining Cr(VI) after the biosorption process was determined [33] using an Atomic Absorption Spectrophotometer (Buck Scientific Accusys 211 series, Norwalk, Connecticut, USA) by an air/acetylene flame system.

2.3. Preparation of Biosorbants

2.3.1. Fungus and Melanin

Aureobasidium pullulans AKW (GenBank accession number OP924554) was obtained in an earlier study [9]. The medium of melanin biosynthesis [34] consisted of potato sucrose broth. The medium was prepared by mixing potato infusion (200 g of sliced, and unpeeled potatoes were boiled in 1 L of distilled water for 30 min, then filtered to remove debris of potato) and sucrose (50 g/L of potato infusion). The medium was adjusted to pH 6.0. Flasks containing 100 sterilized medium (15 min at 121 °C) were inoculated by fungal seeds (3×10^7 cells/mL) in a proportion of 5%. The culture was incubated at 30 °C for 10 days with a rotation speed of 200 rpm.

2.3.2. Separation and Purification of Melanin

The separation and purification of melanin was carried out following The protocol of El-Gamal et al. [35] and Müjdeci [36] with minor modifications. Briefly, melanin was separated from the microbial cell pellets by centrifugation (15 min, 3000× *g*). Melanin in the cell-free filtrate was precipitated by reducing the pH down to 2.0 by HCl (6 M) and kept for 4 h under cooling. The precipitate was separated at 7000× *g* for 15 min. The recovered melanin was washed with distilled water. The melanin purification process was repeated four times followed by lyophilization, then stored (at −20 °C).

2.4. Cr(VI) Ions Biosorption Process

The batch adsorption unit contained 0.01 g of the dried fungal biomass or the biosynthesized melanin and 10 mg/L Cr(VI). 100 mL solution was then agitated (200 rpm at room temperature) in an orbital shaker. Solutions were gathered at a definite time and then centrifuged (12,000 rpm, 10 min). Cr(VI) residual concentrations were analyzed using ICP-OES [37] and the amount of metal ions biosorbed onto the used biosorbents was estimated (Equation (1)).

$$q_e = \frac{(C_0 - C_e)v}{w} \quad (1)$$

where q_e is the quantity of Cr(VI) ions absorbed at a specific period per unit mass of the used materials (mg/g), C_0 is the initial Cr(VI) level (mg/L), C_e is the residual Cr(VI) in solution at equilibrium (mg L^{-1}), V is the volume (L) and W is the amount of the used biosorbent particles (g). The Cr(VI) ions removal percentage was estimated by Equation (2):

$$\%\text{ of Cr(VI) ions removal} = \frac{C_0 - C_e}{C_0} \times 100 \quad (2)$$

where C_o and C_e (mg/L) are the initial and equilibrium levels of Cr(VI) ions in the solution.

The impacts of numerous processing factors, i.e., initial Cr(VI) ions concentration, interaction time, solution pH, and weight of biosorbed material on the Cr(VI) biosorption process were optimized and studied.

2.4.1. Influence of Contact Time

A weighing 0.01 g of the fungal biomass or the biosynthesized melanin particles was added to a 100 mL volume of Cr(VI) solution (10 ppm). Then, the solutions were shaken for diverse intervals of time (10, 20, 30, 60, 120, 180, 240, 280, and 300 min).

2.4.2. Effect of Initial Cr(VI) Concentration

A weight of 0.01 g of the used fungal biomass or the biosynthesized melanin particles was added to a series of initial Cr(VI) solution concentrations (5, 10, 25, 50, 100, and 200 mg/L), and the tests were accomplished for the equilibrium.

2.4.3. Effect of Thresholds of Fungal Biomass and Its Melanin

Different amounts of the used fungal biomass and the biosynthesized melanin particles (0.005, 0.01, 0.05, 0.1, and 0.2 g) were added to each 100 mL of Cr(VI) aqueous solution and shaken for a constant biosorption time.

2.4.4. Solution pH vis. Cr(VI) Sorbtion

The pH effect of the Cr(VI) solution was inspected at diverse pH values (2, 4, 5, 7, 9, and 11) using a pH meter (ADWA, AD1000, Romania), and it was controlled by adding (1 N) NaOH or HCl solutions. The bio-sorbent particles (0.01 g) were added to each 100 mL of Cr(VI) solution using 10 ppm as an initial concentration and were shaken for a constant bio-sorption time.

2.5. Biosorption Isotherm Determination

To study the biosorption isotherms (equilibrium modeling), the impact of the initial Cr(VI) ions level on the biosorption practicability was conducted through batch mode studies by contacting 0.01 g of the used bio-sorbents within a range from 5 to 200 mg/L Cr(VI).

Equilibrium isotherm models defined the adsorbent and adsorbate interactions. The shape of isotherms is determined by many factors such as Cr(VI) level in the solution, their biosorbent capabilities, and the grade of rivalry between Cr(VI) ions to bind to the active sites of biosorbent. The biosorption behavior of Cr(VI) ions onto the used bio-sorbent particles and the optimal terms of biosorption attitude can be appraised from the equilibrium isothermal factors. The Langmuir, Freundlich, and Temkin adsorption isotherm models were applied to explain the equilibrium discrimination of Cr(VI) ions onto the biosorbents [38]. Langmuir isotherm assumes the homogeneity of the biosorption process, for instance, biosorption uniform energy with no transmigration of Cr(VI) ions in the plane of the used bio-sorbent particles surface, with a monolayer coverage, and uniform biosorption energies. The linear isotherm form of Langmuir is presented in Table 1.

Table 1. Equations of Langmuir, Freundlich, and Temkin models for biosorption of Cr(VI) onto the used fungal biomass and the biosynthesized melanin [39].

Adsorption Models	Equation	Parameter
Langmuir	$\frac{C_e}{q_e} = \frac{1}{q_m K} + \frac{C_e}{q_m}$	q_e is the amount of Cr(VI) ions biosorbent at equilibrium (mg/g), q_m is the supreme monolayer coverage aptitudes (mg/g), K is the Langmuir constant (L/mg), and C_e is the equilibrium concentration of Cr(VI) ions (mg/L).
Freundlich	$ln\ q_e = ln\ K_f + \frac{1}{n_f}\ ln\ C_e$	q_e is the Cr (VI) ions amount biosorbent at equilibrium (mg/g); C_e is the Cr (VI) ions equilibrium concentration (mg/L); and K_F and n_f are Freundlich constants related to the biosorption aptitude and biosorption intensity, respectively
Temkin	$q_e = B\ ln\ KT + B\ ln\ C_e$	K_T is the Temkin constant referring to equilibrium maximum binding energy and B is the Temkin constant interrelated to bio-sorption heat.

The model of Freundlich biosorption isotherm suggested that the Cr(VI) ions uptake takes place on a heterogeneous surface by multilayer biosorption with the lateral interaction between biosorbent Cr(VI) ions on the surface of used biosorbent particles surface. The linear form of Freundlich isotherm is publicized in Table 1. The Temkin isotherm transacts with the indirect impacts of interactions between the biosorbent and biosorbent Table 1. The heat of biosorption of all molecule layers decreases linearly with coverage owing to the interactions between bio-sorbent, and biosorbate [40].

2.6. Biosorption Kinetic Studies

The biosorption kinetic study is a crucial step in highlighting the biosorption process's mechanism. An amount of 0.01 g of the used fungal biomass and the biosynthesized melanin particles were mixed with 10 ppm Cr(VI) metal ions at 200 rpm for diverse intervals of time (10–300 min). The final Cr(VI) concentration was measured after each interval time and the amount of biosorbent metal ion (q_t, mg/g) was plotted versus time (t, min) for kinetic modeling. Four prevalent kinetic models (pseudo-first-order, pseudo-second-order, intra-particle Diffusion, and Elovich) were searched to find out the biosorption of Cr(VI) onto the used biosorbent particles. Lagergren's pseudo-first-order supposes that the biosorption process relies merely upon the level of Cr(VI) ions in the aqueous phase and the biosorbent's accessible binding sites at any time [41]. The pseudo-first-order model Equation is presented in Table 2.

The pseudo-second-order kinetic model proposed that chemical biosorption is the rate-controlling step of a biosorption process (Table 2). An intra-particle diffusion model is a mechanistic sequence used to clarify biosorption kinetics. Generally, the biosorption process is controlled by either external, pores, or surface diffusion adsorption on the surface of the pore or a mixture of them. It is a function of adsorption aptitude and time that is calculated by the Equation given in Table 2 as reported by Weber and Morris [42].

Table 2. Pseudo-first-order, Pseudo-second-order, Elovich, and Intra-particle diffusion kinetics models equations for biosorption of Cr(VI) ions onto the used fungal biomass and the biosynthesized melanin at 25 °C [38,43].

Adsorption Kinetic Models	Equation	Parameter
Pseudo-first-order	$ln\ (q_e - q_t) = \ln q_e - K_1 t$	q_t and q_e are the biosorbent Cr(VI) ions amount at time t and equilibrium (mg/g), respectively. k_1 (min^{-1}) is the first-order reaction rate constant
Pseudo-second-order	$\frac{t}{q_t} = \frac{1}{(K_2 q_e)^2}$	qt and q_e are the biosorbent Cr(VI) ions amount at time t and equilibrium (mg/g), respectively, and k_2 is the second-order reaction rate equilibrium constant (g/mg/min).
Elovich	$q_t = \grave{a} + ß\ ln\ t$	à is the initial sorption rate (mg/g/min) and ß is the extent of surface coverage and activation energy for chemisorption (g/mg)
Intra-particle diffusion	$q_t = k_i t^{0.5} + c_i$	k_i is the intra-particle diffusion rate constant, and c_i gives a prediction about the boundary layer thickness

2.7. Decision Tree Learning Algorithm

Two parametric machine learning regression and classification trees were constructed [9] to find out the probable answers for maximizing Cr(VI) removal using the 4 tested variables that were used during equilibrium isotherms and kinetic studies, i.e., contact time, initial Cr(VI) concentration, pH, and fungal biomass or melanin doses. The relative importance of the 4 input continuous predictor variables that mitigate the noise resulting from the experiential procedures was tested. The target response variable was Cr(VI) removal. The 4 continuous predictors in the data were used to generate DT based on the least squared errors rule as a node-splitting process.

The DT statistics for training and testing to select and evaluate the optimal tree, where the maximum correlation coefficients (R^2) and lowest model error (root mean squared error (RMSE), mean squared error (MSE), mean absolute deviation (MAD), and mean percent error (MAPE), and standard deviation, were used judgment statistics.

For both fungal biomass and melanin, validation was carried out at a random fraction of 0.3, where 30% of data were used for testing and 70% for training. A total of 75 data values, representing three repeated experiments of the equilibrium isotherms and kinetic studies, were used. This procedure was performed to evaluate the prediction precision and avoid overfitting.

During the learning process of DT, the nodes continuously split till the terminal nodes could not be further split. Although deeper trees have more precise predictions, the deepness of the DT was stopped when further modifications did not decrease MSE.

2.7.1. Model Validation of DTs

The predicted conditions at various nodes for both DT models were experimentally validated using three different combinations of the 4 parameters. The predicted and observed values were emulated to judge the model's precision in maximizing Cr(VI) removal.

2.7.2. Software and Statistical Procedure

The data retrieved from the equilibrium isotherms and kinetic data were used as historical data for machine learning. Experiments of Cr(VI) removal percentage by fungal biomass or melanin were repeated thrice, and presented as the average with standard deviation (±SD). The algorithm of DT and the statistical calculations were implemented using Minitab software (version 21, Minitab Inc., State College, PA, USA).

2.8. Characterization of the Biosorption Process

Information about the surface morphology and topography of the used fungal biomass and its extracellular melanin particles, as well as their surface chemical composition, were established [44] by scanning electron microscopy (SEM), and energy dispersive x-ray (EDX) on JSM6380, JEOL, Japan. The samples were covered with a thin layer of gold to avoid charging during imaging.

The basic functional groups, and surface chemistry of the fungal biomass, and melanin were scrutinized by Fourier-transform infrared spectroscopy (FT-IR), in which the spectral analysis was run at Thermo-Fisher Nicolet IS10, USA Spectrophotometer at a resolution of four cm^{-1} in the wavenumber ranges of 500–4000 cm^{-1} in transmission approach to identify the influence of the functional groups in the adsorption of Cr(VI) ions on the surface of fungi and melanin. In this route, 20 mL of 1000 mg/L of Cr(VI) solution was mixed with 0.1 g of each solid melanin and fungi biomass at 45 °C. Centrifugation at 200 rpm of the solution was achieved after incubation for 24 h to separate the melanin particles and powdered fungi. The separated pellets were washed with ethanol and dried in air. The solid melanin and fungi biomass were immersed in distilled water, separated following the matching aforementioned conditions, and used as a control.

3. Results and Discussion

Melanin was produced extracellularly by *A. pullulans*. The resulting melanin after fermentation was harvested, purified, dried, and ground before being used in the biosorption process (Figure 1).

The technological significance of heavy metals has been recognized, since the outbreaks of metals in a lot of industrial processes, such as electrical equipment, paints, treated wood, and lead-acid batteries. Subsequently, a massive quantity of heavy metals has been discharged into the ecosystem [45]. As the metals are discharged at doses beyond accepted concentration, human life is threatened [46].

The biosorption process has several advantages compared to other ones e.g., less sludge, cheap running, and installation cost, and flexible design process [46,47]. Likewise, fungal cells play a major role in the biosorption process of heavy metals, because of their content of functional groups that attract and sequester metals on their surface [5]. Wherein, the fungal biomass of *A. pullulans* and its extracellular melanin were comparatively studied as biosorbent agents for Cr(VI) eradication from aqueous solution.

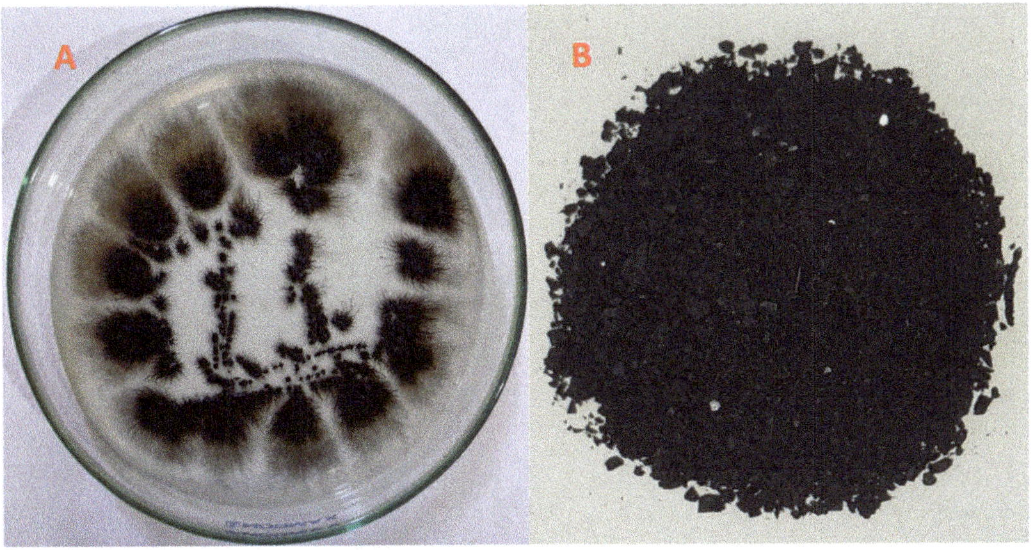

Figure 1. A close-up view of the *A. pullulans* biomass on growth medium (**A**), as well as the biosynthesized melanin by the fungus (**B**).

3.1. Effect of Contact Time

The estimation and design of a biosorbent for the adsorption of Cr(VI) from an aqueous solution depend heavily on the reduction rate. The impact of contact time between Cr(VI) ions, and fungal biomass, or extracellular melanin particles was investigated during reaction intervals up to 300 min) (Figure 2A). The figured data showed that the removal percentage (R%) of Cr(VI) ions increased instantaneously with the augmenting of contact time in the first few minutes, then a slow rate of biosorption was perceived with no notable biosorption rate beyond 240 min (equilibrium time). The biosorption of Cr(VI) ions (R%) increased from 13.46 up to 71.80% and from 9.83 up to 90.06% with increasing the contact time from 10–300 min, by fungal biomass and its extracellular melanin particles, respectively. The biosorption process of Cr(VI) ions is accomplished in two steps. The first step was a fast approach that persisted for a short time. The second stage involved a more gradual procedure that persisted until equilibrium was established. Further increases in contact time after the equilibrium time did not result in any additional removal percentage. The difficulties that Cr(VI) ions encounter in occupying the residual free active surface sites might be the cause of the reduced biosorption rate in the later stage. This is attributed to the forces between the Cr(VI) ions in the solid and bulk phases. The decline in Cr(VI) ions R% with increasing contact time could be a result of the biosorption progression being dominated by the intraparticle diffusion process [48].

Furthermore, metal biosorption is a biphasic process that involves the prompt metal ions sorption on the active groups located on the surface of biomass in the first phase, accompanied by the diffusion of metal throughout the biomass' interior binding sites in the second phase [49]. Moreover, biosynthesized melanin was found to have a higher affinity for adsorbing Cr(VI) than the native fungal biomass. The difference in removal percentage between the two was up to 10%. The literature has reported a similar tendency of chromium adsorption utilizing diverse biomasses as a function of contact time, for example, *Spirulina* sp. biomass [50], *Chlorella* species biomass [51], and dead fungal biomass of *Phanerochaete chrysosporium* [52].

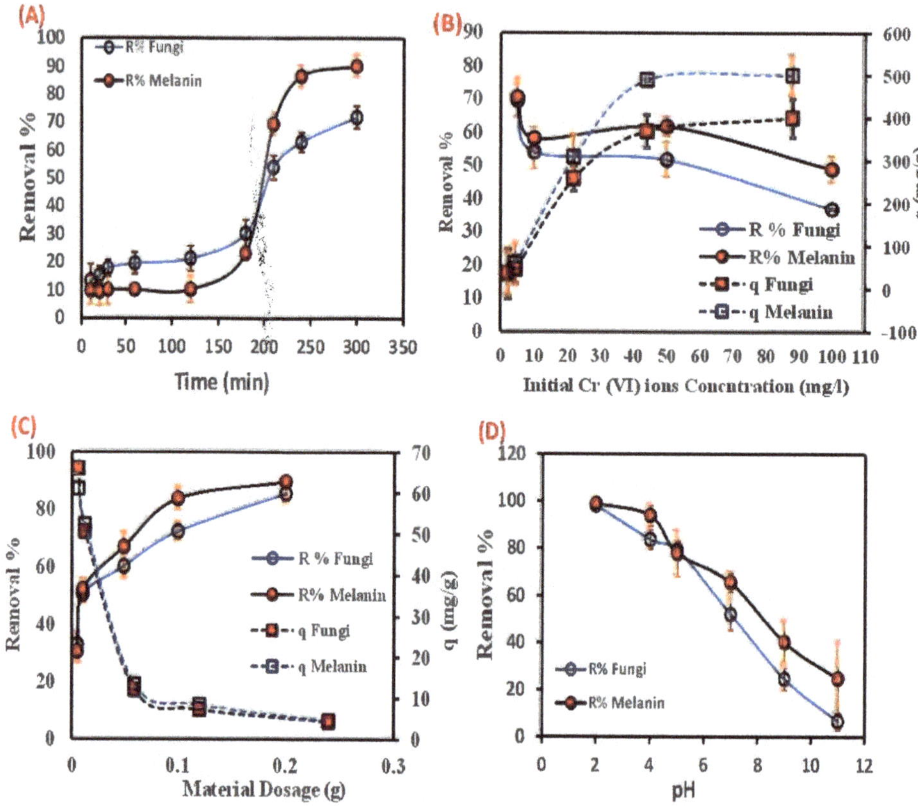

Figure 2. Effect of batch biosorption parameters (**A**) contact time (min), (**B**) initial Cr(VI) metal ions concentration (ppm), (**C**) material dosage (g), and (**D**) pH on the Cr(VI) biosorption onto the used fungal biomass or the biosynthesized melanin.

3.2. *Effect of Initial Concentration*

The impact of the initial concentration of Cr(VI) ions on its R % by the fungal biomass and its extracellular melanin was investigated (Figure 2B). As depicted by the data, there is a reduction in R (%) of Cr(VI) ions correlated with increasing initial concentration and vice versa with the biosorption aptitude. The increase in the concentration of Cr(VI) ions from 5 up to 200 ppm led to a notable decrease in the R (%) of Cr(VI) ions from 69.493 to 20.000%, and 70.373 to 25.000%, by fungal biomass and its extracellular melanin, respectively.

While the biosorption aptitude of Cr(VI) ions increased from 34.747 to 400.000 mg/g, and from 35.187 to 500.000 mg/g onto fungal biomass and extracellular melanin, respectively. This phenomenon might be due to the fast depletion of the biosorption sites [53]. Moreover, the increase in the biosorption aptitude might be ascribed to the presence of Cr(VI) molecules in the aqueous phase, which causes the rapid movement of the molecules to the used bio-sorbent particle surface at higher concentrations, leading to an elevation in the coefficient of mass transfer [54]. This result is consistent with other reported studies [55]. The results signify that the biosynthesized melanin has a great potential for reducing Cr(VI) at higher initial concentrations than the used fungal biomass (~100 mg/g).

Concerning the obtained results, it could be deduced that the used fungal biomass and the extracellular melanin could eliminate a higher Cr(VI) amount per gram of biosorbent.

3.3. Dosage of Biosobants Virsus Cr(VI) Ions Biosorption

The biosorption of Cr(VI) ions was profoundly influenced by the biomass dosage during the process. Figure 2C depicted that the removal percentage (R%) of Cr(VI) ions increased with increasing the biosorbents' dosage. However, the biosorption talent (q) decreased with increasing the biosorbents' dosage. The removal percentage of Cr(VI) ions increased from 33.033% to 85.557% and from 30.550% to 89.763% when the dosage of the fungal biomass and the extracellular melanin increased from 0.005 g to 0.2 g, respectively. However, the biosorption capability decreased from 66.067 mg/g to 4.278 mg/g and from 61.100 mg/g to 4.488 mg/g, respectively, with increasing dosages of fungal biomass and extracellular melanin. Thus, the accessible increase in the active sites on the surface of the biosorbent is a result of the increased amount of the biosorbent in the aqueous phase [56].

Accordingly, a direct relationship between the maximum concentration of the biosorbent capacity and the loaded biomass, for instance, the lowest amount of the used biosorbent particles was already higher than the maximum amount. This inverse relationship is attributed to the agglomeration or coagulation of biomass particles at high biomass loadings. This aggregation of particles reduces the number of accessible sites for metal ion biosorption which could be attributed to the decrease in the metal-to-biosorbent ratio as the dosage was increased [57]. Similar findings have been reported in the recent literature [58–60].

3.4. Effect of Solution pH on the Cr(VI) Biosorption

The adsorption of metallic ions from aqueous solutions could be affected by pH-values, (2–11), where the Cr(VI) ions biosorption by the two used biosorbents has been investigated. The maximum removal percentage of Cr(VI)ions by the fungal biomass and the extracellular melanin 98.147% and 98.880%, respectively, was observed at pH 2 (Figure 2D). Obtained results revealed that the increasing pH up to 5 resulted in a Cr(VI) removal percentage of only 79.120% and 77.800% for fungal biomass and the extracellular melanin, respectively. The increase in Cr(VI) removal could be attributed to the pronated surface hydroxyl groups ($-OH_2^+$) from melanin. Subsequently, these protonated hydroxyl groups enable the used biosorbent particles to electrostatically interact with negatively charged $Cr_2O_7^{2-}$ [61,62]. Similar findings of decreased Cr(VI) biosorption percentage and biosorption capacity with increasing pH have been reported in other reports for a variety of biosorbents such as *Sinorhizobium* sp. [63], and *Pleurotus ostreatus* [64]. On the other hand, it could be observed that melanin particles have a higher removal percentage and biosorption capacity than native *A. pullulans* biomass even at higher pH values. This could be explained by the fact that melanin is a polymer of amino acids that contains indole rings, which are highly electronegative. Additionally, melanin has a strong affinity for positively charged Cr(VI) metal ions [65].

The results presented above clearly demonstrate that extracellular melanin has a higher adsorption capacity than fungus biomass. Sometimes the difference in removal percentage was approximately 20%, and the difference in biosorption capacity was 100 mg/g, due to its unique chemical structure. Melanin is a polymer of amino acids that contains indole rings, which give it its dark color. These indole rings are highly electronegative, which means that they have a strong affinity for positively charged metal ions. Fungus biomass, on the other hand, is composed of a variety of organic compounds, including carbohydrates, proteins, and lipids. These compounds do not have the same strong affinity for positively charged metal ions as indole rings. In addition, melanin is a very porous material, which means that it has a large surface area. This large surface area allows melanin to interact with more metal ions, which results in a higher adsorption capacity.

3.5. Biosorption Isotherms

Adsorption isotherms are mathematical equations that demonstrate how the concentration of a metal in a solution at equilibrium relates to how much metal is adsorbed onto a biosorbent. They are used to determine the spreading and interface of metals in the biphasic system at equilibrium. Adsorption isotherms are defined by several parameters, including the initial level of the adsorbate in the aqueous phase, the amount of biosorbent, the relative capabilities of the metal's adsorption, and competition among the solutes. These parameters are also identified by specific values that reflect the surface properties and ability of a biosorbent for the adsorption of heavy metal ions [65,66].

In this study, the linear Langmuir, Freundlich, and Temkin models were applied to the experimental data to find the best-fitted model. The linear graphs of the three applied models are illustrated in Figure 3, and their corresponding parameters are tabulated in Table 3.

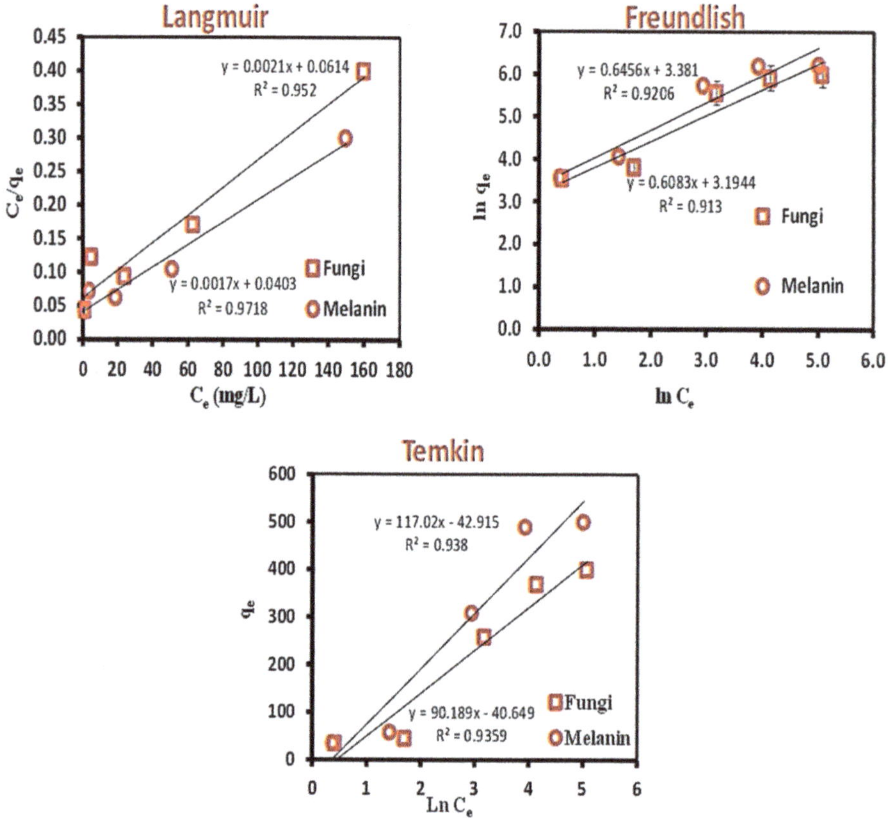

Figure 3. Equilibrium isotherm models; Langmuir, Freundlich, and Temkin for the biosorption of Cr(VI) ions onto the used fungal biomass and the biosynthesized melanin.

The relevance of the isotherm equations is evaluated by R^2. The Langmuir isotherm was further documented by analyzing the Langmuir equilibrium parameter (R_L), as outlined in Equation (3).

$$R_L = \frac{1}{1 + K_{L}\, C_0} \quad (3)$$

The calculated R_L value was found to be 0.749 and 0.706 for the used fungal biomass and the extracellular melanin, respectively. The biosorption process's nature is distinguished by the R_L value, e.g., unfavorable $R_L > 1$); linear ($R_L = 1$); favorable ($0 < R_L < 1$); irreversible ($R_L = 0$) [67,68].

Table 3. The biosorption of Cr(VI) onto the used fungal biomass or the biosynthesized melanin at 25 °C based on Langmuir, Freundlich, and Temkin parameters.

Isotherm Parameters	Fungal Biomass	Melanin
Langmuir		
q_m (mg g^{-1}) calculated	485.747	595.974
K_L (mg L^{-1})	0.034	0.042
R^2	0.976	0.986
Freundlich		
K_F (mgL^{-1}/n L1/n g^{-1})	24.395	1.907
Nf	1.64	1.55
R^2	0.956	0.959
Temkin		
B (mg L^{-1})	90.189	117.017
K_T (KJ mol^{-1})	0.637	0.693
R^2	0.967	0.969

The obtained R_L value indicated that the Cr(VI) biosorption onto the used fungal biomass or the extracellular melanin was a favorable process. The value of 1/n of the Freundlich isotherm model illustrates the probability of the isotherm, e.g., irreversible (1/n = 0); favorable (0 < 1/n < 1); and unfavorable (1/n > 1). The calculated 1/n value (0.608 and 0.646) favored the Freundlich isotherm model. The value of R^2 was applied to evaluate the fit of both isotherm models. The values were attained to be above 0.90 for both models, identifying that the data of isotherm interpretations fit well for both models. However, the fit was more favorable to the Langmuir model than the Freundlich model [69]. The Temkin isotherm model (Table 3) takes into account the influence of adsorbent-adsorbate interactions on the adsorption process. In accordance with the R^2 value, the Temkin isotherm model is well-fitted with the investigational data specifying the energetic homogeneity of the biosorption sites and chemisorption process (RT/b lnKT = 0.637 and 0.693 while B = 90.189 and 117.017 mg L^{-1} for the used fungal biomass and the biosynthesized melanin, respectively). The melanin particles have given the best biosorption rate of Cr(VI) metal ion removal.

3.6. Biosorption Kinetics

It is important to characterize metal ion biosorption in kinetic terms to understand the behavior involved in biosorption, such as mass transfer and chemical interactions, where, the greatest amount of metal is absorbed in the first minutes to hours owing to the accessibility of free active sites of the biosorbent surface [70–73]. In this work, four of the most extensively benefitted kinetic models were applied to define the mechanism of Cr(VI) ions onto the treated fungal biomass and the biosynthesized melanin particles pseudo-first-order, pseudo-second-order, Elovich model, and intraparticle diffusion model. The graphical demonstrations of these models and the kinetic data are publicized in Figure 4 and Table 4.

The first-order Lagergren model assumes that the adsorption rate is positively related to the amount of vacant adsorption sites [74]. Previous studies [75,76] have shown that the first-order Lagergren model is not always relevant to all practical data across the biosorption process. Herein, the model is appropriate for depicting the kinetic data of the Cr(VI) ions biosorption, although the correlation coefficient (R^2 = 0.774 and 0.707 for the used fungal biomass and the biosynthesized melanin, respectively) which is lower than 0.845. Nevertheless, the considered uptake capacity values are respectably matched with the experimental values, also suggesting the suitability of a pseudo-first-order model for describing the kinetics of Cr(VI) ions biosorption onto the used fungal biomass and the biosynthesized melanin, which indicates the physical nature of the adsorption process. These findings are consistent with the previous literature [74].

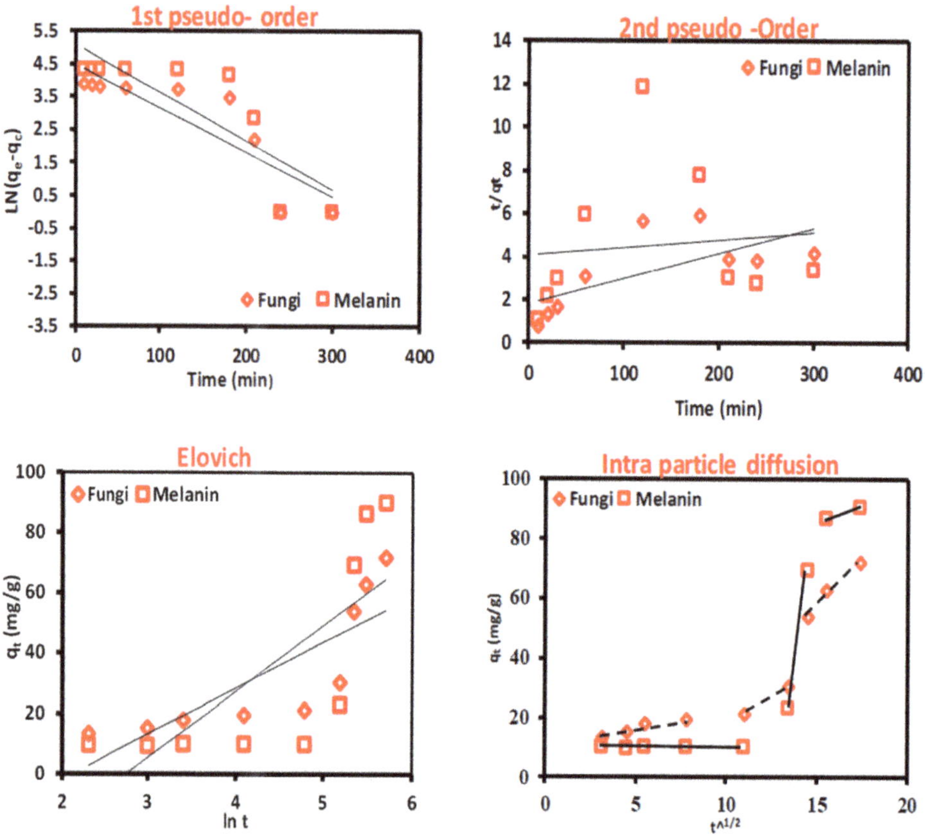

Figure 4. Biosorption kinetics models of Cr(VI) ions onto the used fungal biomass and the biosynthesized melanin particles.

Table 4. Kinetic models for biosorption of Cr(VI) ions onto the used fungal biomass and the biosynthesized melanin particles.

Kinetic Model	Fungal Biomass		Melanin
Pseudo-first-order	q_e (mg/g) Calculated	69.4148	163.531
	q_e (mg/g) Experimental	62.9	86.400
	k_1 (min^{-1})	−0.015	0.015
	R^2	0.774	0.707
Pseudo-second-order	q_e (mg/g) Calculated	85.825	107.704
	q_e (mg/g) Experimental	62.9	86.400
	k_2 (g/mg min)	0.01	0.0048
	R^2	0.678	0.235
Elovich	ß (g/mg)	15.137	21.853
	à (mg/g min)	−32.062	−60.041
	R^2	0.820	0.692
Intra-particle diffusion	$K_{1,}$	3.888	5.748
	C_1	−5.933	−23.697
	R^2	0.900	0.844

The pseudo-second-order model posits that the biosorption capacity is directly proportional to the amount of occupied active sites on the biosorbent surface and that the rate-limiting step in biosorption may involve valence forces through sharing or exchanging electrons between the biosorbent and the adsorbate [76]. The low correlation coefficient (R^2) and the significant difference between the calculated biosorption capacity values ($q_{calculated}$) and the experimental ones ($q_{experimental}$) specify that this model is not appropriate to describe the kinetics of Cr(VI) ions biosorption onto the used biosorbent particles. Elovich kinetic model can be used to identify heterogeneous adsorbents and their role in the adsorption process, but it does not predict any specific adsorption mechanism [42]. The Elovich kinetic model has low R^2 values, which makes sense because the kinetic data were well-fitted to the pseudo-first-order model. This indicates that the biosorption process is physical, while the Elovich model is more suited to describe chemosorption processes.

The steps that control the biosorption rate were elucidated using the intraparticle diffusion model. The three separate regions with varied slopes (K_{id}) and intercepts (C_i) that did not pass through the origin (Figure 4) revealed that the biosorption process of Cr(VI) ions onto the used biosorbent particles was not influenced by the intra-particle diffusion alone but was also affected by more than one process. The intra-particle diffusion model fits the data well, with a high correlation coefficient (0.900 and 0.844) for the used fungal biomass and the biosynthesized melanin, respectively. Data from Table 5 display a comparison of the biosorption capacity of Cr(VI) ions using fungal biomass and melanin.

Table 5. Comparison of the maximum biosorption capacity of Cr(VI) ions onto the used fungal biomass and the extracellular melanin particles with those of previous studies.

Adsorbent	qm (mg/g)	Reference
Fungal biomass and melanin	485.747 and 595.974	Current study
Removal of Cr(VI) by polyethyleneimine-impregnated activated carbon	114	[77]
Biosorption of chromium metal ions onto *Ludwigia stolonifera*	43.478	[39]
Biosorption of Cr(VI) by *Bacillus megaterium* and *Rhodotorula* sp. inactivated biomass	34.80	[56]
Melanin-embedded materials effectively remove Cr(VI)	19.60 and 6.24 for IMB and CMB	[78]
Melanin nano pigment from *Pseudomonas stutzeri*	126.9	[79]
Equilibrium and kinetic studies of copper(II) removal by fungal biomasses	7.74 and 12.08	[74]

3.7. Decision Tree Learning Algorithm

The data from the equilibrium isotherms and kinetic studies (Tables 6 and 7) for both fungal biomass and melanin were used for constructing the DT. For each tree, the data were split into two sets: 70% training and 30% testing. Upon performing the learning process, the predicted and error values mutually with the terminal nodes, were determined for both absorbents (fungal biomass and melanin).

DT begins with a root node and ends up with many branches and roots, i.e., many solutions. The root node is the first node encountered when traversing the tree during the prediction process. From the root node, the decision tree branches into internal nodes (decision nodes) based on different attribute values, and eventually, the leaves (terminal nodes) are reached, providing the final predictions or decisions.

DT creates classifications for the input variables (categorical and/or continuous). As a result, the population can be divided into two or more homogeneous groups. These homogeneous sets are built using the most substantial differentiator on the input variables [27,80]. No information is available about the combination of DT with equilibrium isotherms and kinetic studies for the determination of relative variable importance and modeling the biosorption process. Herein, DT was applied to handle our suggestion regarding identifying the most incredible biosorption parameters and predicting the optimal range of these parameters.

Table 6. The array equilibrium isotherms and kinetic data for maximization of Cr(VI) removal by the biomass of *A. pullulans*, and the predicted and error values as well as terminal nodes of the DT at each data point.

Isotherm or Kinetic Test	Run	Tested Parameter				Cr(VI) Removal, %					
						Actual		Decision Tree			
		Time, min	Cr(VI), mg/mL	Biomass (g/100 mL)	pH	Value	Mean	Type	Fitted	Error	Terminal Node
Contact time	1	10	10	0.010	5	13.44	13.46 ± 0.11	Test	14.64	−1.20	1
	2	20	10	0.010	5	15.41	15.25 ± 0.14	Test	14.64	0.77	1
	3	30	10	0.010	5	17.88	17.87 ± 0.07	Training	17.87	0.01	2
	4	60	10	0.010	5	19.35	19.56 ± 0.19	Test	20.69	−1.34	3
	5	120	10	0.010	5	21.11	21.24 ± 0.15	Training	20.69	0.42	3
	6	180	10	0.010	5	29.23	30.35 ± 1.15	Test	30.31	−1.08	4
	7	240	10	0.010	5	53.00	54.02 ± 1.01	Training	59.48	−6.48	9
	8	280	10	0.010	5	62.50	62.99 ± 0.72	Test	62.65	−0.15	10
	9	300	10	0.010	5	70.95	71.82 ± 1.03	Training	71.25	−0.30	13
Initial Cr(VI)	10	240	5	0.010	5	68.25	69.45 ± 1.05	Training	69.45	−1.19	8
	11	240	10	0.010	5	53.34	54.08 ± 0.79	Training	59.48	−6.15	9
	12	240	50	0.010	5	50.45	51.70 ± 1.21	Test	52.32	−1.86	11
	13	240	100	0.010	5	35.96	36.87 ± 1.03	Test	25.24	10.72	16
	14	240	200	0.010	5	19.43	20.03 ± 0.86	Training	25.24	−5.81	16
Melanin	15	240	10	0.005	5	33.04	33.05 ± 0.99	Training	33.05	−0.01	5
	16	240	10	0.010	5	50.84	50.61 ± 0.65	Training	59.48	−8.64	9
	17	240	10	0.050	5	59.54	60.26 ± 0.81	Training	59.48	0.06	9
	18	240	10	0.100	5	71.59	72.33 ± 0.92	Training	72.33	−0.74	14
	19	240	10	0.200	5	85.32	85.58 ± 0.86	Training	85.58	−0.26	15
pH	20	240	10	0.010	2	98.25	98.16 ± 0.94	Test	99.04	−0.78	6
	21	240	10	0.010	4	82.88	83.67 ± 0.69	Training	83.67	−0.79	7
	22	240	10	0.010	5	79.32	79.16 ± 1.06	Test	59.48	19.84	9
	23	240	10	0.010	7	52.32	52.19 ± 0.89	Test	52.13	0.19	12
	24	240	10	0.010	9	23.87	24.61 ± 0.73	Test	24.65	−0.78	17
	25	240	10	0.010	11	6.53	6.65 ± 0.30	Training	6.65	−0.11	18
Contact time	1	10	10	0.010	5	13.36		Test	14.64	−1.28	1
	2	20	10	0.010	5	15.15		Training	14.64	0.51	1
	3	30	10	0.010	5	17.94		Training	17.87	0.07	2
	4	60	10	0.010	5	19.73		Training	20.69	−0.96	3
	5	120	10	0.010	5	21.22		Training	20.69	0.53	3
	6	180	10	0.010	5	31.52		Test	30.31	1.21	4
	7	240	10	0.010	5	54.04		Test	59.48	−5.44	9
	8	280	10	0.010	5	62.65		Training	62.65	0.00	10
	9	300	10	0.010	5	72.97		Test	71.25	1.71	13
Initial Cr(VI)	10	240	5	0.010	5	69.88		Training	69.45	0.43	8
	11	240	10	0.010	5	54.92		Training	59.48	−4.56	9
	12	240	50	0.010	5	51.77		Training	52.32	−0.55	11
	13	240	100	0.010	5	37.99		Test	25.24	12.74	16
	14	240	200	0.010	5	19.64		Training	25.24	−5.60	16
Melanin	15	240	10	0.005	5	32.07		Training	33.05	−0.99	5
	16	240	10	0.010	5	49.88		Training	59.48	−9.61	9
	17	240	10	0.050	5	60.09		Test	59.48	0.61	9
	18	240	10	0.100	5	72.03		Training	72.33	−0.29	14
	19	240	10	0.200	5	84.88		Training	85.58	−0.70	15

Table 6. *Cont.*

Isotherm or Kinetic Test	Run	Tested Parameter				Cr(VI) Removal, %					
						Actual		Decision Tree			
		Time, min	Cr(VI), mg/mL	Biomass (g/100 mL)	pH	Value	Mean	Type	Fitted	Error	Terminal Node
pH	20	240	10	0.010	2	97.17		Test	99.04	−1.86	6
	21	240	10	0.010	4	83.99		Training	83.67	0.32	7
	22	240	10	0.010	5	78.03		Training	59.48	18.55	9
	23	240	10	0.010	7	51.25		Training	52.13	−0.88	12
	24	240	10	0.010	9	24.65		Training	24.65	0.00	17
	25	240	10	0.010	11	6.42		Training	6.65	−0.22	18
Contact time	1	10	10	0.010	5	13.58		Training	14.64	−1.06	1
	2	20	10	0.010	5	15.20		Training	14.64	0.56	1
	3	30	10	0.010	5	17.79		Training	17.87	−0.07	2
	4	60	10	0.010	5	19.60		Test	20.69	−1.09	3
	5	120	10	0.010	5	21.40		Test	20.69	0.71	3
	6	180	10	0.010	5	30.31		Training	30.31	0.00	4
	7	240	10	0.010	5	55.02		Test	59.48	−4.46	9
	8	280	10	0.010	5	63.81		Test	62.65	1.16	10
	9	300	10	0.010	5	71.55		Training	71.25	0.30	13
Initial Cr(VI)	10	240	5	0.010	5	70.21		Training	69.45	0.77	8
	11	240	10	0.010	5	54.00		Training	59.48	−5.48	9
	12	240	50	0.010	5	52.86		Training	52.32	0.55	11
	13	240	100	0.010	5	36.65		Training	25.24	11.41	16
	14	240	200	0.010	5	21.02		Test	25.24	−4.22	16
Melanin	15	240	10	0.005	5	34.04		Training	33.05	0.99	5
	16	240	10	0.010	5	51.12		Test	59.48	−8.36	9
	17	240	10	0.050	5	61.13		Training	59.48	1.65	9
	18	240	10	0.100	5	73.35		Training	72.33	1.03	14
	19	240	10	0.200	5	86.54		Training	85.58	0.96	15
pH	20	240	10	0.010	2	99.04		Training	99.04	0.00	6
	21	240	10	0.010	4	84.14		Training	83.67	0.47	7
	22	240	10	0.010	5	80.13		Training	59.48	20.65	9
	23	240	10	0.010	7	53.02		Training	52.13	0.88	12
	24	240	10	0.010	9	25.32		Test	24.65	0.67	17
	25	240	10	0.010	11	6.99		Training	6.65	0.34	18

Table 7. The array equilibrium isotherms and kinetic data for maximization of Cr(VI) removal by the melanin of *A. pullulans*, and the predicted and error values, as well as terminal nodes of the DT at each data point.

Isotherm or Kinetic Test	Run	Tested Parameters				Cr(VI) Removal, %					
						Actual		Decision Tree			
	No.	Time, min	Cr(VI), mg/mL	Melanin (g/100 mL)	pH	Value	Mean	Type	Fitted	Error	Terminal Node
Contact time	1	10	10	0.010	5	9.95	9.84 ± 0.14	Test	10.05	−0.10	1
	2	20	10	0.010	5	9.32	9.46 ± 0.24	Test	10.05	−0.73	1
	3	30	10	0.010	5	10.45	10.09 ± 0.45	Training	10.05	0.40	1
	4	60	10	0.010	5	10.36	10.13 ± 0.49	Test	10.05	0.31	1
	5	120	10	0.010	5	10.73	10.11 ± 0.59	Training	10.05	0.68	1
	6	180	10	0.010	5	22.21	23.14 ± 0.93	Test	23.14	−0.92	2
	7	240	10	0.010	5	68.23	69.40 ± 1.16	Training	64.91	3.32	5
	8	280	10	0.010	5	84.55	86.41 ± 1.79	Test	89.07	−4.52	6
	9	300	10	0.010	5	89.05	90.04 ± 0.99	Training	89.07	−0.02	6

Table 7. *Cont.*

Isotherm or Kinetic Test	Run	Tested Parameters				Cr(VI) Removal, %					
						Actual		Decision Tree			
	No.	Time, min	Cr(VI), mg/mL	Melanin (g/100 mL)	pH	Value	Mean	Type	Fitted	Error	Terminal Node
Initial Cr(VI)	10	240	5	0.010	5	70.12	70.39 ± 0.95	Training	64.91	5.21	5
	11	240	10	0.010	5	57.18	58.14 ± 0.97	Training	64.91	−7.73	5
	12	240	50	0.010	5	61.49	61.72 ± 0.88	Test	64.91	−3.42	5
	13	240	100	0.010	5	48.88	48.91 ± 0.94	Test	33.33	15.55	8
	14	240	200	0.010	5	26.00	25.05 ± 0.95	Training	33.33	−7.33	8
Melanin	15	240	10	0.005	5	30.20	30.56 ± 1.21	Training	30.56	−0.36	3
	16	240	10	0.010	5	52.22	52.28 ± 0.91	Training	64.91	−12.69	5
	17	240	10	0.050	5	66.95	66.96 ± 1.00	Training	64.91	2.04	5
	18	240	10	0.100	5	83.90	83.77 ± 0.95	Training	86.75	−2.85	7
	19	240	10	0.200	5	89.65	89.73 ± 0.89	Training	86.75	2.90	7
pH	20	240	10	0.010	2	98.76	98.83 ± 0.98	Test	95.36	3.40	4
	21	240	10	0.010	4	93.88	93.87 ± 0.90	Training	95.36	−1.49	4
	22	240	10	0.010	5	77.69	77.82 ± 0.92	Test	64.91	12.78	5
	23	240	10	0.010	7	65.12	65.59 ± 1.02	Test	64.91	0.21	5
	24	240	10	0.010	9	40.25	40.21 ± 0.98	Test	41.18	−0.93	9
	25	240	10	0.010	11	24.85	24.83 ± 1.06	Training	24.83	0.02	10
Contact time	1	10	10	0.010	5	9.68		Test	10.05	−0.37	1
	2	20	10	0.010	5	9.73		Training	10.05	−0.32	1
	3	30	10	0.010	5	10.23		Training	10.05	0.18	1
	4	60	10	0.010	5	10.45		Training	10.05	0.40	1
	5	120	10	0.010	5	10.06		Training	10.05	0.01	1
	6	180	10	0.010	5	24.07		Test	23.14	0.94	2
	7	240	10	0.010	5	69.42		Test	64.91	4.51	5
	8	280	10	0.010	5	88.12		Training	89.07	−0.95	6
	9	300	10	0.010	5	91.03		Test	89.07	1.96	6
Initial Cr(VI)	10	240	5	0.010	5	71.46		Training	64.91	6.55	5
	11	240	10	0.010	5	58.13		Training	64.91	−6.78	5
	12	240	50	0.010	5	60.98		Training	64.91	−3.93	5
	13	240	100	0.010	5	47.99		Test	33.33	14.66	8
	14	240	200	0.010	5	24.11		Training	33.33	−9.22	8
Melanin	15	240	10	0.005	5	31.91		Training	30.56	1.35	3
	16	240	10	0.010	5	51.40		Training	64.91	−13.51	5
	17	240	10	0.050	5	65.97		Test	64.91	1.06	5
	18	240	10	0.100	5	82.77		Training	86.75	−3.98	7
	19	240	10	0.200	5	88.88		Training	86.75	2.13	7
pH	20	240	10	0.010	2	97.89		Test	95.36	2.53	4
	21	240	10	0.010	4	92.97		Training	95.36	−2.40	4
	22	240	10	0.010	5	78.79		Training	64.91	13.88	5
	23	240	10	0.010	7	66.77		Training	64.91	1.86	5
	24	240	10	0.010	9	41.18		Training	41.18	0.00	9
	25	240	10	0.010	11	25.88		Training	24.83	1.04	10

Table 7. *Cont.*

Isotherm or Kinetic Test	Run	Tested Parameters				Cr(VI) Removal, %					
						Actual		Decision Tree			
	No.	Time, min	Cr(VI), mg/mL	Melanin (g/100 mL)	pH	Value	Mean	Type	Fitted	Error	Terminal Node
Contact time	1	10	10	0.010	5	9.88		Training	10.05	−0.18	1
	2	20	10	0.010	5	9.33		Training	10.05	−0.72	1
	3	30	10	0.010	5	9.58		Training	10.05	−0.47	1
	4	60	10	0.010	5	9.57		Test	10.05	−0.49	1
	5	120	10	0.010	5	9.54		Test	10.05	−0.51	1
	6	180	10	0.010	5	23.14		Training	23.14	0.00	2
	7	240	10	0.010	5	70.54		Test	64.91	5.63	5
	8	280	10	0.010	5	86.54		Test	89.07	−2.53	6
	9	300	10	0.010	5	90.05		Training	89.07	0.97	6
Initial Cr(VI)	10	240	5	0.010	5	69.61		Training	64.91	4.70	5
	11	240	10	0.010	5	59.12		Training	64.91	−5.79	5
	12	240	50	0.010	5	62.69		Training	64.91	−2.22	5
	13	240	100	0.010	5	49.87		Training	33.33	16.55	8
	14	240	200	0.010	5	25.04		Test	33.33	−8.29	8
Melanin	15	240	10	0.005	5	29.56		Training	30.56	−0.99	3
	16	240	10	0.010	5	53.22		Test	64.91	−11.69	5
	17	240	10	0.050	5	67.97		Training	64.91	3.06	5
	18	240	10	0.100	5	84.65		Training	86.75	−2.10	7
	19	240	10	0.200	5	90.65		Training	86.75	3.90	7
pH	20	240	10	0.010	2	99.84		Training	95.36	4.48	4
	21	240	10	0.010	4	94.76		Training	95.36	−0.60	4
	22	240	10	0.010	5	76.97		Training	64.91	12.06	5
	23	240	10	0.010	7	64.89		Training	64.91	−0.02	5
	24	240	10	0.010	9	39.22		Test	41.18	−1.96	9
	25	240	10	0.010	11	23.77		Training	24.83	−1.07	10

3.7.1. Selection of DT

The terminal nodes were plotted versus R^2 (Figure 5) for fungal biomass, and melanin, to identify the smallest regression tree that maximizes the R^2 value. Regarding Cr(VI) removal by the fungal biomass, 18 terminal nodes represent 18 trees that were produced from the validation samples, whereas for melanin as absorbent, there are 10 terminal nodes and 10 trees. The R^2 values of the validation samples for both adsorbents exemplary level off as the tree grows. The maximized R^2 values for the optimal regression tree were 96.26, and 96.70% (for training) and 94.60 and 95.90% (for testing), for fungal biomass and melanin, respectively. These trees had a maximum R^2 value.

Accordingly, the tree was built and the distinction of terminal nodes on the diagram was explored. The DT diagrams for the two absorbents (Figures 6 and 7) depict all 75 cases from the whole data set. The goal was to discover the greatest Cr(VI) removal means with the fewest standard deviations.

For the fungal biomass, the terminal node No. 6 was the best node, achieving 99.07% Cr(VI) removal, whereas, for melanin, the terminal node No. 4 achieved 95.36% Cr(VI) removal, but this node was not selected because it had large standard deviation, being 2.66, instead, the terminal node No. 6 (89.07% Cr(VI) removal) was selected as the best terminal node of the DT. These nodes had a maximum R^2 value with a standard deviation of less than 1.

Regarding the absorption of Cr(VI) by fungal biomass, the first node started with 50 cases and split at two branches of contact time, i.e., $\leq$210 and >210 min. The latter was the best, and split to pH > 8 and $\leq$ 8, the latter was also split based on initial Cr(VI) into the terminal node of Cr(VI) > 75 mg/mL, and node 6 Cr(VI) ≤ 75 mg/mL, which subdivided into terminal node 4 of fungal biomass ≤ 0.0075 g, and node 7 of fungal biomass > 0.0075 g, leading finally to optimum terminal node 5, the node rules were contacting time > 210 min, initial Cr(VI) ≤ 75 mg/mL, fungal biomass > 0.0075 g, and pH ≤ 3 with the highest Cr(VI) removal (99.04%, with lowest SD value = 0.5393), followed by the terminal node 15 (85.58%, with SD value = 0.7046). The other terminal nodes did not show further improvements. Alternatively, terminal node 1 has the smallest Cr(VI) removal (14.64%), suggesting that the data in the terminal node are probably skewed.

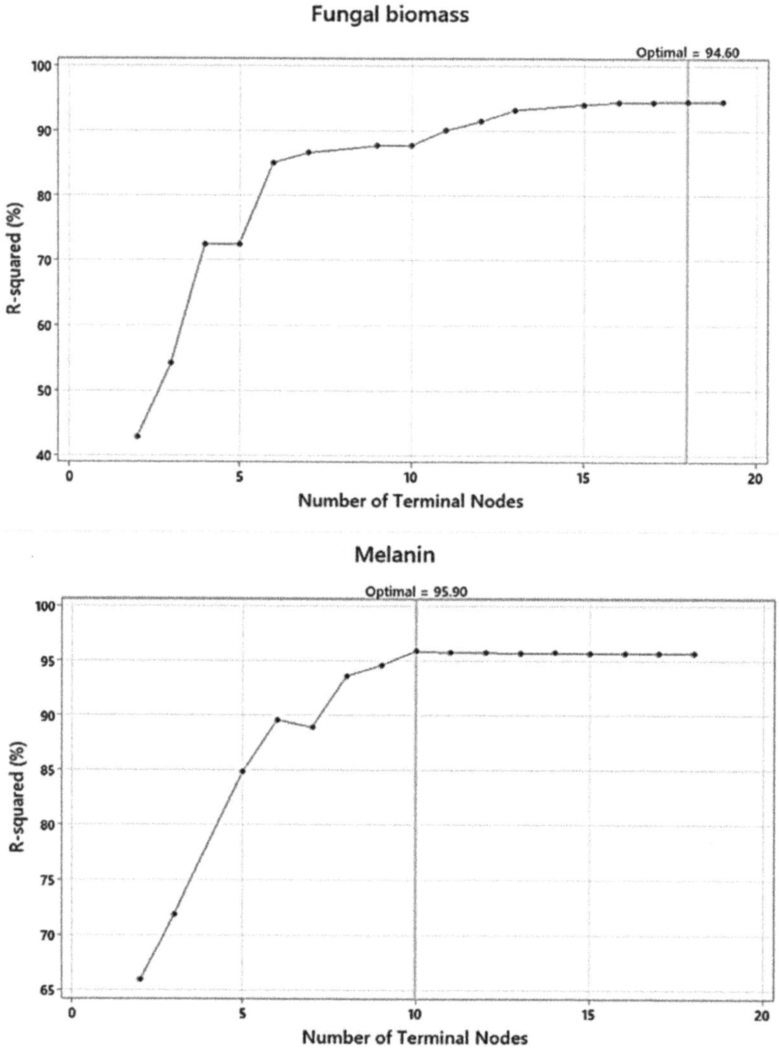

Figure 5. The plots of the number of terminal nodes versus R^2 generated by a DT for Cr(VI) removal by fungal biomass, and melanin.

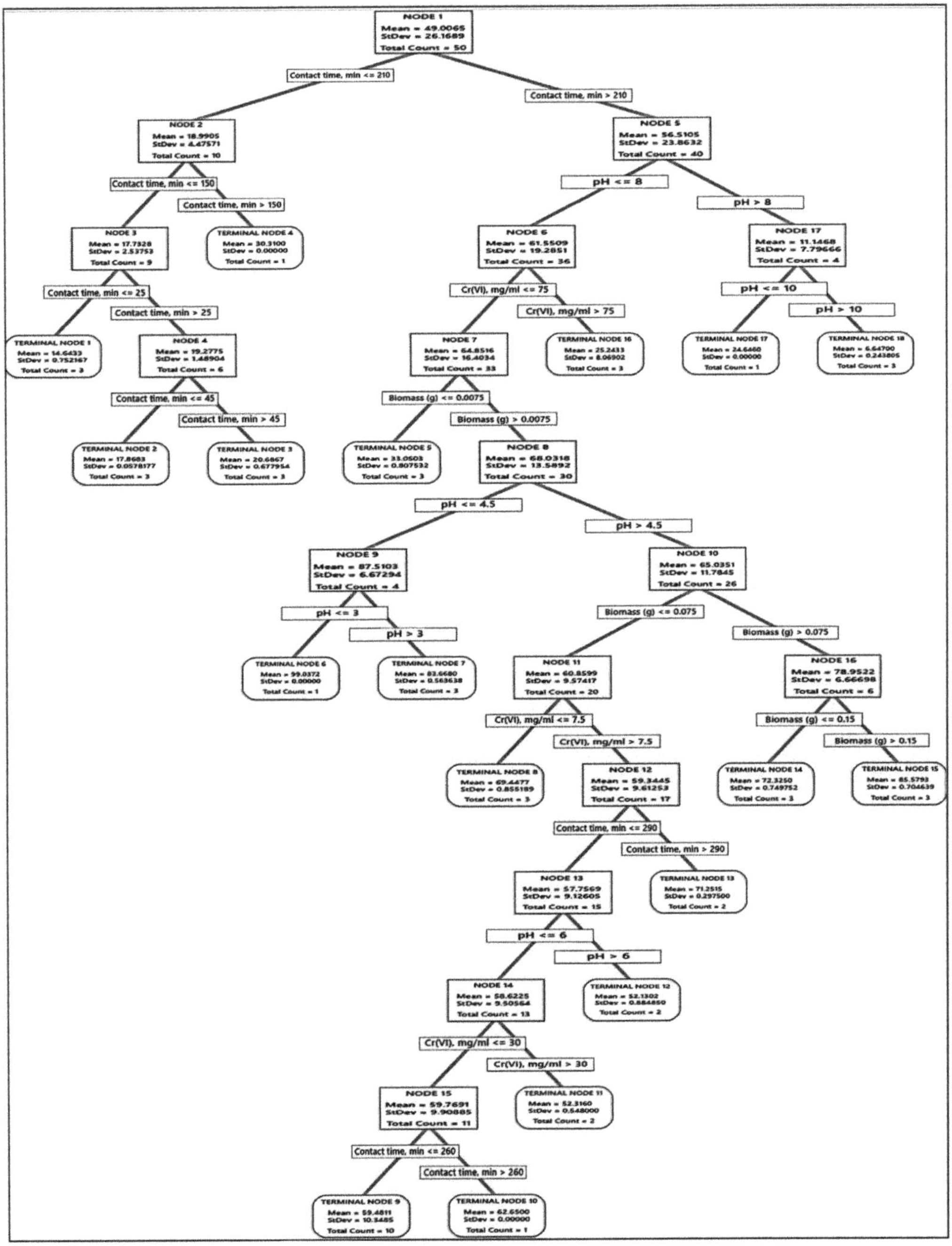

Figure 6. The optimal decision tree diagram of the Cr(VI) removal (%) using fungal biomass. A total of 75 equilibrium isotherms and kinetic data points were employed in the DT to predict the optimum conditions for Cr(VI) removal. The squares contain the data utilized for node selection.

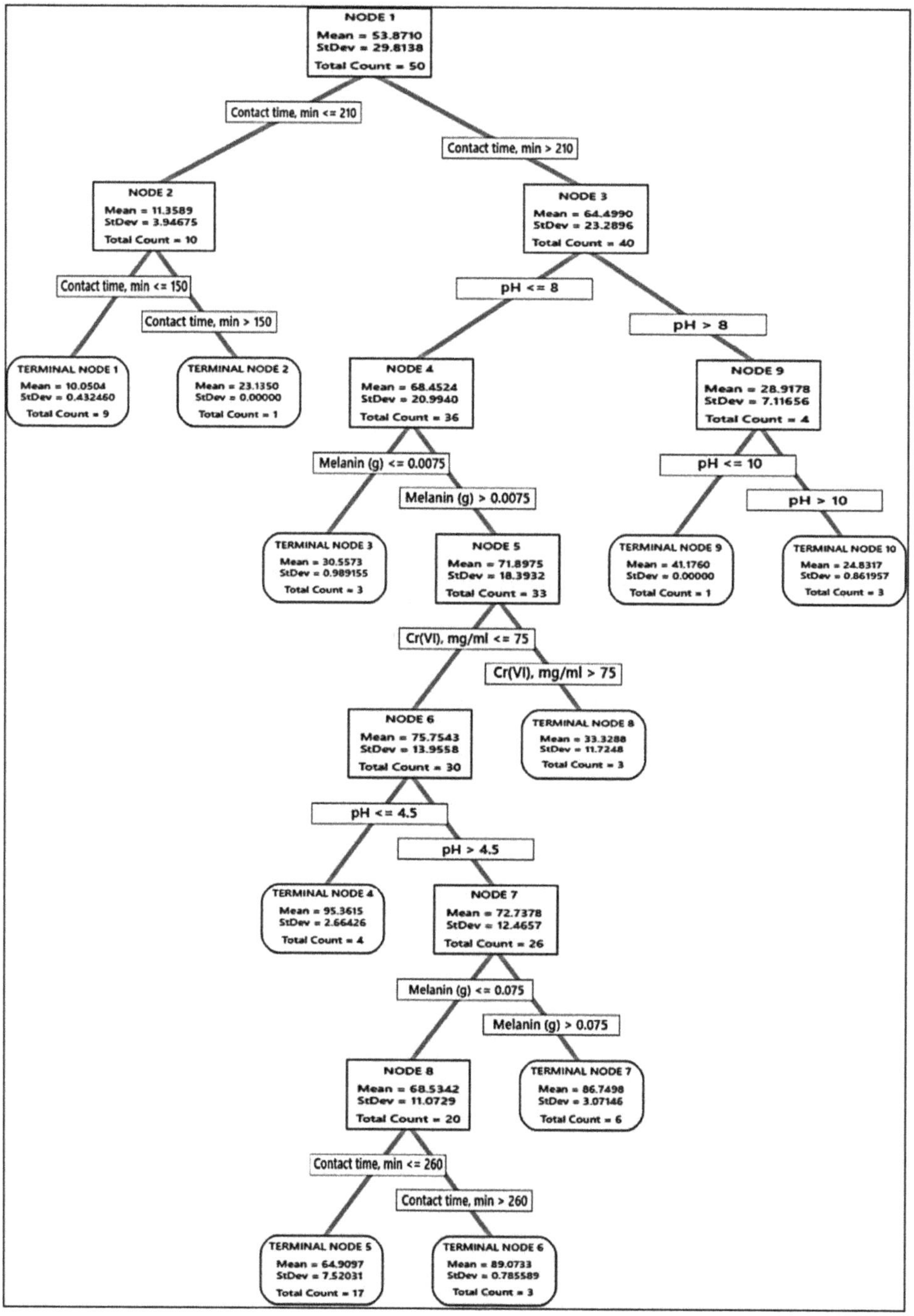

Figure 7. The optimal decision tree diagram of the Cr(VI) removal (%) using melanin. A total of 75 equilibrium isotherms and kinetic data points were employed in the DT to predict the optimum conditions for Cr(VI) removal. The squares contain the data utilized for node selection.

Regarding the absorption of Cr(VI) by melanin, such as fungal biomass DT, the first node uses 50 cases and is split at two levels of contact time, i.e., ≤210 and >210 min. The latter was split into pH > 8 and ≤8, and the latter was split into several nodes and terminal nodes, where terminal node 4 was the best in Cr(VI) removal (95.36%) of the entire DT. However, this point was not a suitable one since it recorded a high value of SD (2.6643). Node 6 has been the second, recording (89.04%), with a SD value of 0.7856, which is less than the overall SD. The role of the terminal node was contact time, min > 260, Cr(VI) ≤ 75 mg/mL, 0.0075 < melanin (g) ≤ 0.075, 4.5 < pH <= 8. No further improvements could be observed on the DT terminal nodes.

3.7.2. Evaluation of the DT Models

The error statistics of both DT models were calculated (Table 8). The accuracy of the resulting two trees was specified with high R-squared, and low values of the error measurements (RMSE, MSE, MAD, and MAPE) for training and testing processes. However, it is obvious that the four tested parameters in the two DTs were important predictors, furthermore, the DT of fungal biomass has a lower error rate in comparison to the DT of melanin, this conclusion is conversely to that obtained by the equilibrium isotherms and kinetics data, suggesting that DT can help improve the accuracy of Cr(VI) removal by discovering the optimum overall situation sittings that maximize the biosorption process.

Table 8. Model summary structure and statistics of decision tree supervised machine learning algorithm for Cr(VI) removal using fungal biomass, and melanin.

		Fungal Biomass		**Melanin**	
Model Summary	Total predictor	4		4	
	Important predictor	4		4	
	Terminal node	18		10	
	Minimum terminal node size	1		1	
Statistics		Training	Test	Training	Test
	R^2, %	96.26	94.60	96.70	95.90
	Root mean squared error	5.0599	5.7665	5.4176	6.1535
	Mean squared error	25.6025	33.2528	29.3509	37.8658
	Mean absolute deviation	2.4584	3.3704	3.5080	3.9997
	Mean absolute percent error	0.0557	0.0832	0.0700	0.0855
	Standard deviation	26.4345	25.3274	30.1164	31.0030
	Number of cases	50	25	50	25

3.7.3. Relative Importance of the Variable

The relative importance of the variables for both DTs was computed as the increasing percentage relative to the leading variable. Figure 8 shows that the four predictors in both DTs are important to the tree. Contact time was the leading predictor in both trees (fungal biomass and melanin) since it has a relative importance of 100% contribution in Cr(VI) removal. The other variables are standardized concerning the most important predictor. As a result, the significance of each variable can be simply interpreted. The leading variable is often identified as the one with the maximum improvement score (and hence regarded as 100% critical), and the remaining variables are sorted consequently [9]. Surrogate or primary splitters in the DT are considered crucial variables. The most essential variable is always given 100% relative importance, while the non-important variable is not represented in the tree. Regarding fungal biomass DT, relative importance in descending order was contact time (100%) > pH (96.6%) > fungal biomass dose (43.4) > initial Cr(VI) concentration

(39.6%), whereas for melanin DT was contact time (100%) > pH (31.4%) > melanin dose (25.7%) > initial Cr(VI) concentration (20.3%).

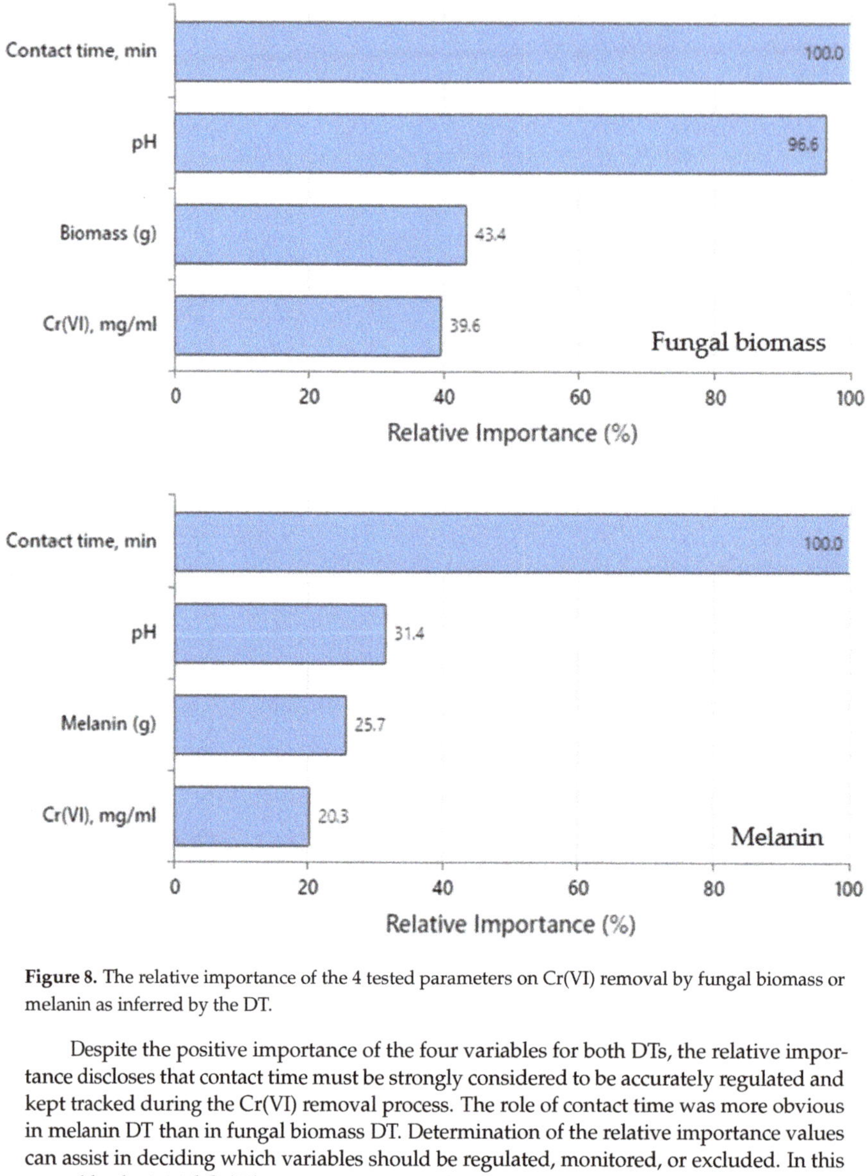

Figure 8. The relative importance of the 4 tested parameters on Cr(VI) removal by fungal biomass or melanin as inferred by the DT.

Despite the positive importance of the four variables for both DTs, the relative importance discloses that contact time must be strongly considered to be accurately regulated and kept tracked during the Cr(VI) removal process. The role of contact time was more obvious in melanin DT than in fungal biomass DT. Determination of the relative importance values can assist in deciding which variables should be regulated, monitored, or excluded. In this assembly, the initial Cr(VI) concentration in melanin DT has the lowest relatively important value (20.3%) concerning contact time. This conclusion comes in line with the goal of the present work, in which the current DTs show efficient accuracy in the detection of these kinds of hidden relationships among tested parameters compared to the equilibrium isotherms and kinetic studies.

3.7.4. Validation of DT Models

Both DT models were experimentally validated. The forecast levels of each of the four tested parameters were determined at various nodes (Table 9). The laboratory validation experiments were carried out in triplicate to validate the procedure conditions estimated from both DT models. When compared to projected values, the experimental Cr(VI) removal date corroborated and validated the fitness of both models. At various nodes, the Cr(VI) removal by both models was confirmed. The speculative values of the ideal levels that maximize Cr(VI) removal by the four variables were also validated. The only exception was nodes number 4 (86.04%), and 7 (80.21%) in the DT model of melanin, which showed shifting by −9, and −6.5% in comparison to the predicted levels (95.36 and 86.75%), respectively. This is expected and can be attributed to the high value of the standard deviation of the node (Figure 7). Interestingly, DT modeling of fungal biomass was found to be better than melanin in the Cr(VI) removal process. One other positive outcome of the current validation is that both DTs effectively introduce more alternative solutions for the Cr(VI) removal process, in comparison to the equilibrium isotherms and kinetic data.

Table 9. The expected variables conditions that were projected based on the DT models for fungal biomass, and melanin as well as the expected and actual values of Cr(VI) removal.

DT Model	Investigated Parameters					Cr(VI) Removal, %		Terminal Node
	Contact Time, min	Initial Cr(VI), mg/mL	Fungal Biomass (g)	Melanin (g)	pH	Predicted	Actual	
Fungal biomass	220	70	0.100	-	3.0	99.04	98.28 ± 0.24	6
	230	75	0.050	-	4.0	83.67	86.00 ± 0.56	7
	240	75	0.120	-	5.0	85.33	84.35 ± 0.80	15
Melanin	240	75	-	0.100	4.5	95.36	86.04 ± 0.50	4
	280	70	-	0.070	5.0	89.07	91.22 ± 0.29	6
	220	70	-	0.100	5.0	86.75	80.21 ± 0.28	7

3.8. *Surface Topology, and Chemistry of the Biosorbents*

3.8.1. FT-IR Spectral Analysis

The FT-IR analysis was applied to explore the vibrations of the characteristic functional groups of the fungi biomass, and melanin as well as chromium ions adsorbed on the surface of fungi biomass, and melanin [79]. The results of FT-IR spectral analysis (Figure 9 and Table S1) for the pure fungi powder (F) revealed the appearance of an absorption band ascribed to a medium N-H or OH stretching group identified at $\nu = 3272.99$ cm^{-1} with a shift recorded in the value of the same group for chromium adsorbed on fungi biomass (F-Cr) at $\nu = 3281.85$ cm^{-1}. The stretching absorption bands of OH or N-H stretching group in the case of melanin (M) and chromium adsorbed on melanin (M-Cr) were specified at wavenumbers (ν) 3289.57, and 3311.59 cm^{-1}, respectively. A previous study supported that this range of wavenumbers is related to OH groups [79].

The low to high shift in values of peak wavenumbers is ascribed to the binding of chromium to the -OH and –NH functional groups [81]. The wavenumber value in the case of M-Cr was shifted from the value of the same group in the analysis of melanin (M). The strong broad absorption band due to C-H stretching appeared in the analysis of all samples with a slight shift in the case of chromium ions adsorbed on fungi, and melanin surfaces within $\nu = 2921.32$–2923.82 cm^{-1}. The shifted intensity of the C-H stretching peak specified the binding of Cr(VI) to the C-H group of melanin, and fungi biomass. The FT-IR data verified the distinctive absorption bands attributed to the sp3 C-H stretching group at $\nu = 2851.54$ cm^{-1} (F), attended by a slight shift for F-Cr at $\nu = 2853.71$ cm^{-1}, $\nu = 2853.50$ cm^{-1} (M), and $\nu = 2852.89$ cm^{-1} (M-Cr). Specifically, a strong stretching absorption band was noticed in the FT-IR analysis of melanin at $\nu = 2323.61$ cm^{-1} is ascribed to the O=C=O group, and disappeared in the analysis of M-Cr. This result supported the participation of the O=C=O group in the adsorption process of chromium ions on the melanin surface.

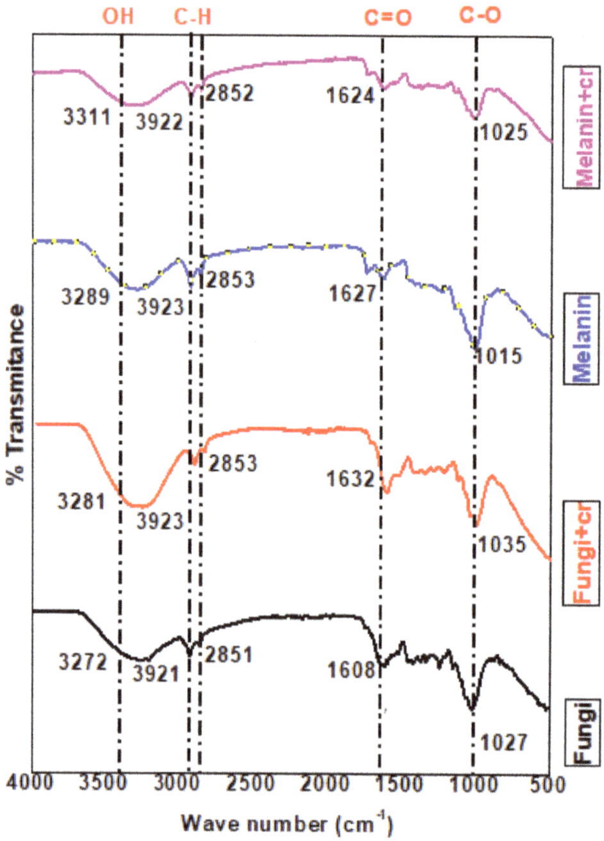

Figure 9. The FT-IR spectral chart of the used fungal biomass and its extracellular melanin before and after Cr(VI) biosorption.

Similarly, absorption bands ascribed to the stretching C-H group appeared in the analysis of melanin within ν = 2170.03–2011.0 cm^{-1} along with the disappearance of these absorption bands in the analysis of M-Cr. The data of the melanin sample referred to the presence of a stretching carbonyl ester group at ν = 1740.66 cm^{-1} with a slight shift in the value of this band at ν = 1740.82 cm^{-1} in the analysis of M-Cr. The fact that the shifted values of carbonyl groups reinforced the contribution of this group in the absorption of chromium ions. Alternatively, a strong absorption band related to stretching amide carbonyl group was noticed in all samples at ν = 1608.82 cm^{-1} (F), ν = 1632.11 cm^{-1} (F-Cr), ν = 1627.26 cm^{-1} (M), and ν = 1624.12 cm^{-1} (M-Cr). Predominantly, the FT-IR data of fungi sample indicated the presence of absorption bands at ν = 1589.78, and 1513.92 cm^{-1} owing to medium bending N-H groups. The disappearance of absorption bands in these regions in the analysis of F-Cr verified the adsorption of chromium ions on the fungi surface with the aid of N-H groups. The FT-IR analyses of melanin, and Cr-adsorbed on melanin surface revealed the absence of the characteristic absorption bands in the regions of N-H groups [78]. In addition, bending O-H groups appeared in the analysis of fungi within ν = 1416.81–1330.35 cm^{-1}, while these values were intensively shifted in the analysis of F-Cr (ν = 1443.78–1315.49 cm^{-1}). The analysis of melanin revealed medium absorption bands at ν = 1413.49, and 1370.51 cm^{-1} attributed to the bending O-H group, although one of these bands disappeared, and the other absorption band was shifted at ν = 1374.93 cm^{-1} in the analysis of M-Cr. The stretching C-N groups were documented for all samples with a slightly shifted value at ν = 1244.81 (F), ν = 1248.68 (F-Cr), ν = 1232.96 (M), and ν = 1238.20 cm^{-1} (M-Cr). The shift of peak to higher wavenumber values indicated the chemisorption of Cr(VI) to the functional groups. Our results are in accordance with the previous report concerning the interpretation of the C-N group, and the increased shift in intensity [79]. Correspondingly, stretching absorption bands assigned for Cr=O groups appeared at ν = 1035.11 cm^{-1} (F-Cr), and for M-Cr at ν = 1025.67 cm^{-1} indicating the adsorption of chromium as dichromate anions on the surface of fungi biomass and melanin [82,83]. An absorption band at ν= 840.55 cm^{-1} was established for strong bending vibration of the C-H group in the FT-IR analysis of melanin, this absorption band disappeared in the analysis of F-Cr. The disappearance of the C-H group is related to the involvement of this group in the adsorption of chromium ions on the fungi biomass. The spectral of FT-IR revealed absorption bands at ν = 525.47, 471.52 (F), ν = 486.78 (F-Cr), ν = 482.99, and 425.76 (M) owing to phenyl rings, while this band was not recorded for the sample of M-Cr. Cr(VI) ions predominantly bind to fungi biomass, and melanin as negatively charged $HCrO_4{}^{-1}$ and $Cr_2O_4{}^{-2}$ groups to positively charged sites or functional groups of fungi, and melanin. The introduction of oxygen and hydrogen atoms in the adsorbed complex of Cr(VI) might be the cause for increased perceived % transmittance [84]. The interpretations of the FT-IR verified that the adsorbent has active sites for Cr(VI) adsorption.

3.8.2. SEM Investigation

The SEM image of biomass cells of *A. pullulans* AKW, and its melanin particles were investigated against Cr(VI) ions using two different magnification powers. The Cr(VI)-unloaded biomass cells of *A. pullulans* Akw, showed to be characterized by the regular and homogenous shape of the cells (Figure 11(A1,A2)). In the treated fungal biomass with Cr(VI), detectable alterations in morphology (cell shape, size, and surface features), and also indentation in cells, clear deformation, shrinking of cells, flighting, and some distortion, with enlargement, could be noticed (Figure 11(B1,B2)). Our results are in harmony with the previous findings that pointed out that the uptake and accumulation of heavy metals by fungal cells, caused morphological alteration, physiological damage, molecular disturbance, ultrastructural changes, and inhibition of the antioxidant system [85,86]. Similarly, the SEM images of extracellular melanin particles before and after the adsorption process of Cr(VI) were investigated. The controlled sample of melanin particles (Figure 11(C1,C2)) appeared to be more amorphic and opaque, while, the treated melanin (Figure 11(D1,D2)) showed flourished amorphous particles as a result of Cr(VI) ions. The image showed also changes in the aggregation pattern and the surface appearance of the melanin particles, including texture, shape, and aggregation of the particles. Similar conclusions were reported during the examination of the structure of natural and synthetic eumelanin using SEM [87].

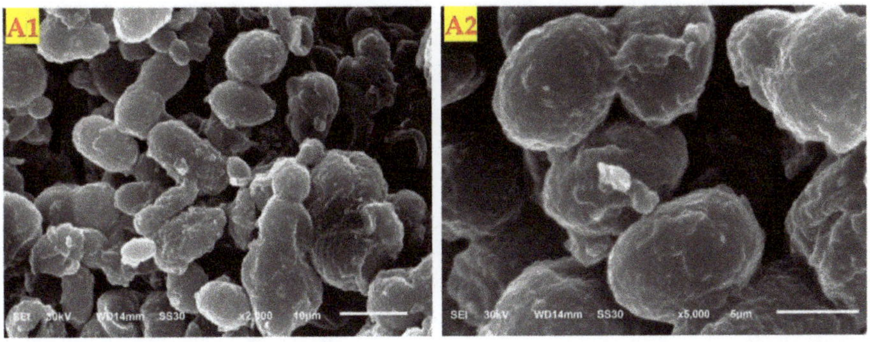

Figure 10. *Cont.*

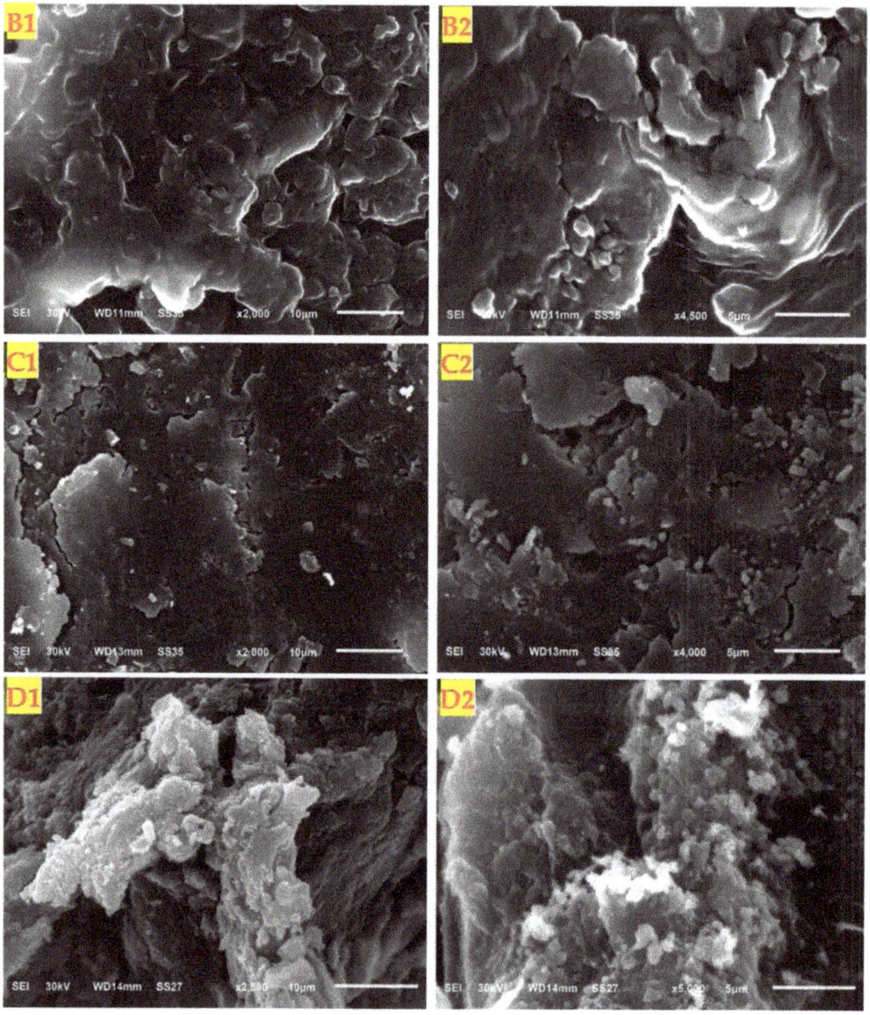

Figure 11. SEM photographs of the fungi biomass and melanin before and after Cr(VI) biosorption, showing fungal biomass before (**A1**,**A2**) and after (**B1**,**B2**) adsorption, as well as melanin particles before (**C1**,**C2**), after (**D1**,**D2**) adsorption.

Melanin and fungal biomass have better morphology to adsorb Cr(VI), and their surfaces are rich with porous which acquire Cr(VI). This provides favorable conditions for the bioadsorption process of organic pollutants [88]. The adsorbent surface of melanin and fungal biomass profoundly adsorbed Cr(VI) ions in the inner walls.

3.8.3. EDX Analyses

The analysis of fungal biomass revealed the presence of carbon (75.97%), and oxygen (23.32%) as the main elements, along with rare elements such as chlorine, potassium, calcium, copper, and zinc. The analysis of Cr(VI) adsorbed on fungal biomass indicated the adsorption of Cr(VI) with 1.39% of weight %, and 0.35% of atomic percentage. Hereafter, it is established that the treatment of melanin, and fungal biomass with Cr(VI) solution does not destroy any of the basic functional groups of the adsorbent [89].

The images of the EDX analyses of melanin, fungal biomass, and Cr(VI) ions adsorbed on their surfaces are shown in Figure 12. The results of melanin indicated the presence of carbon (68.44%), and oxygen (30.24%) as the major elements with rare atomic percentages for potassium, copper, and zinc. The melanin-adsorbed Cr(VI) displayed the presence of Cr(VI) with 1.34% of the weight and 0.34% of atomic percentages. Similarly, the analyses were applied for fungi biomass, and melanin as well as chromium ions adsorbed on their surface to examine the possible adsorption of Cr(VI) ions on their surfaces [79]. Our results are in line with the results obtained by [90].

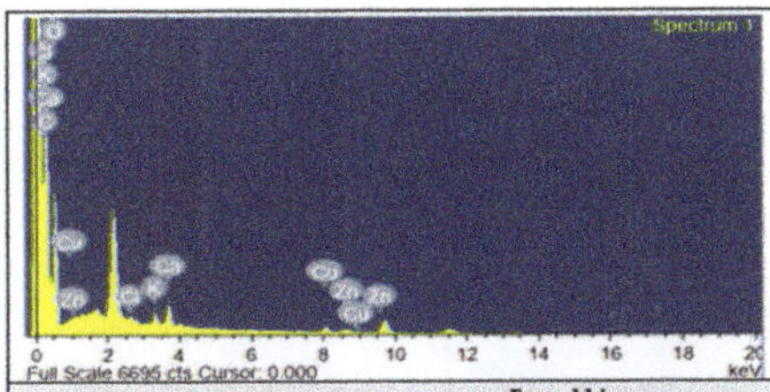

Element	Weight %	Atomic %
C K	69.10	75.97
O K	28.26	23.32
Cl K	0.17	0.06
K K	0.35	0.12
Ca K	0.72	0.24
Cu K	0.79	0.16
Zn K	0.62	0.13

Fungal biomass

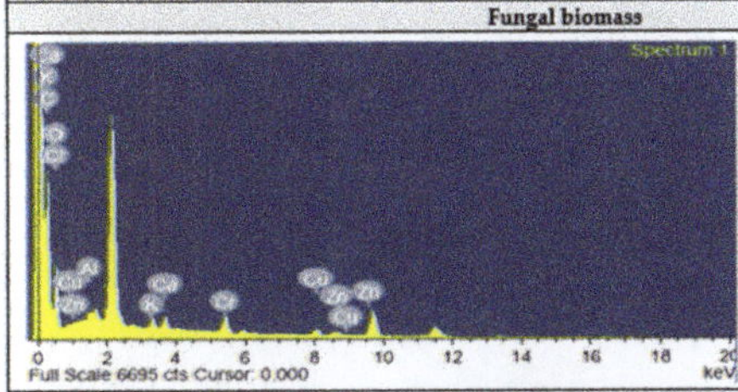

Element	Weight %	Atomic %
C K	79.74	86.08
O K	15.82	12.82
Al K	0.06	0.03
K K	0.42	0.14
Ca K	0.53	0.17
Cr K	1.39	0.35
Cu K	1.29	0.26
Zn K	0.74	0.15

Fungi biomass + Cr(VI)

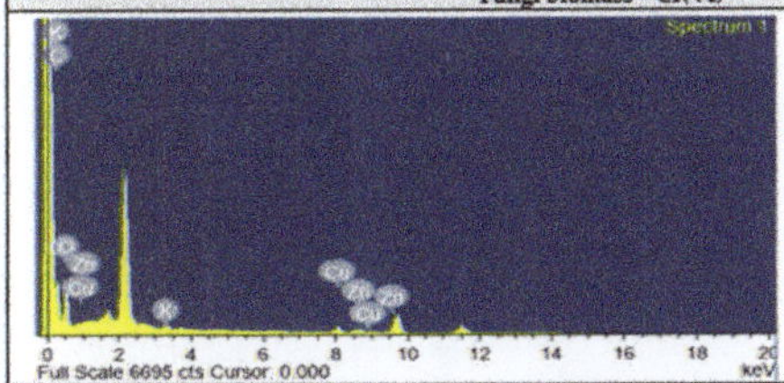

Element	Weight %	Atomic %
C K	59.22	68.44
O K	34.86	30.24
K K	0.32	0.11
Cu K	3.40	0.74
Zn K	2.20	0.47

Melanin

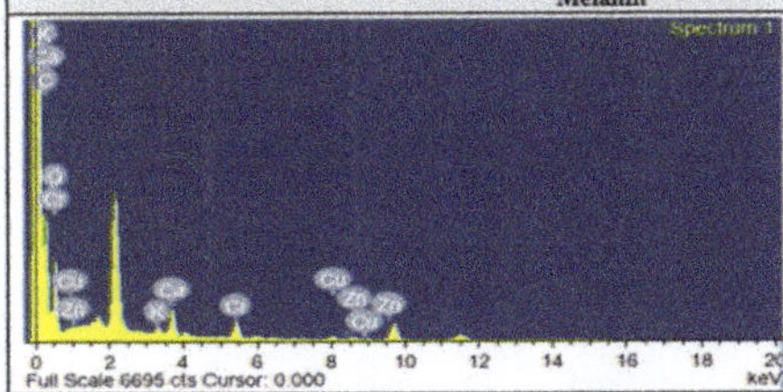

Element	Weight %	Atomic %
C K	76.81	83.26
O K	19.30	15.70
K K	0.31	0.10
Ca K	1.18	0.38
Cr K	1.34	0.34
Cu K	0.69	0.14
Zn K	0.37	0.07

Melanin + Cr(VI)

Figure 12. EDX-analyses of the fungi biomass and melanin before and after Cr(VI) biosorption.

Based on the current studies, it is supposed that Cr(VI) may interact with the surface of the biomass or melanin through a variety of mechanisms, including hydrophobic interactions, electrostatic interactions, and hydrogen bonding.

3.9. Interaction Mechanism of Cr(VI) Ions onto the Biosorbent Particles

The biosorption of Cr(VI) is affected by the nature of the functional groups in the metal ion, the textural and surface properties of the used biosorbent materials, the diffusion of the metal ion molecules to the used biosorbent, and how the metal ion molecules interact with the used material particles. The surface of the used biosorbent particles has pronated hydroxyl groups ($-OH_2^+$) and ($-NH_2$), which enable them to electrostatically interact with negatively charged $Cr_2O_7^{2-}$. Moreover, the abundant hydroxyl groups are highly capable of forming hydrogen bonds with CrO_4^{2-}, and $HCrO_4^-$. Taken collectively, the combination of hydrogen bonding and electrostatic attraction could maximize the biosorption of Cr(VI) onto the used biosorbent particles. The removal percentage of Cr(VI) ions increased under acidic conditions, while decreasing under basic conditions, indicating that electrostatic attraction plays a critical role in the biosorption of negative Cr(VI) ions onto the used biosorbent particles surface (Figure 13).

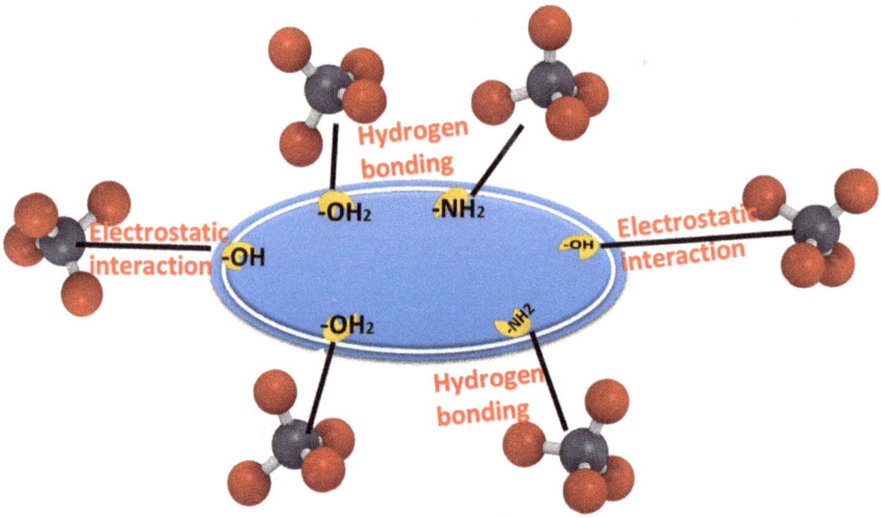

Figure 13. Binding mechanism of Cr(VI) ions onto the used biosorbent particles.

4. Conclusions

The comparative study involving fungus, *A. pullulans* AKW biomass and its extracellular melanin particles as biosorbent agents for Cr(VI) ions was conducted. The parameters of contact time, initial concentration of Cr(VI), biosorbent dosages (fungal biomass or melanin), and pH with their influence on the biosorption process of Cr(VI) ions were investigated. Equilibrium isotherms were applied for experimental data. Langmuir model was shown to be more favorite for the biosorption process compared to the Freundlich isotherm. As well as, the Temkin isotherm fit well with the experimental data, indicating the energetic homogeneity of the biosorption sites and chemisorption process for fungus biomass and melanin. Regarding the kinetic studies, the biosorption process obeyed pseudo-first-order rather than pseudo-second-order. The DTs of both fungal biomass and melanin were constructed. The DT of fungal biomass was a lower error compared to melanin. Dts have efficiency in optimizing Cr(VI)removal compared to equilibrium isotherms and kinetic study. The FT-IR spectra of melanin, and fungal biomass verified the adsorption of Cr(VI) on their surface as appeared from the obtained data with a notable shift in the values of the stretching vibrations of most of the functional groups. The C-H stretching, stretching of the N-H bond, and "O-H" hydroxyl groups produced the shifting of absorption bands in the FT-IR spectra attributed to the biosorption of Cr(VI). The SEM and EDX analyses supported the reality that melanin and fungal biomass can remove heavy metals leading to detoxification. Summing up, the extracellular melanin was found to be more efficient in the biosorption process of Cr(VI) ions compared to fungus biomass, it could be applicable on a large scale since its feasibility of production by *A. pullulans* AKW, using a cheap medium.

Based on the data of the current work, further research can focus on applying the DT algorithm modeling Cr(VI) ions removal using a combination of fungal biomass and melanin in biosorption strategies. Synergistic effects between different biosorbents may enhance the overall efficiency and versatility of the process. Likewise, combining another approach of machine learning such as artificial neural networks could lead to enhanced efficiency and uptake capacity of the biosorbents, making them more attractive for practical applications. Furthermore, the generalization of the modeling process in the current study to include other sorbent/absorbate interactions. Moreover, the scenario of the performance of fungal biomass and melanin under real-world conditions will provide valuable information for their implementation. However, it is recommended to assess the toxicity and potential environmental impact of using fungal biomass and/or melanin for Cr(VI) removal during the biosorption process.

Supplementary Materials: The following supporting information can be downloaded at: https://www.mdpi.com/article/10.3390/polym15183754/s1, Table S1. The adsorption of chromium ions on the melanin surface was evidenced by Fourier Transform Infrared Spectroscopy (FT-IR).

Author Contributions: Conceptualization, H.F., M.S.E.H. and W.I.A.S.; Data curation, H.F.; Formal analysis, H.F., Y.A.H. and M.S.E.H.; Funding acquisition, F.O.A.-O.; Investigation, H.F., A.A.G. and W.I.A.S.; Methodology, A.A.G., F.O.A.-O., Y.A.H., K.M.E. and A.E.; Project administration, W.I.A.S.; Resources, F.O.A.-O., Y.A.H. and K.M.E.; Software, H.F., A.A.G., Y.A.H., M.S.E.H., K.M.E., W.I.A.S. and A.E.; Validation, A.A.G., F.O.A.-O., K.M.E. and A.E.; Writing—original draft, H.F., A.A.G., F.O.A.-O., Y.A.H., K.M.E. and A.E.; Writing—review and editing, M.S.E.H. and W.I.A.S. All authors have read and agreed to the published version of the manuscript.

Funding: The authors extend their appreciation to the Researchers Supporting Project number (RSP2023R114), King Saud University, Riyadh, Saudi Arabia.

Institutional Review Board Statement: Not applicable.

Data Availability Statement: All data generated or analyzed during this study are included in this published article and the Supplementary Materials.

Conflicts of Interest: The authors declare no conflict of interest.

References

1. Tran-Ly, A.N.; Ribera, J.; Schwarze, F.W.; Brunelli, M.; Fortunato, G. Fungal melanin-based electrospun membranes for heavy metal detoxification of water. *Sustain. Mater. Technol.* **2020**, *23*, e00146. [CrossRef]
2. Tchounwou, P.B.; Yedjou, C.G.; Patlolla, A.K.; Sutton, D.J. Heavy metal toxicity and the environment. *Mol. Clin. Environ. Toxicol.* **2012**, *101*, 133–164.
3. Saber, W.I.; El-Naggar, N.E.-A.; El-Hersh, M.S.; El-Khateeb, A.Y.; Elsayed, A.; Eldadamony, N.M.; Ghoniem, A.A. Rotatable central composite design versus artificial neural network for modeling biosorption of Cr^{6+} by the immobilized Pseudomonas alcaliphila NEWG-2. *Sci. Rep.* **2021**, *11*, 1717. [CrossRef] [PubMed]
4. El-Naggar, N.E.-A.; El-Khateeb, A.Y.; Ghoniem, A.A.; El-Hersh, M.S.; Saber, W.I. Innovative low-cost biosorption process of Cr^{6+} by Pseudomonas alcaliphila NEWG-2. *Sci. Rep.* **2020**, *10*, 14043. [CrossRef] [PubMed]
5. Samuel, M.S.; Chidambaram, R.J. Hexavalent chromium biosorption studies using *Penicillium griseofulvum* MSR1 a novel isolate from tannery effluent site: Box–Behnken optimization, equilibrium, kinetics and thermodynamic studies. *J. Taiwan Inst. Chem. Eng.* **2015**, *49*, 156–164.
6. Vafajoo, L.; Cheraghi, R.; Dabbagh, R.; McKay, G. Removal of cobalt (II) ions from aqueous solutions utilizing the pre-treated 2-Hypnea Valentiae algae: Equilibrium, thermodynamic, and dynamic studies. *Chem. Eng. J.* **2018**, *331*, 39–47. [CrossRef]
7. Kapahi, M.; Sachdeva, S.J. Bioremediation options for heavy metal pollution. *J. Health Pollut.* **2019**, *9*, 191203. [CrossRef]
8. Vidal-Cros, A.; Viviani, F.; Labesse, G.; Boccara, M.; Gaudry, M. Polyhydroxynaphthalene reductase involved in melanin biosynthesis in Magnaporthe grisea: Purification, cDNA cloning and sequencing. *Eur. J. Biochem.* **1994**, *219*, 985–992. [CrossRef] [PubMed]
9. Saber, W.I.; Al-Askar, A.A.; Ghoneem, K.M. Exclusive Biosynthesis of Pullulan Using Taguchi's Approach and Decision Tree Learning Algorithm by a Novel Endophytic *Aureobasidium pullulans* Strain. *Polymers* **2023**, *15*, 1419. [CrossRef]
10. Pombeiro-Sponchiado, S.R.; Sousa, G.S.; Andrade, J.C.; Lisboa, H.F.; Gonçalves, R. Production of melanin pigment by fungi and its biotechnological applications. *Melanin* **2017**, *1*, 47–75.
11. Seelam, S.D.; Agsar, D.; Shetty, P.R.; Vemireddy, S.; Reddy, K.M.; Umesh, M.; Rajitha, C. Characterization and photoprotective potentiality of lime dwelling Pseudomonas mediated melanin as sunscreen agent against UV-B radiations. *J. Photochem. Photobiol. B Biol.* **2021**, *216*, 112126. [CrossRef]

12. El-Bialy, H.A.; El-Gamal, M.S.; Elsayed, M.A.; Saudi, H.; Khalifa, M. Microbial melanin physiology under stress conditions and gamma radiation protection studies. *Radiat. Phys. Chem.* **2019**, *162*, 178–186. [CrossRef]
13. Brenner, M.; Hearing, V.J. The protective role of melanin against UV damage in human skin. *Photochem. Photobiol.* **2008**, *84*, 539–549. [CrossRef]
14. Kimura, Y.; Sumiyoshi, M.; Kobayashi, T. Whey peptides prevent chronic ultraviolet B radiation–induced skin aging in melanin-possessing male hairless mice. *J. Nutr.* **2014**, *144*, 27–32. [CrossRef] [PubMed]
15. Ribera, J.; Panzarasa, G.; Stobbe, A.; Osypova, A.; Rupper, P.; Klose, D.; Schwarze, F.W. Scalable biosynthesis of melanin by the basidiomycete Armillaria cepistipes. *J. Agric. Food Chem.* **2018**, *67*, 132–139. [CrossRef] [PubMed]
16. Sáiz-Jiménez, C. The chemical nature of the melanins from Coprinus spp. *Soil Sci.* **1983**, *136*, 65–74. [CrossRef]
17. Froncisz, W.; Sarna, T.; Hyde, J.S. Cu^{2+} probe of metal-ion binding sites in melanin using electron paramagnetic resonance spectroscopy: I. *Synthetic melanins. Arch. Biochem. Biophys.* **1980**, *202*, 289–303. [CrossRef]
18. Saiz-Jimenez, C.; Shafizadeh, F. Iron and copper binding by fungal phenolic polymers: An electron spin resonance study. *Curr. Microbiol.* **1984**, *10*, 281–285. [CrossRef]
19. d'Ischia, M.; Wakamatsu, K.; Napolitano, A.; Briganti, S.; Garcia-Borron, J.C.; Kovacs, D.; Meredith, P.; Pezzella, A.; Picardo, M.; Sarna, T.; et al. Melanins and melanogenesis: Methods, standards, protocols. *Pigment. Cell Melanoma Res.* **2013**, *26*, 616–633. [CrossRef]
20. Hong, L.; Simon, J.D. Current understanding of the binding sites, capacity, affinity, and biological significance of metals in melanin. *J. Phys. Chem. B* **2007**, *111*, 7938–7947. [CrossRef]
21. Felix, C.; Hyde, J.; Sarna, T.; Sealy, R.J. Interactions of melanin with metal ions. Electron. spin resonance evidence for chelate complexes of metal ions with free radicals. *J. Am. Chem. Soc.* **1978**, *100*, 3922–3926. [CrossRef]
22. Buszman, E.; Pilawa, B.; Zdybel, M.; Wilczyński, S.; Gondzik, A.; Witoszyńska, T.; Wilczok, T. EPR examination of Zn^{2+} and Cu^{2+} binding by pigmented soil fungi *Cladosporium cladosporioides. Sci. Total Environ.* **2006**, *363*, 195–205. [CrossRef] [PubMed]
23. Oh, J.-J.; Kim, J.Y.; Kim, Y.J.; Kim, S.; Kim, G.-H. Utilization of extracellular fungal melanin as an eco-friendly biosorbent for treatment of metal-contaminated effluents. *Chemosphere* **2021**, *272*, 129884. [CrossRef] [PubMed]
24. Ghosh, S.; Rusyn, I.; Dmytruk, O.V.; Dmytruk, K.V.; Onyeaka, H.; Gryzenhout, M.; Gafforov, Y. Filamentous fungi for sustainable remediation of pharmaceutical compounds, heavy metal and oil hydrocarbons. *Front. Bioeng. Biotechnol.* **2023**, *11*, 1106973. [CrossRef]
25. Dusengemungu, L.; Kasali, G.; Gwanama, C.; Ouma, K.O. Recent Advances in Biosorption of Copper and Cobalt by Filamentous Fungi. *Front. Microbiol.* **2020**, *11*, 582016. [CrossRef]
26. Goodman, K.E.; Lessler, J.; Cosgrove, S.E.; Harris, A.D.; Lautenbach, E.; Han, J.H.; Milstone, A.M.; Massey, C.J.; Tamma, P.D. A clinical decision tree to predict whether a bacteremic patient is infected with an extended-spectrum β-lactamase–producing organism. *Clin. Infect. Dis.* **2016**, *63*, 896–903. [CrossRef]
27. Ashino, K.; Sugano, K.; Amagasa, T.; Ying, B.-W. Predicting the decision making chemicals used for bacterial growth. *Sci. Rep.* **2019**, *9*, 7251. [CrossRef]
28. Giri, D.D.; Shah, M.; Srivastava, N.; Hashem, A.; Abd_Allah, E.F.; Pal, D.B. Sustainable chromium recovery from wastewater using mango and jackfruit seed kernel bio-adsorbents. *Front. Microbiol.* **2021**, *12*, 717848. [CrossRef]
29. He, H.; Lu, Q.; Huang, H.; Xue, F.; Lin, W.; Zhou, H.; Wei, W. Biomass bagasse-based hyperbranched adsorbent for the complete removal of low-level Cr (VI). *Cellulose* **2020**, *27*, 8121–8134. [CrossRef]
30. Rai, R.; Karki, D.R.; Bhattarai, K.P.; Pahari, B.; Shrestha, N.; Adhikari, S.; Gautam, S.K.; Poudel, B.R. Recent advances in biomass-based waste materials for the removal of chromium (VI) from wastewater: A review. *Amrit Res. J.* **2021**, *2*, 37–50. [CrossRef]
31. Li, Q.; Huang, Q.; Pan, X.Y.; Yu, H.; Zhao, Z.T. Adsorption behavior of Cr (VI) by biomass-based adsorbent functionalized with deep eutectic solvents (DESs). *BMC Chem.* **2022**, *16*, 41. [CrossRef]
32. Blaga, A.C.; Zaharia, C.; Suteu, D. Polysaccharides as support for microbial biomass-based adsorbents with applications in removal of heavy metals and dyes. *Polymers* **2021**, *13*, 2893. [CrossRef] [PubMed]
33. Allen, S. *Chemical Analysis of Ecological Material*; Blackwell Scientific Publications: Oxford, UK, 1974.
34. Saber, W.I.; Ghoniem, A.A.; Al-Otibi, F.O.; El-Hersh, M.S.; Eldadamony, N.M.; Menaa, F.; Elattar, K.M. A comparative study using response surface methodology and artificial neural network towards optimized production of melanin by Aureobasidium pullulans AKW. *Sci. Rep.* **2023**, *13*, 13545. [CrossRef]
35. El-Gamal, M.; El-Bialy, H.; Elsayed, M.; Khalifa, M. Isolation and characterization of melanized yeast form of Aureobasidium pullulans and physiological studies on the melanization process. *J. Nucl. Sci. Technol.* **2017**, *5*, 57–72.
36. Müjdeci, G.N. Experimental modeling and optimization of melanin production by Aureobasidium pullulans NBRC 100716 in carrot peel extract. *Environ. Prog. Sustain. Energy* **2022**, *41*, e13919. [CrossRef]
37. Şahin, Y.; Öztürk, A. Biosorption of chromium (VI) ions from aqueous solution by the bacterium Bacillus thuringiensis. *Process Biochem.* **2005**, *40*, 1895–1901. [CrossRef]
38. Fakhry, H.; Hassan, H.M.; El-Aassar, M.; Alsohaimi, I.H.; Hussein, M.F.; ALQahtani, M.M.; El-Amier, Y.A. A treatment of wastewater containing Safranin O using immobilized Myriophyllum spicatum L. onto polyacrylonitrile/polyvinylpyrrodlidone biosorbent. *J. Inorg. Organomet. Polym. Mater.* **2022**, *32*, 3181–3195. [CrossRef]

39. El-Amier, Y.A.; Elsayed, A.; El-Esawi, M.A.; Noureldeen, A.; Darwish, H.; Fakhry, H. Optimizing the biosorption behavior of Ludwigia stolonifera in the removal of lead and chromium metal ions from synthetic wastewater. *Sustainability* **2021**, *13*, 6390. [CrossRef]
40. Cheng, Q.; Fang, Z.; Yi, X.S.; An, X.; Tang, B.; Xu, Y. "Ex situ" concept for toughening the RTMable BMI matrix composites, Part I: Improving the interlaminar fracture toughness. *J. Appl. Polym. Sci.* **2008**, *109*, 1625–1634. [CrossRef]
41. Ho, Y.-S.; McKay, G. Pseudo-second order model for sorption processes. *Process Biochem.* **1999**, *34*, 451–465. [CrossRef]
42. Priya, A.; Yogeshwaran, V.; Rajendran, S.; Hoang, T.K.; Soto-Moscoso, M.; Ghfar, A.A.; Bathula, C. Investigation of mechanism of heavy metals (Cr^{6+}, Pb^{2+} & Zn^{2+}) adsorption from aqueous medium using rice husk ash: Kinetic and thermodynamic approach. *Chemosphere* **2022**, *286*, 131796.
43. Fakhry, H.; El-Sonbati, M.; Omar, B.; El-Henawy, R.; Zhang, Y.; Marwa, E.-K. Novel fabricated low-cost hybrid polyacrylonitrile/polyvinylpyrrolidone coated polyurethane foam (PAN/PVP@ PUF) membrane for the decolorization of cationic and anionic dyes. *J. Environ. Manag.* **2022**, *315*, 115128. [CrossRef] [PubMed]
44. Sivashankar, R.; Sathya, A.; Kanimozhi, J.; Deepanraj, B. Characterization of the biosorption process. In *Biosorption for Wastewater Contaminants*; Wiley: Hoboken, NJ, USA, 2022; pp. 102–116.
45. Wang, H.; Ren, Z.J. Bioelectrochemical metal recovery from wastewater: A review. *Water Res.* **2014**, *66*, 219–232. [CrossRef]
46. Bolisetty, S.; Peydayesh, M.; Mezzenga, R. Sustainable technologies for water purification from heavy metals: Review and analysis. *Chem. Soc. Rev.* **2019**, *48*, 463–487. [CrossRef]
47. Hong, L.; Liu, Y.; Simon, J.D. Binding of metal ions to melanin and their effects on the aerobic reactivity. *Photochem. Photobiol.* **2004**, *80*, 477–481. [CrossRef] [PubMed]
48. Zubair, A.; Bhatti, H.N.; Hanif, M.A.; Shafqat, F. Kinetic and equilibrium modeling for Cr (III) and Cr (VI) removal from aqueous solutions by Citrus reticulata waste biomass. *Water Air Soil Pollut.* **2008**, *191*, 305–318. [CrossRef]
49. Sibi, G. Biosorption of chromium from electroplating and galvanizing industrial effluents under extreme conditions using Chlorella vulgaris. *Green. Energy Environ.* **2016**, *1*, 172–177. [CrossRef]
50. Rezaei, H. Biosorption of chromium by using *Spirulina* sp. *Arab. J. Chem.* **2016**, *9*, 846–853. [CrossRef]
51. Kanchana, S.; Jeyanthi, J.; Kumar, R.D. Equilibrium and kinetic studies on biosorption of chromium (VI) on to Chlorella species. *Eur. J. Sci. Res.* **2011**, *63*, 255–262.
52. Marandi, R. Biosorption of hexavalent chromium from aqueous solution by dead fungal biomass of Phanerochaete crysosporium: Batch and fixed bed studies. *Can. J. Chem. Eng. Technol.* **2011**, *2*, 8–22.
53. Suganya, E.; Saranya, N.; Patra, C.; Varghese, L.A.; Selvaraju, N. Biosorption potential of Gliricidia sepium leaf powder to sequester hexavalent chromium from synthetic aqueous solution. *J. Environ. Chem. Eng.* **2019**, *7*, 103112.
54. Nakkeeran, E.; Patra, C.; Shahnaz, T.; Rangabhashiyam, S.; Selvaraju, N. Continuous biosorption assessment for the removal of hexavalent chromium from aqueous solutions using Strychnos nux vomica fruit shell. *Bioresour. Technol. Rep.* **2018**, *3*, 256–260. [CrossRef]
55. Mangwandi, C.; Kurniawan, T.A.; Albadarin, A.B. Comparative biosorption of chromium (VI) using chemically modified date pits (CM-DP) and olive stone (CM-OS): Kinetics, isotherms and influence of co-existing ions. *Chem. Eng. Res. Des.* **2020**, *156*, 251–262. [CrossRef]
56. Roșca, M.; Silva, B.; Tavares, T.; Gavrilescu, M. Biosorption of Hexavalent Chromium by *Bacillus megaterium* and *Rhodotorula* sp. *Inact. Biomass. Process.* **2023**, *11*, 179.
57. Bhatti, H.N.; Khalid, R.; Hanif, M.A. Dynamic biosorption of Zn (II) and Cu (II) using pretreated Rosa gruss an teplitz (red rose) distillation sludge. *Chem. Eng. J.* **2009**, *148*, 434–443. [CrossRef]
58. Wang, G.; Zhang, S.; Yao, P.; Chen, Y.; Xu, X.; Li, T.; Gong, G. Removal of Pb (II) from aqueous solutions by *Phytolacca americana* L. biomass as a low cost biosorbent. *Arab. J. Chem.* **2018**, *11*, 99–110.
59. Naseem, K.; Farooqi, Z.H.; Begum, R.; Rehman, M.Z.U.; Shahbaz, A.; Farooq, U.; Ali, M.; Rahman, H.M.A.U.; Irfan, A.; Al-Sehemi, A.G. Removal of cadmium (II) from aqueous medium using vigna radiata leave biomass: Equilibrium isotherms, kinetics and thermodynamics. *Z. Für Phys. Chem.* **2019**, *233*, 669–690. [CrossRef]
60. Ume, O.L.; Ekeoma, B.C.; Yusuf, M.; Al-Kahtani, A.A.; Ubaidullah, M.; Sillanpää, M. Batch studies of hexavalent chromium biosorption from mining wastewater using Aspergillus niger. *Results Chem.* **2022**, *4*, 100490. [CrossRef]
61. Miretzky, P.; Cirelli, A.F. Cr(VI) and Cr(III) removal from aqueous solution by raw and modified lignocellulosic materials: A review. *J. Hazard. Mater.* **2010**, *180*, 1–19. [CrossRef]
62. Krishna Kumar, A.S.; Warchol, J.; Matusik, J.; Tseng, W.-L.; Rajesh, N.; Bajda, T. Heavy metal and organic dye removal via a hybrid porous hexagonal boron nitride-based magnetic aerogel. *NPJ Clean. Water* **2022**, *5*, 24. [CrossRef]
63. Jobby, R.; Jha, P.; Gupta, A.; Gupte, A.; Desai, N. Biotransformation of chromium by root nodule bacteria *Sinorhizobium* sp. SAR1. *PLoS ONE* **2019**, *14*, e0219387. [CrossRef]
64. da Rocha Ferreira, G.L.; Vendruscolo, F.; Antoniosi Filho, N.R. Biosorption of hexavalent chromium by *Pleurotus ostreatus*. *Heliyon* **2019**, *5*, e01450. [CrossRef] [PubMed]
65. Pradhan, D.; Sukla, L.B.; Mishra, B.B.; Devi, N. Biosorption for removal of hexavalent chromium using microalgae *Scenedesmus* sp. *J. Clean. Prod.* **2019**, *209*, 617–629. [CrossRef]
66. Manzoor, Q.; Nadeem, R.; Iqbal, M.; Saeed, R.; Ansari, T.M. Organic acids pretreatment effect on Rosa bourbonia phyto-biomass for removal of Pb (II) and Cu (II) from aqueous media. *Bioresour. Technol.* **2013**, *132*, 446–452. [CrossRef] [PubMed]

67. El-Aassar, M.; Ibrahim, O.M.; Omar, B.M.; El-Hamid, H.T.A.; Alsohaim, I.H.; Hassan, H.M.; Althobaiti, I.O.; El-Sayed, M.Y.; Goher, M.E.; Fakhry, H. Hybrid Beads of Poly (Acrylonitrile-co-Styrene/Pyrrole)@ Poly Vinyl Pyrrolidone for Removing Carcinogenic Methylene Blue Dye Water Pollutant. *J. Polym. Environ.* **2023**, *31*, 2912–2929. [CrossRef]
68. Omar, B.M.; Abdelgalil, S.A.; Fakhry, H.; Tamer, T.M.; El-Sonbati, M.A. Wheat husk-based sorbent as an economical solution for removal of oil spills from sea water. *Sci. Rep.* **2023**, *13*, 2575. [CrossRef]
69. Gupta, V.; Rastogi, A. Biosorption of hexavalent chromium by raw and acid-treated green alga *Oedogonium hatei* from aqueous solutions. *J. Hazard. Mater.* **2009**, *163*, 396–402. [CrossRef]
70. Bulgariu, L.; Bulgariu, D.; Rusu, C. Marine algae biomass for removal of heavy metal ions. In *Springer Handbook of Marine Biotechnology*; Springer: Berlin/Heidelberg, Germany, 2015; pp. 611–648.
71. Shamim, S. Biosorption of heavy metals. *Biosorption* **2018**, *2*, 21–49.
72. Pagnanelli, F. Equilibrium, kinetic and dynamic modelling of biosorption processes. In *Microbial Biosorption of Metals*; Springer: Dordrecht, The Netherlands, 2011; pp. 59–120.
73. Roşca, M.; Hlihor, R.M.; Cozma, P.; Simion, I.M.; Filote, C.; Grecu, C.; Stoleru, V.; Gavrilescu, M. Scaling-Up Strategies of Heavy Metals Microbial Bioremediation. In Proceedings of the 2021 International Conference on e-Health and Bioengineering (EHB), Iasi, Romania, 18–19 November 2021; pp. 1–4.
74. Li, X.; Xu, Q.; Han, G.; Zhu, W.; Chen, Z.; He, X.; Tian, X. Equilibrium and kinetic studies of copper (II) removal by three species of dead fungal biomasses. *J. Hazard. Mater.* **2009**, *165*, 469–474. [CrossRef]
75. Taty-Costodes, V.C.; Fauduet, H.; Porte, C.; Delacroix, A. Removal of Cd (II) and Pb (II) ions, from aqueous solutions, by adsorption onto sawdust of Pinus sylvestris. *J. Hazard. Mater.* **2003**, *105*, 121–142. [CrossRef]
76. Yan, C.-Z.; Wang, S.-R.; Zeng, A.-Y.; Jin, X.-C.; Xu, Q.-J.; Zhao, J.-Z. Equilibrium and kinetics of copper (II) biosorption by *Myriophyllum spicatum* L. *J. Environ. Sci.* **2005**, *17*, 1025–1029.
77. Masinga, T.; Moyo, M.; Pakade, V.E. Removal of hexavalent chromium by polyethyleneimine impregnated activated carbon: Intra-particle diffusion, kinetics and isotherms. *J. Mater. Res. Technol.* **2022**, *18*, 1333–1344. [CrossRef]
78. Cuong, A.M.; Le Na, N.T.; Thang, P.N.; Diep, T.N.; Thuy, L.B.; Thanh, N.L.; Thang, N.D. Melanin-embedded materials effectively remove hexavalent chromium (CrVI) from aqueous solution. *Environ. Health Prev. Med.* **2018**, *23*, 9. [CrossRef] [PubMed]
79. Manirethan, V.; Raval, K.; Rajan, R.; Thaira, H.; Balakrishnan, R.M. Kinetic and thermodynamic studies on the adsorption of heavy metals from aqueous solution by melanin nanopigment obtained from marine source: Pseudomonas stutzeri. *J. Environ. Manag.* **2018**, *214*, 315–324. [CrossRef] [PubMed]
80. Ghoniem, A.A.; Moussa, Z.; Alenzi, A.M.; Alotaibi, A.S.; Fakhry, H.; El-Khateeb, A.Y.; Saber, W.I.; Elsayed, A. Pseudomonas alcaliphila NEWG-2 as biosorbent agent for methylene blue dye: Optimization, equilibrium isotherms, and kinetic processes. *Sci. Rep.* **2023**, *13*, 3678. [CrossRef]
81. Sajjan, S.S.; Anjaneya, O.; Kulkarni, G.B.; Nayak, A.S.; Mashetty, S.B.; Karegoudar, T.B. Properties and Functions of Melanin Pigment from *Klebsiella* sp. GSK. *Korean J. Microbiol. Biotechnol.* **2013**, *41*, 60–69. [CrossRef]
82. Kalidhasan, S.; Gupta, P.A.; Cholleti, V.R.; Kumar, A.S.K.; Rajesh, V.; Rajesh, N. Microwave assisted solvent free green preparation and physicochemical characterization of surfactant-anchored cellulose and its relevance toward the effective adsorption of chromium. *J. Colloid Interface Sci.* **2012**, *372*, 88–98. [CrossRef] [PubMed]
83. Kalidhasan, S.; Kumar, A.S.K.; Rajesh, V.; Rajesh, N. Enhanced adsorption of hexavalent chromium arising out of an admirable interaction between a synthetic polymer and an ionic liquid. *Chem. Eng. J.* **2013**, *222*, 454–463. [CrossRef]
84. Liao, L.F.; Lien, C.F.; Shieh, D.L.; Chen, M.T.; Lin, J.L. FTIR Study of Adsorption and Photoassisted Oxygen Isotopic Exchange of Carbon Monoxide, Carbon Dioxide, Carbonate, and Formate on TiO_2. *J. Phys. Chem. B* **2002**, *106*, 11240–11245. [CrossRef]
85. Chen, A.; Zeng, G.; Chen, G.; Liu, L.; Shang, C.; Hu, X.; Lu, L.; Chen, M.; Zhou, Y.; Zhang, Q. Plasma membrane behavior, oxidative damage, and defense mechanism in Phanerochaete chrysosporium under cadmium stress. *Process Biochem.* **2014**, *49*, 589–598. [CrossRef]
86. Gururajan, K.; Belur, P.D. Screening and selection of indigenous metal tolerant fungal isolates for heavy metal removal. *Environ. Technol. Innov.* **2018**, *9*, 91–99. [CrossRef]
87. Simon, J.D.; Peles, D.; Wakamatsu, K.; Ito, S. Current challenges in understanding melanogenesis: Bridging chemistry, biological control, morphology, and function. *Pigment Cell Melanoma Res.* **2009**, *22*, 563–579. [CrossRef] [PubMed]
88. Biswas, S.; Mishra, U. Continuous Fixed-Bed Column Study and Adsorption Modeling: Removal of Lead Ion from Aqueous Solution by Charcoal Originated from Chemical Carbonization of Rubber Wood Sawdust. *J. Chem.* **2015**, *2015*, 907379. [CrossRef]
89. Manirethan, V.; Gupta, N.; Balakrishnan, R.M.; Raval, K. Batch and continuous studies on the removal of heavy metals from aqueous solution using biosynthesised melanin-coated PVDF membranes. *Environ. Sci. Pollut. Res.* **2020**, *27*, 24723–24737. [CrossRef]
90. Rizvi, A.; Ahmed, B.; Zaidi, A.; Khan, M.S. Bioreduction of toxicity influenced by bioactive molecules secreted under metal stress by Azotobacter chroococcum. *Ecotoxicology* **2019**, *28*, 302–322. [CrossRef]

polymers

MDPI

Article

Application of Ionic Liquid Crosslinked Hydrogel for Removing Heavy Metal Ions from Water: Different Concentration Ranges with Different Adsorption Mechanisms

Jian Sun [1], Ziqi Jin [1], Jiyang Wang [1], Hong Wang [1], Qian Zhang [2], Huajing Gao [1], Zhaohui Jin [1,*], Jianlin Zhang [3] and Zhiwei Wang [4]

1 Institute of Petrochemical Technology, Jilin Institute of Chemical Technology, Jilin 132022, China; sunjian@jlict.edu.cn (J.S.)
2 Shandong Chambroad HoldingGroup Co., Ltd., Binzhou 256599, China
3 Shandong Efirm Biochemistry and Environmental Protection Co., Ltd., Binzhou 256500, China
4 Key Laboratory of Clean Pulp & Papermaking and Pollution Control of Guangxi, College of Light Industrial and Food Engineering, Guangxi University, Nanning 530004, China
* Correspondence: jinzhaohui@jlict.edu.cn

Abstract: Heavy metal wastewater poses a significant environmental challenge due to its harmful effect on organisms and difficult biodegradation. To address this issue, hydrogel has been used as a promising solution for the adsorption of heavy metal ions in water, offering advantages such as low cost, simple design, and environmental friendliness. In this study, we synthetized a novel polyacrylamide/acrylic acid/vinyl imidazole bromide (PAM/AA/[Vim]Br_2) hydrogel as an effective adsorbent for the removal of Ni^{II}, Cu^{II}, Zn^{II}, and Cr^{III} from water. The structure of the hydrogel was characterized by using techniques such as Fourier transform infrared spectroscopy (FTIR) and scanning electron microscopy (SEM). By exploring various parameters such as monomer ratio, neutralization degree, crosslinking agent addition amount, and initiator addition amount, the highest swelling ratio of the PAM/AA/[Vim]Br_2 hydrogel reached 40,012%. One of the notable aspects of this study lay in the investigation of the adsorption behavior of the hydrogel towards heavy metal ions at different concentrations. The adsorption isotherm calculations and X-ray photoelectron spectroscopy (XPS) analysis revealed distinct adsorption mechanisms. At low concentrations, the hydrogel exhibits a multilayer physical adsorption mechanism, with heavy metal ion removal rates exceeding 80%; while at high concentrations, it demonstrates a monolayer chemical adsorption mechanism, with heavy metal ion removal rates above 90%. This dual mechanism approach distinguishes our study from previous reports on the removal of heavy metal ions using hydrogels and shows good ion adsorption efficiency at both high and low concentrations. To the best of our knowledge, this is the first report to explore the removal of heavy metal ions from water using hydrogels with such intriguing dual mechanisms. Overall, the utilization of the PAM/PAA/[Vim]Br_2 hydrogel as an adsorbent for heavy metal ion removal presents a promising and innovative approach, contributing to the development of environmentally friendly solutions for heavy metal wastewater treatment.

Keywords: heavy metal ions; hydrogels; ionic liquid; crosslinking agent

1. Introduction

With the rapid development of industries, the problem of water pollution has become increasingly severe [1–4]. Among the various types of water pollutants, heavy metal ion pollution poses significant threats to ecosystems, human health, production, and the daily life of individuals [5–7]. Currently, several main treatment methods for removing heavy metal ions from water are employed, including filtration [8], ion exchange [9], oxidation–reduction [10,11], and chemical precipitation [12]. However, these treatment methods have inherent limitations, such as complicating the composition of wastewater, low removal efficiency, high costs, generation of large amounts of sludge, and potential secondary pollution. Activated carbon, a widely used porous material for water treatment through adsorption [13,14], has certain limitations that restrict its broad application, such as high experimental costs, difficulties in separation, and long adsorption time. Therefore, there is a need to develop environmentally friendly adsorbents with high removal efficiency, simple structural design, and low cost to address the treatment of wastewater containing heavy metal ions.

Hydrogels possess an extremely hydrophilic three-dimensional network structure, which can rapidly swell in water and retain a large volume of water without dissolving [15–18]. Due to their structural design, low cost, good water permeability, and biodegradability, hydrogels find wide applications in adsorbing heavy metal ions from wastewater [19–23]. Zahra [24] synthesized magnetic hydrogel beads based on poly(vinyl alcohol)/carboxymethyl starch-g-poly(vinylimidazole) for the removal of CuII and CdII, achieving removal rates (RR%) of 93.2% and 62.5%, respectively. The hydrogel exhibited a high RR (%) for heavy metal ions at a low concentration of 20 ppm. However, as the concentration of metal ions increased, the RR (%) gradually decreased. Shah [25] synthesized a PAA/PAM superabsorbent polymer hydrogel for the removal of CdII, NiII, and CuII from aqueous samples. The RR (%) for CdII, NiII, and CuII exceeded 75% across the entire concentration range. When the metal ion concentration was high (100 ppm), the hydrogel showed a high RR (%); however, the RR (%) was extremely low at low concentrations. Previous studies have also reported that chitosan/polyethyleneimine hydrogels exhibited removal rates were lower than 75% for Pb^{2+}, Ni^{2+}, and Cu^{2+} after multiple cycles of use at high concentration (100 ppm) [26], and the composite hydrogels exhibited a removal efficiency of 80% for Cu^{2+} at a concentration of 100 ppm [27]. Although these hydrogels demonstrated high removal efficiency for heavy metal ion adsorption at low or high concentrations, there have been no reports of hydrogels exhibiting high RR for heavy metal ions in both low and high concentration ranges.

Ionic liquids, also known as low-temperature molten salts, are a class of compounds composed entirely of ions and exhibit a liquid state at room temperature. Different from traditional high-temperature molten salt such as NaCl, due to the good symmetry of anion and cation and small ion radius, they can be firmly combined by electrostatic force [28–30]. In addition, ionic liquids show great potential in catalysis, separation, and electrochemistry due to their high thermal stability, wide electrochemical window, and structural designability [31–35]. However, the use of ionic liquids as crosslinking agents of hydrogels in the synthesis for adsorption of heavy metal ions in water has not been reported.

In this work, we use ionic liquid $[Vim]Br_2$ as a crosslinking agent to synthesize PAM/AA/$[Vim]Br_2$ hydrogel, which was mainly used to adsorb heavy metal ions such as Ni^{II}, Cu^{II}, Zn^{II}, and Cr^{III} from water. The optimal conditions for hydrogel synthesis were optimized by response surface methodology. The swelling ratio of PAM/AA/[Vim]Br2 hydrogel could reach 40,012%. The effects of temperature, pH, initial concentration of heavy metal ions, and the amount of hydrogel added are discussed. The adsorption isotherm and XPS analysis indicated that the hydrogel exhibits different adsorption mechanisms when adsorbing heavy metal ions with different concentrations. At low concentrations, the adsorption of the hydrogel was attributed to multilayer physical adsorption, and the enhanced removal rates of heavy metal ions could be attributed to the electrostatic interaction of the bromide ions from the crosslinking ionic liquid agent in the hydrogel structure. On the one hand, this is the first report of achieving removal rates exceeding 80% at low concentrations using hydrogels. On the other hand, at high concentrations, the adsorption of the hydrogel follows a monolayer chemical adsorption mechanism, and the removal rates of heavy metal ions above 90% could be attributed to the chelation interaction between the active sites and the metal ions.

2. Experimental Methods

2.1. Materials

Acrylic acid (AA), acrylamide (AM), ammonium persulfate (APS), 1-vinylimidazole (Vim), 1,2-dibromoethane, nickel(II) chloridehexahydrate ($NiCl_2{\cdot}6H_2O$), copper(II) chloride hexahydrate ($CuCl_2{\cdot}6H_2O$), zinc(II) acetate dihydrate ($Zn[CH_3COO]_2{\cdot}2H_2O$), chromium(III) chloride hexahydrate ($CrCl_3{\cdot}6H_2O$), sodium hydroxide (NaOH), ether, and methanol were purchased from Energy-Chemical (Shanghai, China). All chemicals were of analytical grade and used without any purification. The experimental water was ultrapure water prepared by an ELGA CLXXDM2 ultrapure water instrument ($\geq$18.2 MΩ·cm).

2.2. Synthesis of [Vim]Br_2

Under N_2 protection, 1,2-dibromoethane (1.88 g, 10 mmol) was dissolved in 10 mL methanol, meanwhile, 1-vinylimidazole (1.88 g, 20 mmol) was dripped. The reaction mixture was stirred and heated to 50 °C for 24 h. The methanol solution was removed by vacuum distillation and yellow powder product was obtained. The crude product was washed 3 times with ether and dried under vacuum at 50 °C for 2 h. Finally, the yellow powdery product was obtained (Scheme 1).

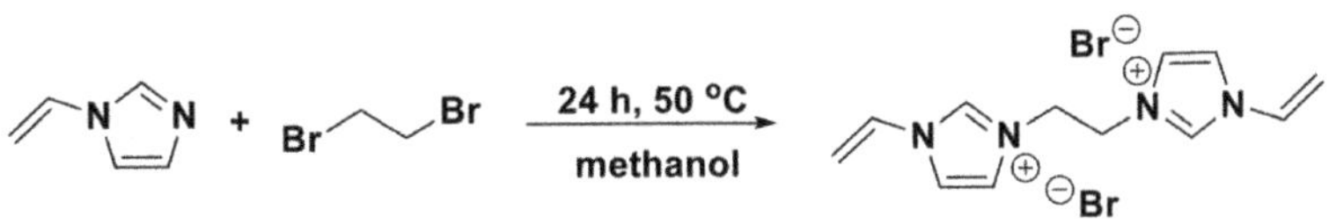

Scheme 1. The synthesis of [Vim]Br_2.

The yield of Bis1-vinylimidazole ethyl bromide ([Vim]Br_2) was 89%. ^{1}HNMR (400 MHz, D_2O), δ:4.8 (m, 4H), 5.43 (m, 2H), 5.78 (m, 2H), 7.1 (m, 2H), 7.56(m, 2H), 7.8 (m, 2H), 9.11 (s, 2H).

2.3. Preparation of PAM/AA/[Vim]Br_2 Hydrogel

Appropriate amounts of acrylic acid (AA), acrylamide (AM), NaOH, and distilled water were mixed. The mixture was stirred and cooled to room temperature. Then, a certain amount of crosslinking agent [Vim]Br_2 was added and stirred magnetically at room temperature for 30 min. N_2 was introduced to drain the oxygen in the flask and APS was added to seal the mixture. The polymerization was carried out in a water bath at 60 °C for 6 h. The hydrogels were washed with distilled water several times to remove unreactive monomers, dried at 60 °C until constant weight, crushed, screened, and reserved. Figure 1 shows the synthetic process of hydrogel and its adsorption mechanisms for heavy metal ions.

2.4. Hydrogel Swelling

The swelling behavior of PAM/AA/[Vim]Br_2 hydrogels was investigated by immersion of 0.1 g of the SPH in 100 mL Milli-Q distilled water at room temperature and the hydrogels reached equilibrium swelling for 12 h. The influence of pH on the swelling behavior was tested using HCl and NaOH. Equation (1) calculates the percentage of hydration.

$$\text{Swelling} = \frac{m_w - m_d}{m_d} \times 100\% \quad (1)$$

where m_w is the mass of the swollen sample at time t and m_d is the weight of the dry sample.

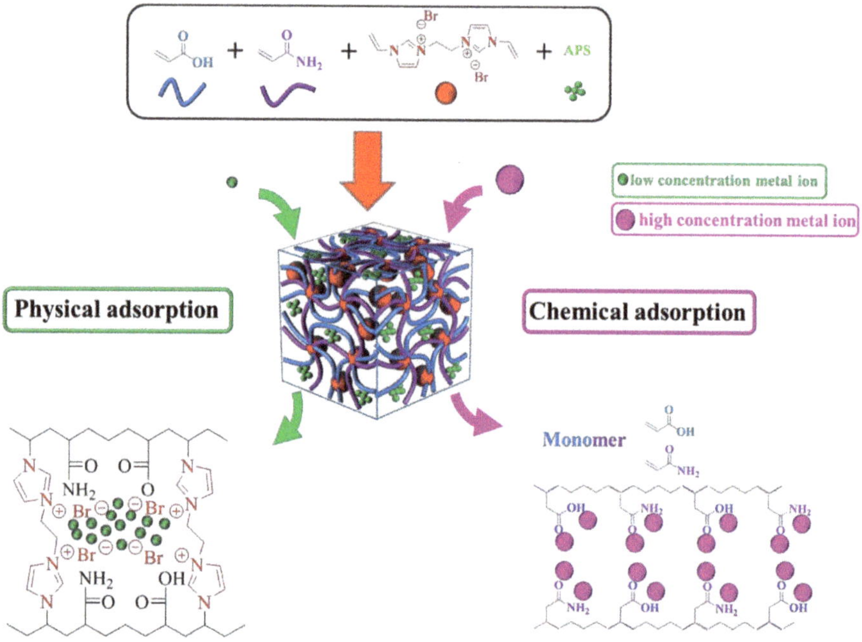

Figure 1. The synthetic process of hydrogel and its adsorption mechanisms for heavy metal ions.

2.5. Adsorption Experiments

First, 0.1 g of dry hydrogel powder was added to a 200 mL Erlenmeyer flask, 100 mL of heavy metal solutions (Ni^{II}, Cu^{II}, Zn^{II}, Cr^{III}) was added at different concentrations, and it was shaken at 25 °C for 12 h. After reaching the adsorption equilibrium, an atomic absorption spectrometer was used to detect the metal ion content in the remaining solution. Q_e and RR (%) were calculated using the following Equations (2) and (3):

$$Q_e = \frac{(C_0 - C_e) \times V}{M} \tag{2}$$

$$RR\% = \frac{C_0 - C_e}{C_0} \times 100\% \tag{3}$$

Q_e (mg/g) represents the equilibrium removal efficiency of hydrogel. C_0(mg/L) and C_e (mg/L) are the initial and the equilibrium concentrations of metal ions in liquid phase, respectively. V(L) is the volume of metal solution and M(g) is the weight of dried hydrogel.

2.6. Design of Response Surface Experiment

The response surface method was used to design the experiment, and the factors affecting the swelling performance of the hydrogel were evaluated. There are four main variables: monomer ratio (A, 0.7–0.9%), neutralizing (B, 60–80), initiator (C, 0.2–0.6), and the crosslinking agent (D, 0.2–0.8). The analysis of variance of the results was carried out using the following quadratic model [36]:

$$Y = \beta_0 + \beta_1 A + \beta_2 B + \beta_3 C + \beta_4 D + \beta_{12} AB + \beta_{13} AC + \beta_{14} AD + \beta_{23} BC + \beta_{24} BD + \beta_{34} CD + \beta_{11} A^2 + \beta_{22} B^2 + \beta_{33} C^2 + \beta_{44} D^2$$

where Y is the percentage of hydrogel swelling rate responses, β_n is the linear regression coefficients, A, B, C, and D are the studied factors (shown in Table 1). Values of "prob>F" which are less than 0.05 indicate that model terms are significant [37]. As shown in Figure 2, the response surface diagram shows that the swelling rate of the gelatin is affected by the interaction of four factors, rather than a single linear relationship. Optimization by four factors indicates: monomer ratio: 70:30, neutralizing: 60%, initiator: 0.4 wt%, crosslinking agent: 0.8 wt%, the maximum swelling rate of hydrogels reached 40,012%.

Table 1. Independent variables and coded level of the experiment.

Variable	Code	Level		
Monomer ratio	A	0.7	0.8	0.9
% Neutralizing	B	60	70	80
% Initiator	C	0.2	0.4	0.6
% Crosslinking	D	0.2	0.5	0.8

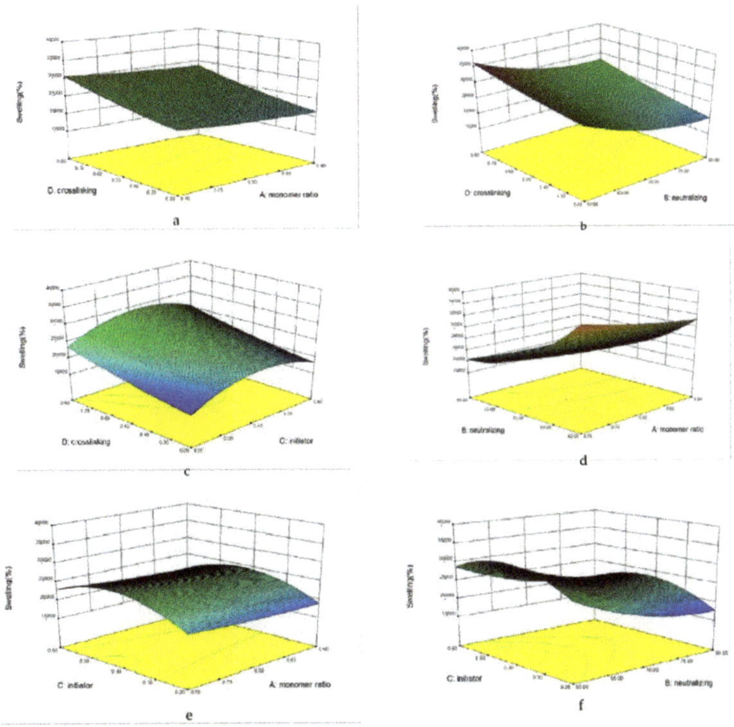

Figure 2. The influence of various factors on the response value in the response surface three-dimensional graph. The z-axis coordinate of six graphs is swelling (%). The horizontal and vertical coordinates are variables: (**a**) monomer ratio and crosslinking; (**b**) neutralizing and crosslinking; (**c**) initiator and crosslinking; (**d**) monomer ratio and neutralizing; (**e**) monomer ratio and initiator; (**f**) crosslinking and initiator.

2.7. Characterization

^{1}H NMR spectra were recorded at 400 MHz on a Bruker Ascend 400 spectrometer (Bruker Daltonics Inc., Billerica, MA, USA) with tetramethylsilane as the internal standard. The Fourier transform infrared (FTIR) spectroscopy was performed on a Nicolet iS5FTIR spectrometer equipped with an attenuated total reflectance (ATR) accessory. The samples were first mixed with dried KBr before analysis and the spectrum of each sample was obtained in the range of 4000–500 cm^{-1}. The surface morphology of hydrogels was observed with a scanning electron microscope JSM-6490LV (JEOL, Tokyo, Japan). The

XPS measurements were conducted on an ESCALAB 250Xi spectrometer. An atomic absorption spectrometer (AAS) was applied for the determination of the metal ions in the aqueous medium.

3. Results and Discussion

3.1. FTIR Analysis

In the infrared spectrum of AM (Figure 3a), the strong absorption bands of amide groups are observed at 3372 and 3195 cm^{-1}. In the case of PAM/AA/[Vim]Br_2 hydrogel (Figure 3b), the strong absorption peaks of amide groups have shifted to 3419 and 3232 cm^{-1}. Additionally, the C=O stretching vibration absorption peak at 1674 cm^{-1} in the AM infrared spectrum corresponds to the C=O stretching vibration absorption peak at 1652 cm^{-1} in the PAM/AA/[Vim]Br_2 hydrogel. The peaks observed at 1562 and 1448 cm^{-1} in the hydrogel correspond to the C=O stretching vibration peaks of the carboxyl anion (-COO^-) in AA. These data indicate the successful copolymerization of AA and AM [38], resulting in the synthesis of PAM/AA/[Vim]Br_2 hydrogel.

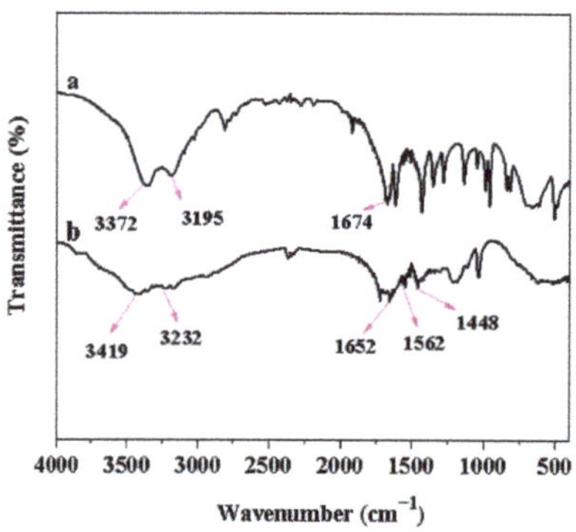

Figure 3. (**a**) Infrared spectrum of acrylamide (AM). (**b**) PAM/AA/[Vim]Br_2 hydrogel.

3.2. SEM Analysis of PAM/AA/[Vim]Br_2 Hydrogel

As shown in Figure 4a, after freezing intervention, the prepared PAM/AA/[Vim]Br_2 hydrogel has a clearer layered structure, indicating that the internal molecular structure of the prepared hydrogel is more evenly distributed. As shown in Figure 4b, the internal pore size of the hydrogel is uniformly distributed, and it is a polymer material with a three-dimensional network structure. As shown in Figure 4c,d, the hydrogel has a large pore structure. This large loose pore structure further increases the contact area between the hydrogel and heavy metal ions, which is conducive to hydrogel adsorption of heavy metal ions.

3.3. Effect of Initial Concentration of Ni^{II}, Cu^{II}, Zn^{II}, Cr^{III} on RR (%)

The effects of the initial concentrations of Ni^{II}, Cu^{II}, Zn^{II}, and Cr^{III} on the removal rate (RR%) are depicted in Figure 5. Heavy metal solutions with initial concentrations of 20, 40, 60, 80, 100, 120, and 150 mg/L were selected for the experimental study. The four adsorption curves exhibited a consistent pattern and could be divided into two stages. In the first stage, within the low concentration range (<40 ppm), the adsorption capacities of the hydrogels for heavy metal ions increased as the initial concentration of metal ions rose. This could be attributed to the electrostatic attraction between anions, such as Br- and Ni^{II}, towards heavy metal ions. However, as the metal ion concentration continued to increase (<60 ppm), the electrostatic force weakened, resulting in a decline in the removal rate. In the second stage, within the high concentration range (<100 ppm), the RR% of heavy metal ions increased as the concentration of metal ions rose. This could be attributed to the increased contact probability between the active sites in the hydrogel and the heavy metal ions, enhancing the coordination between the functional groups of the hydrogel and the metal ions. However, once the concentration surpassed a threshold value (>100 ppm), the adsorption capacity of the hydrogel approached saturation, and the RR% started to decrease. This decrease was due to the saturation of active sites on the hydrogel, which limited the coordination between the functional groups in the hydrogel and the metal ions. Given the generally consistent trends of the four metal ions in both the low and high concentration ranges, our subsequent study primarily focused on Ni^{II}.

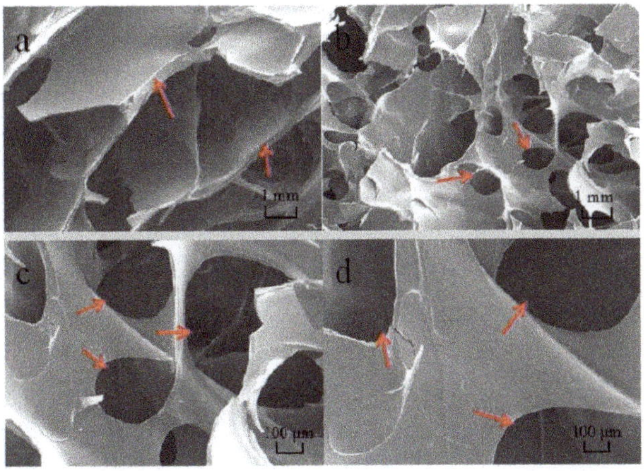

Figure 4. SEM images of PAM/AA/[Vim]Br$_2$ hydrogel. (**a**,**b**) are 1 mm structural illustrations, (**c**,**d**) are 100 μm structural illustrations. The red arrows points to the internal pores of the hydrogel.

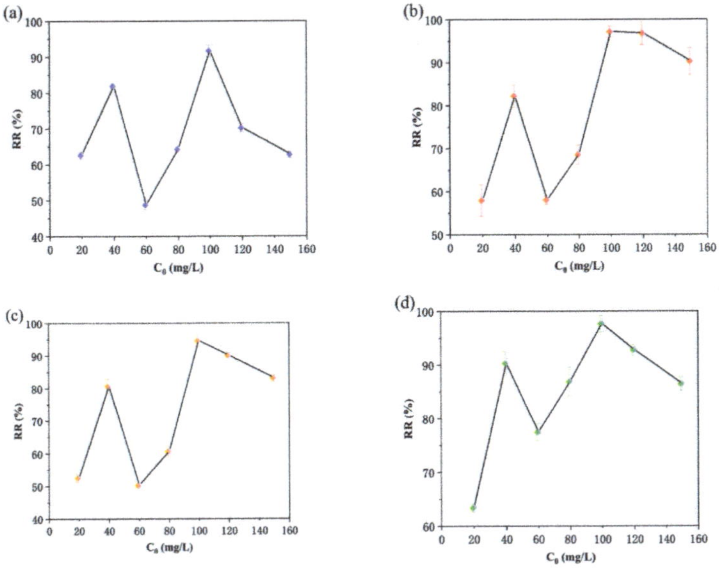

Figure 5. (**a**) Effect of initial concentration of Ni^{II} on RR (%). (**b**) Cu^{II} RR (%). (**c**) Zn^{II} RR (%). (**d**) Cr^{III} RR (%).

3.4. Effect of Hydrogel Dosage on RR (%)

The effect of the added amount of hydrogel on RR (%) is shown in Figure 6. The results showed that at low concentration (40 ppm) or high concentration (100 ppm), RR (%) increased sharply with the increase in hydrogel dosage, which mainly increased the surface area and active sites of hydrogel. When the hydrogel dosage was 2 g/L, the RR (%) value reached 86.4% at a low concentration. When the concentration was high, the RR (%) value reached 91.8%. However, when the hydrogel dosage was 1 g/L, the saturation phenomenon appeared, and the RR (%) increased slowly with the increase in hydrogel addition. Therefore, 1 g/L was selected as the best dosage for further adsorption experiments.

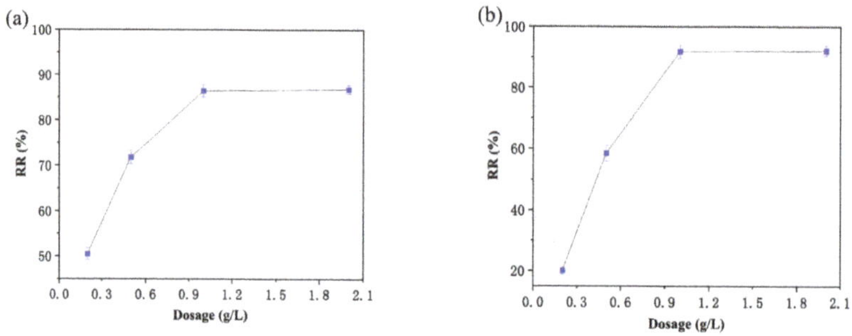

Figure 6. (**a**) Effect of hydrogel addition RR (%) (40 ppm); (**b**) 100 ppm.

3.5. Effect of Temperature on RR (%)

The changes in the adsorption capacity of the hydrogel for heavy ions at different temperatures are shown in Figure 7. As the temperature increased from 15 °C to 55 °C, the hydrogel's removal rates (RRs) significantly increased. When the temperature exceeded 55 °C, there was little alteration observed in the removal efficacy for heavy metal ions. This phenomenon indicates that with increasing temperature, the activity of heavy metal ions is enhanced, leading to the disruption of hydrogen bonds between the hydrogel and water. This exposes more active functional groups for complexation with heavy metal ions [38]. However, when the adsorption capacity approaches saturation, further increasing the temperature does not significantly increase the removal efficiency of the hydrogel.

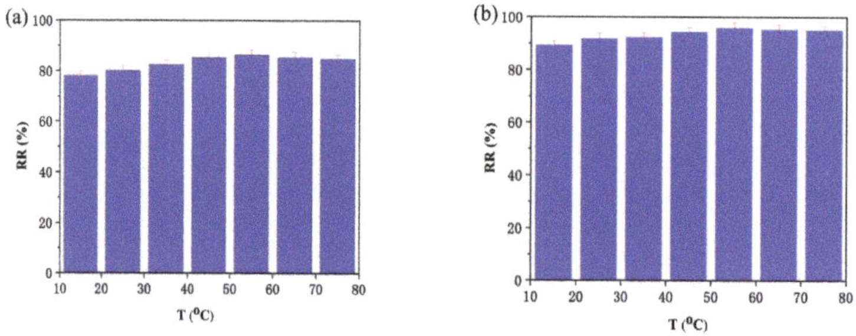

Figure 7. (**a**) Effect of temperature on RR (%) (40 ppm); (**b**) 100 ppm.

3.6. Effect of pH on RR (%)

Figure 8 illustrated the RR (%) of PAM/PAA/[Vim]Br2 hydrogel (40 ppm, 100 ppm) for heavy metal ions (NiII, CuII, ZnII, CrIII) in solutions with different pH values. The RR (%) peaked at pH 7, while it significantly decreased in strong acid (pH = 3) or strong alkaline (pH = 11) environments compared to the neutral environment. This was because, in acidic conditions, H+ protonated the -COO^- and -NH_2 groups on the hydrogel structure, forming

-COOH and $-NH_3^+$. The H^+ ions also competed with heavy metal ions for adsorption sites. As the solution pH gradually increased to 7, the competitive effect of H^+ ions weakened, allowing a large number of $-NH_2$ and $-COO^-$ groups to re-coordinate with heavy metal ions, resulting in an increase in the RR (%) [39]. However, when the solution pH was above 7 and gradually increased to a strong alkaline environment, the increased OH^- ion concentration in the external solution led to an anion shielding effect on the $-COO^-$ and $-NH_2$ groups of the hydrogel structure. This caused a decrease in osmotic pressure inside and outside the hydrogel network, hindering the effective diffusion of heavy metal cations into the hydrogel. As a result, the RR (%) decreased [40].

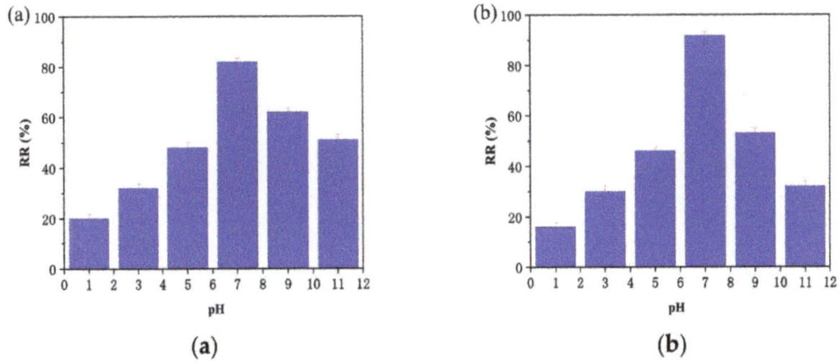

Figure 8. (**a**) Impact of pH on Ni^{II} adsorption onto hydrogel (40 ppm); (**b**) 100 ppm.

3.7. Adsorption Kinetics of Heavy Metal Ions by PAM/AA/[Vim]Br$_2$ Hydrogel

The adsorption effect of PAM/PAA/[Vim]Br2 hydrogel on Ni^{II} with initial concentrations of 40 ppm and 100 ppm at different time points (1, 2, 3, 4, 5, 6, 7, 8, 9, and 10 h) is shown in Figure 9. During the initial adsorption period (0–2 h), the adsorption of heavy metal ions by the hydrogel increased rapidly. After 3 h of adsorption, the efficiency of heavy metal ion adsorption gradually decreased. At 4 h, the adsorption capacity of the hydrogel for heavy metal ions approached saturation. These adsorption behaviors indicated that PAM/PAA/[Vim]Br_2 hydrogel could achieve relatively fast and near-saturation adsorption of heavy metal ions. [41,42].

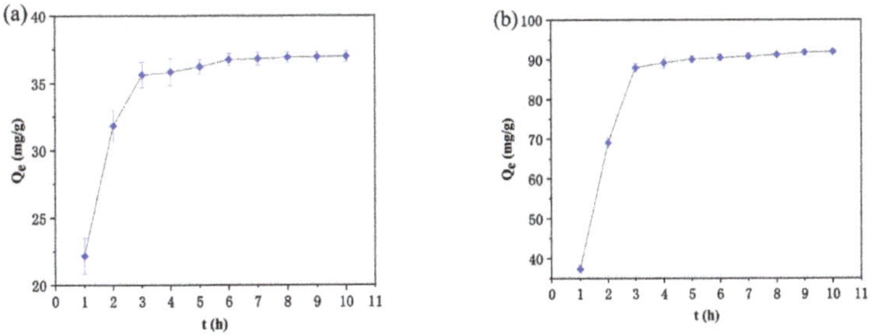

Figure 9. (**a**) Adsorption kinetics curves of PAM/AA/[Vim]Br_2 hydrogels for heavy metal ions (40 ppm); (**b**) 100 ppm.

The adsorption kinetics of PAM/PAA/[Vim]Br_2 hydrogel on Ni^{II} was fitted by a pseudo-first-order kinetics and two-stage kinetic Equations (4) and (5), and the kinetics of the adsorption reaction of PAM/AA/[Vim]Br_2 hydrogel on Ni^{II} was obtained. The equation expression is as follows.

$$\ln(Q_e - Q_t) = \ln Q_e - k_1 t \quad (4)$$

$$\frac{t}{Q_t} = \frac{t}{Q_e} + \frac{1}{k_2 Q_e^2} \quad (5)$$

where t is the adsorption time (min); Q_e and Q_t are, respectively, the adsorption capacity of PAM/AA/[Vim]Br_2 hydrogel for heavy metal ions when adsorbed at equilibrium time and t (mg/g); K_1 and K_2 are quasi-one- and quasi-two-stage adsorption rate constants, respectively.

The results of pseudo-first-order and pseudo-second-order reaction kinetics fitting curves of Ni^{II} at 40 ppm and 100 ppm are shown in Figure 10 and Table 2. The results show that: compared with the pseudo-first-order adsorption kinetics fitting results (40 ppm, R^2 = 0.8156, Figure 10a; 100 ppm, R^2 = 0.953, Figure 10c), the pseudo-second-order adsorption kinetic model fitting is more consistent (40 ppm, R^2 = 0.9983, Figure 10b; 100 ppm, R^2 = 0.9986, Figure 10d). Therefore, the adsorption of heavy metal ions by hydrogel (40 ppm, 100 ppm) is a multistep process. First, the heavy metal ions adhere to the metal surface and then enter the hydrogel through the channel of the hydrogel to further spread.

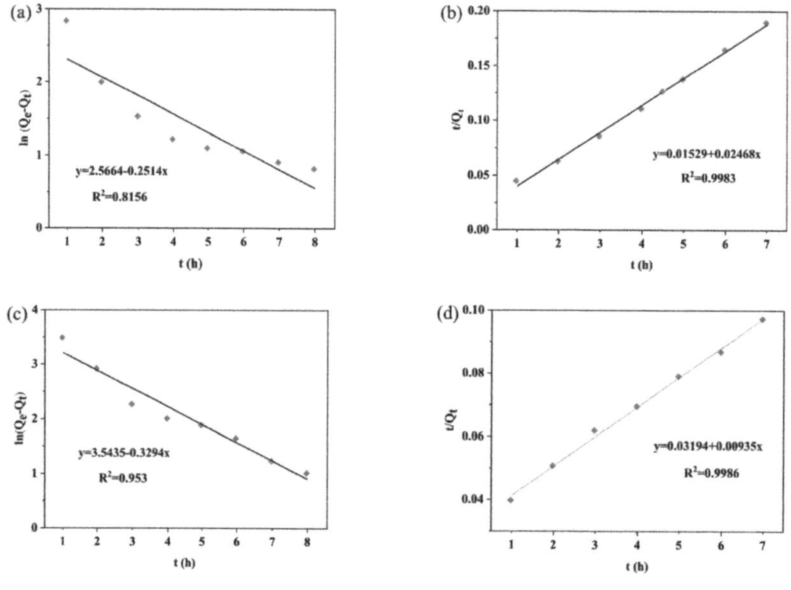

Figure 10. Fitting results of adsorption kinetics of Ni^{II} by PAM/AA/[Vim]Br_2 hydrogel. (**a**,**b**) The first-order kinetics and second-order kinetics fitting results of 40 ppm, respectively; (**c**,**d**) 100 ppm.

Table 2. Fitting parameters of adsorption kinetics model of different concentrations of Ni^{II} for PAM/PAA/[Vim]Br_2 hydrogel.

Ion	Concentration	Pseudo-First-Order			Pseudo-Second-Order		
		$Q_{e,1}$ (mg/g)	K_1	R^2	$Q_{e,2}$ (mg/g)	K_2	R^2
Ni^{2+}	40 ppm	32.23	0.833	0.8156	38.28	0.01152	0.9983
	100 ppm	89.12	0.572	0.953	98.98	0.00418	0.9986

3.8. Adsorption Isotherms of PAM/AA/[Vim]Br_2 Hydrogels for Heavy Metal Ions

The adsorption isotherms of Freundlich and Langmuir were studied. The Freundlich isotherm is a heterogeneous, multilayer adsorption system, and the absorption process takes place on an active heterogeneous surface. The Langmuir isotherm is a homogeneous, single molecular layer adsorption system, each binding site on the absorption surface absorbs the same energy, and each binding site is occupied by only one metal ion. The two models are presented in Equations (6) and (7):

$$lnQ_e = lnK_F + \frac{lnC_e}{n} \quad (6)$$

$$\frac{C_e}{Q_e} = \frac{1}{K_L Q_m} + \frac{C_e}{Q_m} \quad (7)$$

where C_e, Q_e, Q_m were the initial equilibrium concentration (mg/L) of heavy metal ion solution, the adsorption capacity of PAM/AA/[Vim]Br_2 hydrogel to heavy metal ions (mg/g), and the saturated adsorption capacity of PAM/AA/[Vim]Br_2 hydrogen to heavy metal ions (mg/g), K_F and K_L are Freundlich and Langmuir equilibrium constants, respectively, and n is the concentration index.

The Freundlich and Langmuir adsorption isotherm models were fitted to the initial concentration and equilibrium adsorption capacity of the PAM/AA/[Vim]Br_2 hydrogel to adsorb Ni^{II}. The results are shown in Figure 11 and Table 3. When the initial concentration of Ni^{II} was 40 ppm, regarding the adsorption isotherm of Ni^{2+} by the hydrogel, the Freundlich adsorption isotherm (R^2 = 0.9935) (Figure 11a) is better than the Langmuir adsorption isotherm (R^2 = 0.9859) (Figure 11b), which shows that the adsorption isotherm of hydrogel for Ni^{II} is more in line with multilayer physical adsorption. When the initial concentration of Ni^{II} is 100 ppm, regarding the adsorption isotherm of hydrogel for Ni^{II}, the Langmuir adsorption isotherm (R^2 = 0.9954) (Figure 11d) has better fitting results, which shows that the adsorption of Ni^{II} to hydrogel is more in line with multilayer chemical adsorption.

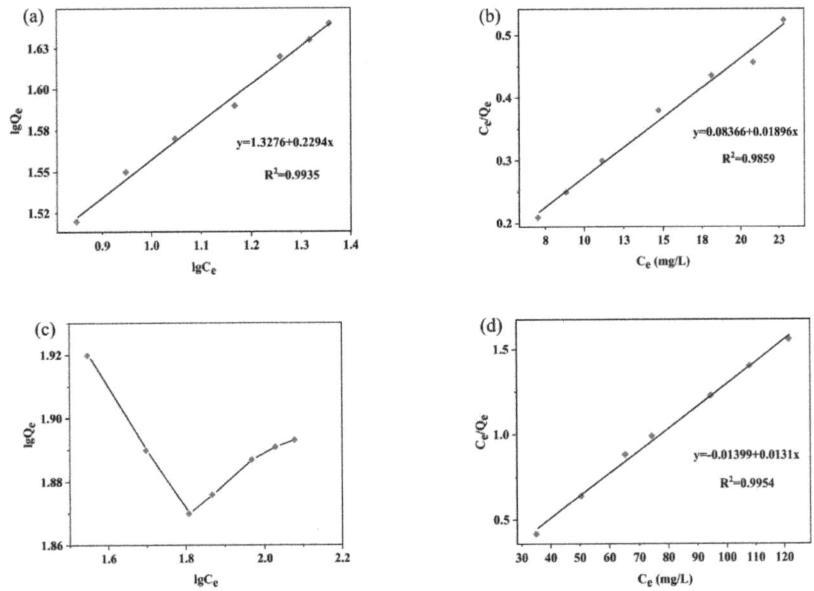

Figure 11. Fitting results of adsorption isotherms of Ni^{II} for hydrogel PAM/AA/[Vim]Br_2 hydrogel. (**a**,**b**) The Freundlich adsorption isotherm and Langmuir adsorption isotherm fitting results of 40 ppm, respectively; (**c**,**d**) 100 ppm.

Table 3. Fitting parameters of adsorption isotherm model of different concentrations of Ni^{II} for PAM/PAA/[Vim]Br_2 hydrogel.

Ion	Concentration	Langmuir			Freundlich		
		K_L	Q_m	R^2	K_F	n	R^2
Ni^{2+}	40 ppm	0.29	49.19	0.9859	816.58	3.33	0.9935
	100 ppm	−0.153	69.06	0.9954	86,377	−6.97	/

3.9. XPS Analysis of PAM/AA/[Vim]Br_2 Hydrogel

To further investigate the adsorption mechanism of hydrogels for heavy metal ions, XPS was used to analyze the binding energy of the hydrogel before and after adsorption, and the results are shown in Figures 12 and 13.

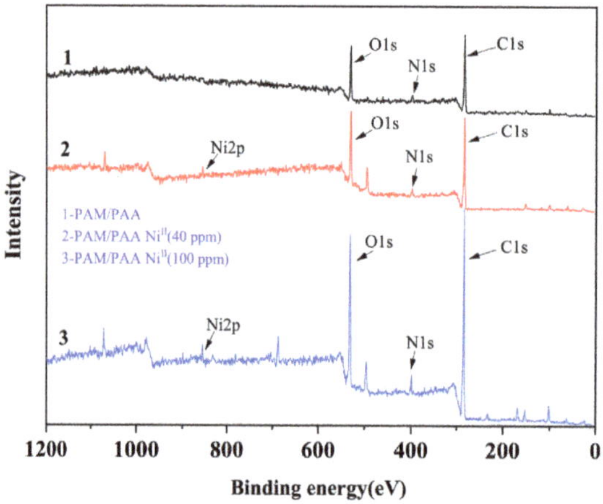

Figure 12. XPS spectrum of PAM/AA/[Vim]Br_2 hydrogel before and after Ni^{II} ion adsorption.

In Figure 12, the main elements of the hydrogel such as the binding energy peaks of C1s, O1s, N1s can be seen. The Figure 12 (2) and (3) curves exhibit that the binding energy peak of Ni2p appears after adsorption, which proves that the Ni^{II} was adsorbed by the hydrogel. As shown in Figure 13a, in the PAM/AA sample before the adsorption of the Ni^{II} ions (40 ppm, 100 ppm), the N1s spectrum showed distinct peaks at ~399.6 eV and ~401.65 eV, corresponding to the -NH_2 or -NH or C-NH^{3+} groups [43]. In the O1s range, the spectrum showed a distinct peak at ~531.75 (Figure 13d) [44,45], corresponding to the oxygen of C=O or C-O. After the adsorption of the Ni^{II} ions, the N1s spectrum shows no significant changes (40 ppm) (Figure 13b), however, the binding energy peaks of the N1s move from 399.65 eV to 399.95 eV and 401.65 eV to 402.45 eV (100 ppm) (Figure 13c). The O1s peak shifted slightly from 531.75 eV to 531.85 eV (40 ppm) (Figure 13e) and the O1s peak shifted strongly from 531.75 eV to 532.2 eV (100 ppm) (Figure 13f). In addition, Figure 13g (40 ppm) and Figure 13h (100 ppm) show the XPS spectrum of Ni2p with a binding energy ranging from 849.1 eV to 886.2 eV. Two major peaks with binding energies of 855.9 eV and 870.6 eV have a significant corresponding relationship to $NiCl_2$, and the peaks of Ni_2O_3 are found at 865.7 eV and 873.1 eV (40 ppm) (864.9 and 874.4 (100 ppm)), which indicates that the PAM/PAA hydrogel can effectively offer O element as a chelating group for removal of Ni^{II} [46]. The other two peaks at 860.5 eV and 879.6 eV (40 ppm) (860.5 and 879.6 (100 ppm)) can be assigned to the corresponding satellite peaks of Ni $2p_{3/2}$ and Ni $2p_{1/2}$ [47]. The XPS spectra indicate that the concentration of heavy metal ions was low (40 ppm), and the adsorption of heavy metal ions by the hydrogel was mainly physical adsorption, which was the electrostatic attraction between ionic liquids and heavy metal ions. When the concentration of heavy metal ions was high (100 ppm), the adsorption of heavy metal ions by the hydrogel was mainly chemical adsorption, which was due to the chelation and coordination reaction of heavy metal ions with the amino, hydroxyl, and carboxyl groups of the hydrogel. The conclusion was consistent with the adsorption isotherm of hydrogel.

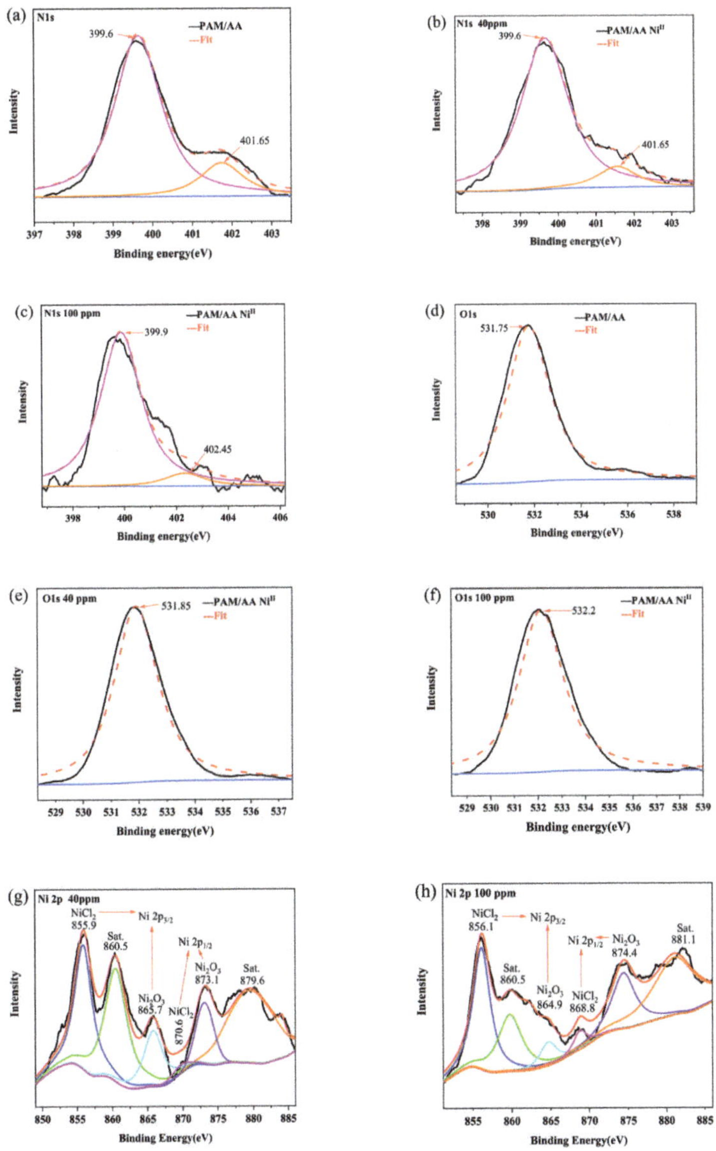

Figure 13. **(a–c)**: XPS spectrum of N1s with PAM/AA hydrogel adsorbing NiII ions (0, 40, 100 ppm). **(d–f)**: XPS spectrum of O1s with PAM/AA hydrogel adsorbing NiII ions (0, 40, 100 ppm). **(g,h)**: XPS spectrum of Ni2p (40, 100 ppm) with a binding energy ranging from 849.1 eV to 886.2 eV (The black line represents the test line, the red line represents the fitting line, and other colors represent reference lines).

4. Conclusions

To our best knowledge, there have been no reports on the use of ionic liquids as crosslinking agents to prepare hydrogels for the adsorption of heavy metal ions in water. The experiment successfully synthesized a PAM/AA/[Vim]Br_2 hydrogel using the ionic liquid [Vim]Br_2 as the crosslinking agent, which was confirmed by FTIR characterization. In our experiment, under near-neutral solution conditions and at a heavy metal ion concentration of 100 ppm, the PAM/AA/[Vim]Br_2 hydrogel demonstrated superior adsorption performance for Ni^{II}, Cu^{II}, Zn^{II}, and Cr^{III}, achieving the removal rates of 91.8%, 97.2%, 95.6%, and 98.1%, respectively, with RR% values all exceeding 90%. This indicates a certain advantage in heavy metal removal rates compared to the reported values. However, further testing in actual wastewater and assessment of the removal efficiency after multiple cycles of use are still required.

The adsorption isotherms and XPS analysis revealed that, at low concentrations (40 ppm), the hydrogel follows the Freundlich isotherm for adsorbing heavy metal ions, primarily through multilayer physical adsorption. One major highlight of this article is the removal rate of heavy metal ions in the low concentration range, which exceeds 80%. This is mainly attributed to the electrostatic interaction between anions in the ionic liquid and the heavy metal ions. At higher concentrations (100 ppm), the hydrogel follows the Langmuir isotherm, indicating monolayer chemical adsorption. This is mainly attributed to the coordination between carboxyl and amino groups within the hydrogel and the metal ions.

Author Contributions: Conceptualization, J.S. and Z.J. (Zhaohui Jin) methodology, J.S. and Z.J. (Ziqi Jin); software, J.W. and Z.J. (Ziqi Jin); validation, Z.J. (Ziqi Jin) and H.W.; formal analysis, Q.Z.; investigation, H.G.; resources, J.Z.; data curation, J.W.; writing—original draft preparation, Z.J. (Ziqi Jin) and J.W.; writing—review and editing, Z.J. (Zhaohui Jin), Z.J. (Ziqi Jin) and Z.W.; visualization, H.W.; supervision, J.S. and H.G.; project administration, J.S.; funding acquisition, J.S. All authors have read and agreed to the published version of the manuscript.

Funding: This research was funded by [Jilin Province Science and Technology Development Plan Project] grant number [20180201019SF], [the Research Fund of the Science and Technology Department of Jilin Province] grant number [20100551], [Special Project of Industrial Technology Research and Development in Jilin Province] grant number [2013779], and [Research Science Institute of Jilin Provincial Department of Education] grant number [JJKH20230307KJ].

Institutional Review Board Statement: Not applicable.

Data Availability Statement: The data presented in this study are available on request from the corresponding author.

Acknowledgments: All authors are aware of the publication of the article and are grateful for the financial support of the fund.

Conflicts of Interest: The authors declare no conflict of interest.

References

1. Liu, S.; Lou, S.; Kuang, C.; Huang, W.; Chen, W.; Zhang, J.; Zhong, G. Water quality assessment by pollution-index method in the coastal waters of Hebei Province in western Bohai Sea, China. *Mar. Pollut. Bull.* **2011**, *62*, 2220–2229. [CrossRef] [PubMed]
2. Foy, R.H.; Lennox, S.D.; Smith, R.V. Assessing the effectiveness of a three-stage on-farm biobed in treating pesticide contaminated wastewater. *Water Res.* **2011**, *181*, 874–882. [CrossRef]
3. Looi, L.J.; Aris, A.Z.; Johari, W.L.W.; Yusoff, F.M.; Hashim, Z. Baseline metals pollution profile of tropical estuaries and coastal waters of the Straits of Malacca. *Mar. Pollut. Bull.* **2013**, *74*, 471–476. [CrossRef] [PubMed]
4. Oudeika, M.S.; Altinoglu, F.F.; Akbay, F.; Aydin, A.B. The use of magnetic susceptibility and chemical analysis data for characterizing heavy metal contamination of topsoil in Denizli city, Turkey. *Appl. Geophys.* **2020**, *183*, 104208. [CrossRef]
5. Li, G.; Zhao, Z.; Liu, J.; Jiang, G.; Haza, J. Effective heavy metal removal from aqueous systems by thiol functionalized magnetic mesoporous silica. *J. Hazard. Mater.* **2011**, *198*, 277–283. [CrossRef] [PubMed]
6. Järup, L. Hazards of heavy metal contamination, British Medical Bulletin. *Br. Med. Bull.* **2003**, *68*, 167–182. [CrossRef] [PubMed]
7. Choi, J.Y.; Jeong, H.; Choi, K.-Y.; Hong, G.H.; Yang, D.B.; Kim, K.; Ra, K. Source identification and implications of heavy metals in urban roads for the coastal pollution in a beach town, Busan, Korea. *Mar. Pollut. Bull.* **2020**, *161*, 111724. [CrossRef]

8. Abri, M.; Dakheel, A.; Tizaoui, C.; Hilal, N. Combined humic substance and heavy metals coagulation, and membrane filtration under saline conditions. *Desalination* **2010**, *253*, 46–50. [CrossRef]
9. Sun, B.; Li, X.; Zhao, R.; Yin, M.; Wang, Z.; Jiang, Z.; Wang, C. Hierarchical aminated PAN/γ–AlOOH electrospun composite nanofibers and their heavy metal ion adsorption performance. *J. Taiwan Inst. Chem. Eng.* **2016**, *62*, 219–227. [CrossRef]
10. Huang, L.L.; Liu, J.F.; Sun, B.; Zhang, N.; Tang, Y.Q.; Feng, Y.J. Experimental Study on Papermaking Wastewater by Advanced Electrochemical Oxidation Method. *Adv. Mater. Res.* **2013**, *726*, 1699–1703. [CrossRef]
11. Lee, Y.-J.; Son, C.-Y.; Lee, C.-G.; Jeong, Y.J.; Cho, I.S.; Park, S.-J.; Lee, J. Enhancement of photocatalytic Cr(VI) reduction using man-darin peel extract as natural sacrificing agent. *Alex. Eng. J.* **2023**, *75*, 151–163. [CrossRef]
12. Kurniawan, T.A.; Chan, G.Y.; Lo, W.-H.; Babel, S. Physico–chemical treatment techniques for wastewater laden with heavy metals. *Chem. Eng. J.* **2006**, *118*, 83–98. [CrossRef]
13. Tan, X.Y.; Zhao, F.; Wen, R.Z.; Dai, Y. Adsorption of Pb(ll) on Activated Carbon Treated by Nitric Acid. *Adv. Mater. Res.* **2012**, *550*, 2190–2193. [CrossRef]
14. Hadi, P.; Xu, M.; Ning, C.; Lin, C.S.K.; McKay, G. Valorization of an Electronic Waste-Derived Aluminosilicate: Surface Functionalization and Porous Structure Tuning. *Chem. Eng. J.* **2015**, *4*, 2980–2989. [CrossRef]
15. Zohuriaan-Mehr, M.J.; Motazedi, Z.; Kabiri, K.; Ershad-Langroudi, A.; Allahdadi, I. Gum arabic–acrylic superabsorbing hydrogel hybrids: Studies on swelling rate and environmental responsiveness. *J. Appl. Polym. Sci.* **2006**, *102*, 5667–5674. [CrossRef]
16. Cha, G.D.; Lee, W.H.; Lim, C.; Choi, M.K.; Kim, D.-H. Materials engineering; processing, and device application of hydrogel nanocomposites. *Nanoscale* **2020**, *12*, 10456–10473. [CrossRef] [PubMed]
17. Xue, B.; Kozlovskaya, V.; Kharlampieva, E. Shaped stimuli-responsive hydrogel particles: Syntheses, properties and biological responses. *J. Mater. Chem. B* **2017**, *5*, 9–35. [CrossRef] [PubMed]
18. Seidi, F.; Zhao, W.; Xiao, H.; Jin, Y.; Saeb, M.R.; Zhao, C. Radical polymerization as a versatile tool for surface grafting of thin hydrogel films. *Polym. Chem.* **2020**, *11*, 4355–4381. [CrossRef]
19. Javed, R.; Shah, L.A.; Sayed, M.; Khan, M.S. Uptake of heavy metal ions from aqueous media by hydrogels and their conversion to nanoparticles for generation of a catalyst system: Two-fold application study. *RSC Adv.* **2018**, *8*, 14787–14797. [CrossRef] [PubMed]
20. Lv, Q.; Hu, X.; Shen, Y.; Sun, G. Polymer hydrogel cross-linked by inorganic nanoparticles for removing trace metal ions. *J. Appl. Polym. Sci.* **2020**, *137*, 49004. [CrossRef]
21. Sinha, V.; Chakma, S. Advances in the preparation of hydrogel for wastewater treatment: A concise review. *Environ. Chem. Eng.* **2019**, *7*, 103295. [CrossRef]
22. Pakdel, P.M.; Peighambardoust, S.J. A review on acrylic based hydrogels and their applications in wastewater treatment. *Environ. Manag.* **2018**, *217*, 123–143. [CrossRef]
23. Halah, A.; Machado, D.; González, N.; Contreras, J.; Carrasquero, F. Use of super absorbent hydrogels derivative from acrylamide with itaconic acid and itaconates to remove metal ions from aqueous solutions. *J. Appl. Polym. Sci.* **2019**, *136*, 46999. [CrossRef]
24. Pour, Z.S.; Ghaemy, M. Removal of dyes and heavy metal ions from water by magnetic hydrogel beads based on poly(vinyl alcohol)/carboxymethyl starch-g-poly(vinyl imidazole). *RSC Adv.* **2015**, *5*, 64106–64118. [CrossRef]
25. Shah, L.A.; Khan, M.; Javed, R.; Sayed, M.; Khan, M.S.; Khan, A.; Ullah, M. Superabsorbent polymer hydrogels with good thermal and mechanical properties for removal of selected heavy metal ions. *J. Clean. Prod.* **2018**, *201*, 78–87. [CrossRef]
26. Chen, Z.; Wang, Y.-F.; Zeng, J.; Zhang, Y.; Zhang, Z.-B.; Ma, S.; Tang, C.-M.; Xu, J.-Q. Chitosan/polyethyleneimine magnetic hydrogels for adsorption of heavy metal ions. *Iran Polym. J.* **2022**, *31*, 1273–1282. [CrossRef]
27. Chen, J.; Zhao, K.; Liu, L.; Gao, Y.; Zheng, L.; Liu, M. Modified kaolin hydrogel for Cu2+ adsorption. *e-Polymers* **2022**, *22*, 986–996. [CrossRef]
28. Chen, S.; Zhang, S.; Liu, X.; Wang, J.; Wang, J.; Dong, K.; Sun, J.; Xu, B. Ionic liquid clusters: Structure; formation mechanism, and effect on the behavior of ionic liquids. *Phys. Chem. Chem. Phys.* **2014**, *16*, 5893–5906. [CrossRef]
29. Dong, K.; Liu, X.; Dong, H.; Zhang, X.; Zhang, S. Multiscale Studies on Ionic Liquids. *Chem. Rev.* **2017**, *117*, 6636–6695. [CrossRef]
30. Zhang, S.; Zhou, G.; Liu, X. Structure, interaction and property of amino-functionalized imidazolium ILs by molecular dynamics simulation and Ab initio calculation. *AICHE J.* **2010**, *53*, 3210–3221. [CrossRef]
31. Yang, S.; Liu, X.; Yao, H.; Xin, J.; Xu, J.; Kang, Y.; Yang, Y.; Cai, G.; Zhang, S. Efficient hydrodeoxygenation of lignin-derived phenols and dimeric ethers with synergistic [Bmim]PF6-Ru/SBA-15 catalysis under acid free conditions. *Green Chem.* **2019**, *21*, 597–605. [CrossRef]
32. Smiglak, M.; Pringle, J.M.; Lu, X.; Han, L.; Zhang, S.; Gao, H.; MacFarlane, D.R.; Rogers, R.D. Ionic liquids for energy, materials, and medicine. *Chem. Commun.* **2014**, *50*, 9228–9250. [CrossRef] [PubMed]
33. Zhang, S.; Zhu, Y.; Lu, X.; Li, S. A promising method for electrodeposition of aluminium on stainless steel in ionic liquid. *Aiche. J.* **2010**, *55*, 783–796. [CrossRef]
34. Wang, H.; Yan, R.; Li, Z.; Zhang, X.; Zhang, S. Fe-containing magnetic ionic liquid as an effective catalyst for the glycolysis of poly(ethylene terephthalate). *Catal. Commun.* **2010**, *11*, 763–767. [CrossRef]
35. Dai, F.; Li, Z.; Chen, X.; He, B.; Liu, R.; Zhang, S. Synthesis of vanadium phosphorus oxide catalysts promoted by iron-based ionic liquids and their catalytic performance in selective oxidation of n-butane. *Catal. Sci. Technol.* **2018**, *8*, 4515–4525. [CrossRef]
36. Sosa, M.; Contreras, A.; Durán, C.; García, R.; Palestino, G. Tuning the pH-responsiveness capability of poly(acrylic acid-co-itaconic acid)/NaOH hydrogel: Design, swelling, and rust removal evaluation. *J. Appl. Polym. Sci.* **2020**, *137*, 48403. [CrossRef]

37. Jiang, C.; Wang, X.; Hou, B.; Hao, C.; Li, X.; Wu, J. Construction of a Lignosulfonate–Lysine Hydrogel for the Adsorption of Heavy Metal Ions. *J. Agric. Food Chem.* **2020**, *68*, 3050–3060. [CrossRef] [PubMed]
38. Tally, M.; Atassi, Y. Synthesis and characterization of pH-sensitive superabsorbent hydrogels based on sodium alginate-g-poly(acrylic acid-co-acrylamide) obtained via an anionic surfactant micelle templating under microwave irradiation. *Polym. Bull.* **2016**, *73*, 3183–3208. [CrossRef]
39. Moradi, O.; Aghaie, M.; Zare, K.; Monajjemi, M.; Aghaie, H. The study of adsorption characteristics Cu2+ and Pb2+ ions onto PHEMA and P(MMA-HEMA) surfaces from aqueous single solution. *J. Hazard. Mater.* **2009**, *170*, 673–679. [CrossRef]
40. Olad, A.; Gharekhani, H.; Mirmohseni, A.; Bybordi, A. Synthesis, characterization, and fertilizer release study of the salt and pH-sensitive NaAlg-g-poly(AA-co-AAm)/RHA superabsorbent nanocomposite. *Polym. Bull.* **2017**, *74*, 3353–3377. [CrossRef]
41. Puspitasari, T.; Pangerteni, D.; Nurfilah, E.; Darwis, D.S. Study of Metal Ions Removal from Aqueous Solution by Using Radiation Crosslinked Chitosan-co-Poly(Acrylamide)-Based Adsorbent. *Macromol. Symp.* **2015**, *353*, 168–177. [CrossRef]
42. Samandari, S.; Samandari, S.; Nezafati, N.; Yahya, K. Efficient removal of lead (II) ions and methylene blue from aqueous solution using chitosan/Fe-hydroxyapatite nanocomposite beads. *Environ. Manag.* **2014**, *146*, 484–490. [CrossRef]
43. Sutirman, Z.A.; Sanagi, M.M.; Karim, K.J.A.; Ibrahim, W.A.W.; Jume, B.H. Equilibrium, kinetic and mechanism studies of Cu(II) and Cd(II) ions adsorption by modified chitosan beads. *Int. J. Biol. Macromol.* **2018**, *116*, 255–263. [CrossRef]
44. Wang, J.; Li, Z. Enhanced selective removal of Cu(II) from aqueous solution by novel polyethylenimine-functionalized ion imprinted hydrogel: Behaviors and mechanisms. *J. Hazard. Mater.* **2015**, *300*, 18–28. [CrossRef] [PubMed]
45. Li, P.; Xing, C.; Qu, S.; Li, B.; Shen, W. Carbon Dioxide Capturing by Nitrogen-Doping Microporous Carbon. *Sustain. Chem. Eng.* **2015**, *3*, 1434–1442. [CrossRef]
46. Li, Y.P.; Shi, L.; Gao, X.P.; Wang, J.Y.; Hu, Y.Y.; Wu, X.W.; Wen, Z.Y. Constructing a charged-state Na-NiCl2 battery with NiCl2/graphene aerogel composite as cathode. *Chem. Eng. J.* **2020**, *421*, 127853. [CrossRef]
47. Ao, X.; Wen, Z.; Hu, Y.; Wu, T.; Wu, X.; He, Q. Enhanced stability performance of nickel nanowire with 3D conducting network for planar sodium-nickel chloride batteries. *J. Power Sources* **2017**, *360*, 345–352. [CrossRef]

polymers

Review

Optimized Polymeric Membranes for Water Treatment: Fabrication, Morphology, and Performance

Avneesh Kumar and Dong Wook Chang *

Department of Industrial Chemistry, ECS Core Research Institute, Pukyong National University, Busan 48513, Republic of Korea; avneesh@pknu.ac.kr
* Correspondence: dwchang@pknu.ac.kr

Abstract: Conventional polymers, endowed with specific functionalities, are extensively utilized for filtering and extracting a diverse set of chemicals, notably metals, from solutions. The main structure of a polymer is an integral part for designing an efficient separating system. However, its chemical functionality further contributes to the selectivity, fabrication process, and resulting product morphology. One example would be a membrane that can be employed to selectively remove a targeted metal ion or chemical from a solution, leaving behind the useful components of the solution. Such membranes or products are highly sought after for purifying polluted water contaminated with toxic and heavy metals. An efficient water-purifying membrane must fulfill several requirements, including a specific morphology attained by the material with a specific chemical functionality and facile fabrication for integration into a purifying module Therefore, the selection of an appropriate polymer and its functionalization become crucial and determining steps. This review highlights the attempts made in functionalizing various polymers (including natural ones) or copolymers with chemical groups decisive for membranes to act as water purifiers. Among these recently developed membrane systems, some of the materials incorporating other macromolecules, e.g., MOFs, COFs, and graphene, have displayed their competence for water treatment. Furthermore, it also summarizes the self-assembly and resulting morphology of the membrane materials as critical for driving the purification mechanism. This comprehensive overview aims to provide readers with a concise and conclusive understanding of these materials for water purification, as well as elucidating further perspectives and challenges.

Keywords: supramolecular assemblies; functional polymers; membranes; water purification; pollutants in water

1. Introduction

Human activities and interventions in ecosystems have led to numerous global challenges [1,2]. Among these, the availability of freshwater for drinking stands out as a significant concern. Scarcity of potable water in many regions worldwide presents a formidable challenge for sustaining life on this planet [3]. Large quantities of chemicals from industrial plants and households are dumped into rivers and other natural water reservoirs [4]. These chemicals or pollutants mainly comprise heavy metal ions, dyes, and other organic contaminants. They accumulate in water bodies, posing risks to living organisms and causing not only water pollution but also potentially life-threatening illnesses, such as cancer [5–8]. Consequently, there is an urgent need for efficient methods to remove or decompose these pollutants, prompting extensive research in this area. Most of these techniques involve incorporating functional materials, such as polymers or hybrid composites, within the performing device [9–12]. While bulk materials have also been utilized for this purpose, there has been a growing emphasis on focusing on nanoscale functional materials. This is because nanomaterials can provide several benefits ranging from control over the morphology to selectivity and fabrication [13–16]. Among these nanomaterials, polymers are of particular interest as they offer multiple features (as mentioned above) suitable for similar applications. In polymers, the introduction of a specific chemical functional group (polar or nonpolar) before or post synthesis can help determine the final morphology and selectivity for a particular pollutant and its removal during the water treatment process [17–24]. The chemical and physical properties of a polymer, characterized by a well-designed backbone and desired functional groups, play a considerable role in the fabrication of water purification devices or units [25–27]. Functional polymers employed in water treatment processes are modified with polar or ionic groups, such as hydroxyl, carboxylic, amine, phosphonic, and sulfonic groups, among others, to remove a specific pollutant (for example, a metal) [28–31]. Therefore, when designing functional polymers, the removal pathway in the water treatment process must be considered. To further understand the holistic approach applied to functional polymers concerning their morphology, selectivity, and role in the water treatment processes, it is essential to summarize and highlight all the recent research efforts aimed at developing efficient functional polymers dedicated to pollutant removal. This review provides a detailed presentation of functional polymers, encompassing their distinctive features, microstructures, fabrication techniques, and water treatment properties. This thorough review aims to provide readers with a comprehensive overview of the advancements in functional polymer-based water purification methods to date. Additionally, this review sheds light on prevalent challenges associated with the performance of these polymers, offering a roadmap to address these concerns. The functional polymers under scrutiny are categorized as water-soluble and insoluble, contingent on their repeating units and chemical functional groups. Both types of polymers have found application in various extraction methods, including liquid–liquid extraction, adsorption, and precipitation, based on the distribution of the two phases [32,33]. These methodologies are directly related to the morphology, functionality, and dimensions of the polymeric phase in a membrane [32,33]. Numerous functional polymers have been specifically designed and fabricated for use in membrane preparation. As shown in Figure 1, as a membrane material the functional polymer is an essential component of a water purifier device. Its morphology also plays a decisive role in selective filtration or removal of certain pollutants.

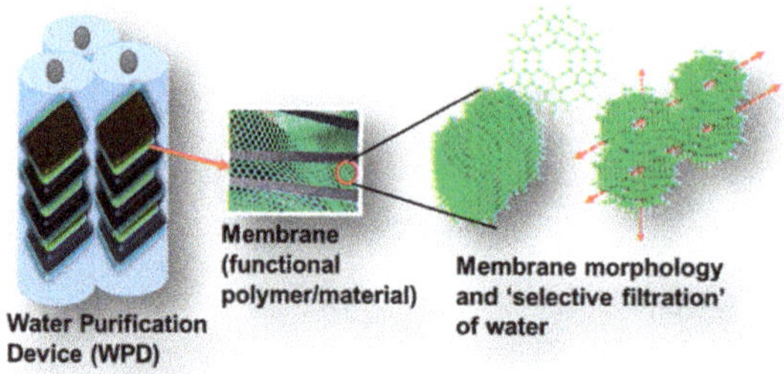

Figure 1. A membrane purifier comprising a functional polymer or material fabricated as a membrane and integrated within a device. The microscopic structure demonstrates the pivotal role of "specific" chemical functionality in governing the water filtration process, selectivity, and flux.

A well-controlled and directed morphology further dictates the formation of nanopores within the membrane. The size and shape of these nanopores are crucial for the pathway of water flux. Figure 2 provides a detailed schematic overview showcasing a few polymers, the membrane casting in a device, and the resultant membrane morphology. Common polymers, including polyacrylonitrile (PAN), polyvinylidene fluoride (PVDF), polyethersulfone (PES), polysulfone (PSF), poly(ether ether ketone) (PEEK), poly(amide) (PA), polyurethane (PU), and poly(styrene), along with their derivatives, such as copolymers, have been extensively utilized for membrane filtration [34–48]. Crucial considerations when casting or processing membranes in a device involves evaluating the chemical structure, morphology, mechanical, and thermal properties of these polymers [43,44,46,47]. These parameters significantly impact the membrane's performance in the device. Membranes produced from these synthetic polymers are categorized into the following groups: (1) isotropic or symmetric and (2) anisotropic or asymmetric. This categorization depends on factors like nonporosity, as well as the presence of charged or uncharged polymeric chains (as fabricated through solvent-based or other techniques) [49,50].

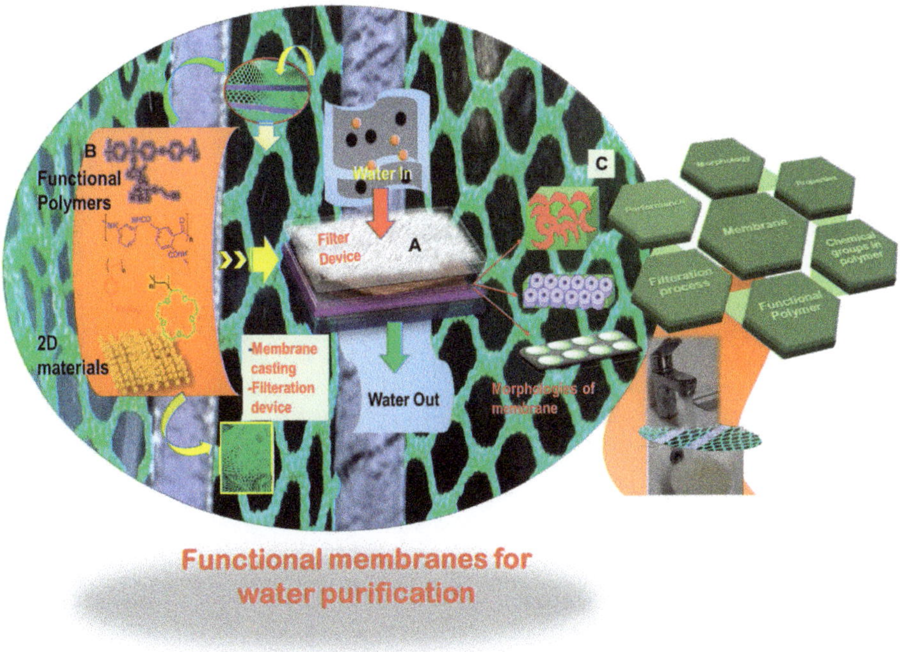

Figure 2. (**A**) A water filter device composed of different components, such as a separating layer, substrate, and a polymeric membrane cast from a relevant functional polymer; (**B**) chemical structures of a few polymers, including their two-dimensional (2D) materials counterparts, highlighting active functional groups crucial for selectivity and filtration, as well as a specific nanostructure suitable for an effective membrane and its processability; and (**C**) various morphologies of polymeric membranes showcasing diverse features, including nanopores, three-dimensional (3D) interconnected channels and homogenous distribution of uniform nanopores, and structure–properties correlation.

In isotropic or symmetric membranes, the porosity and pore size regulate the filtration process, that is, the microfiltration (MF) of molecules or particles with varying sizes or dimensions. Alternatively, nonporous yet dense isotropic membranes have undergone testing for their performance. These membranes allow water to pass through while retaining other "species" or pollutants depending on their solubility and diffusivity. This phenomenon is observed in gas separation, pervaporation, and reverse osmosis (RO) processes. In isotropically charged membranes, such as those used in electrodialysis, the separation process occurs via ionic interactions (regardless of porosity) between the polymeric network's ionic groups in the membrane and the positively/negatively charged molecules of the impurities [51]. Conversely, anisotropic (asymmetric) membranes exhibit structural or chemical heterogeneity in their networks and are classified into two subgroups: (i) Loeb and Sourirajan membranes and (ii) Thin-film composite (TFC) membranes [52–55]. Loeb and Sourirajan membranes are fabricated using polymer-forming layers with indistinguishable pore sizes and porosities. In contrast, TFC membranes are prepared employing two different polymers, creating a dense surface layer responsible for separation and a thick, porous layer for providing mechanical support [52–55]. Several techniques are employed for fabricating polymer-based membranes, serving both research and industrial purposes. Notable methods include interfacial polymerization (IP), sputtering, solution casting, extruding, melt pressing, phase inversion, and electrospinning, all commonly applied for producing polymeric membranes [56–62]. Figure 3 illustrates a graphical depiction of some of these fabrication techniques. The choice of fabrication process largely depends on the physical and chemical properties of the polymer selected for membrane production. Each of these fabrication techniques possess distinct advantages and disadvantages and may require optimization through factors such as choice of solvent (polar/nonpolar), temperature variation, substrate selection, and the utilization of electrostatic forces. In the phase inversion method, the solid phase of the polymer is converted to either a solution or molten state by adjusting specific parameters, as previously mentioned. For instance, this involves selecting an appropriate solvent and controlling the melting temperature. This technique is further subdivided into immersion precipitation and controlled evaporation precipitation (Figure 3A) [63]. Commercially available membranes are typically fabricated using the immersion precipitation technique. This process involves casting a polymeric solution (containing some additives) onto a suitable substrate, normally glass. Subsequently, the supported polymer is immersed in a nonsolvent agent like pure deionized (DI) water. The immersion-precipitation method is known to yield membranes with higher tensile strength, porosity, and improved hydrophilicity. Additionally, this process results in small-diameter pores that help enhance the thermal shock resistance and volumetric power density of the membrane. In controlled evaporation precipitation, a membrane is cast by depositing a polymeric solution onto a porous substrate and subsequently evaporating the solvent under controlled speed and temperature. By carefully regulating the solvent evaporation from the polymeric solution, a denser, thicker membrane with paper-like characteristics can be obtained. Furthermore, in a slightly modified protocol, when the temperature is significantly increased, a fibrous membrane with an interconnected polymer network can be cast from the solution [64,65]. This process, known as thermally induced phase separation (TIPS), offers numerous advantages, including high porosity, simple processing, easy reproducibility, and a reduced occurrence of defects [65]. In this unique method, IP involves the polymerization of a selected monomer at the interface of water and a suitable solvent, utilizing a porous support as the reaction reservoir. This method leads to the formation of an ultrathin membrane that is highly suitable for ultrafiltration (UF) and RO applications (Figure 3B). Through IP, desirable properties such as high performance, ultrahigh permeability, and a high level of membrane rejection can be introduced during the fabrication process. Additionally, to exclude the use of organic solvents and make fabrication environmentally friendly, the extrusion process has gained widespread use, in which a polymer is subjected to stretching under cold and hot conditions (Figure 3C). In this process, the polymer is heated to its melting point and then extruded through specially designed piston–cylinder equipment, producing polymer aligned fibers when used with a substrate. As these fibers align in a particular direction, the membrane produced with this method displays good mechanical properties, controllable pore size, and thickness.

In the advanced method known as electrospinning, polymeric fibers are produced under an electric field, leading to the evaporation of the solvent and the fabrication of a membrane or film with a considerably large surface area (Figure 3D). In this process, a viscous solution of the polymer is placed in the injector. Furthermore, a strong electric field (DC voltage) is applied across the injector [66]. Once the electrostatic repulsion surpasses the surface tension, the viscous solution flows from the needle as fibers, which are then collected on a designated collector. Through electrospinning, polymeric fibers with multiple arrangements and morphological structures can be created [66]. Table 1 summarizes the aforementioned fabrication techniques, outlining the products formed via each relevant technique, along with their specific properties and applications. This review focuses solely on recent examples of optimized membranes with enhanced performance to elucidate their significant milestones and gain deeper insights. The subsequent section delves into a comprehensive discussion of these membranes, highlighting their distinctive features and examining their perspectives. Earlier, a class of hyperbranched macromolecules known as dendrimers and other polymers have also been explored as membrane materials for similar applications. To describe a few, poly(amidoamine) terminated with sodium carboxylate groups (PAMAM-COONa) was studied for its forward osmosis (FO) to draw solute from the sea water with a relatively high water flux of 9 L m^{-2} h^{-1} [67]. In another report, thermoresponsive magnetic nanoparticles (MNPs) incorporated with a copolymer poly(sodium styrene-4-sulfonate)-*co*-poly(*N*-isopropylacrylamide) (PSSS-PNIPAM) have been shown to draw solute and to extract water from brackish water or seawater via forward osmosis (FO) [68]. Polyacrylic acid and its salt derivatives were also proposed previously for similar applications related to water treatment [69]. In this current review, the major focus is on the most recent state-of-the-art fabrication, developments, and performance of the membrane materials, to provide a comprehensive overview so that a further roadmap can be drawn, if necessary.

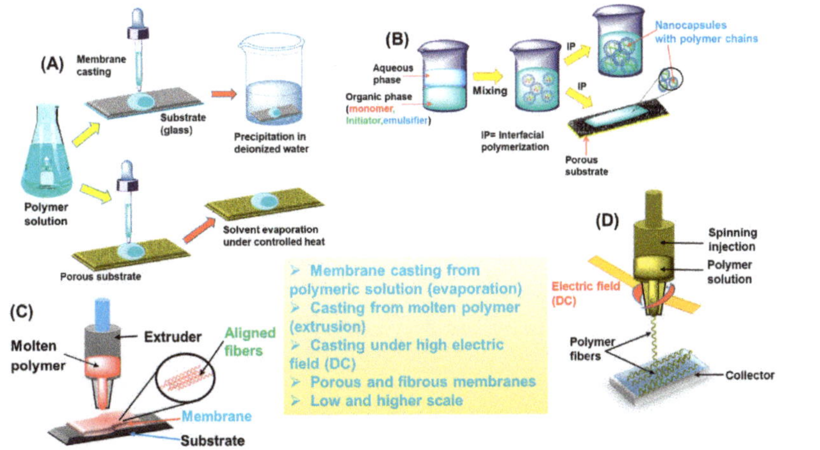

Figure 3. Various techniques for fabricating membranes from polymeric precursors: (**A**) casting a membrane from a polymeric solution on a substrate and evaporating the solvent either via precipitation in a nonsolvent (DI-water) or under controlled heat; (**B**) membrane formation via IP of a monomer at the interface of organic–aqueous phase, resulting in nanocapsules with aligned polymer chains on a porous substrate; (**C**) solvent-free extrusion of a molten polymer, generating aligned fibers in a membrane on a substrate; and (**D**) formation of polymeric fibers and a membrane via electrospinning, achieved by injecting a viscous polymeric solution through needles under electrodynamics or high DC voltage.

Table 1. Various functional materials with tuned properties used for fabricating their respective membrane by using various techniques and their major applications.

S. No.	Material(s)	Membrane Technique	Parameters	Membrane Characteristics	Applications
1.	Poly(arylene sulfide sulfone) (PASS)	Electrospinning	Conc.=0.27 g·mL^{-1}, Voltage = 20 kV, and Speed = 300 rpm	Smallest pore size, highest mechanical property, and best surface wettability	MF, water flux of 747.76 L·m^{-2}·h^{-1} and high separation efficiency of 99.9% to 0.2 μm particles [70]
2.	PES	Phase inversion (solvent evaporation)	9 wt.% in dimethylacetamide (DMAc), ZnO-nanoparticles, and Temperature >100 °C	Asymmetry, pores 20–100 nm, and controlled morphology	UF, water flux of over 5600 L m^{-2} h^{-1}, and high efficiency [71,72]

Table 1. *Cont.*

S. No.	Material(s)	Membrane Technique	Parameters	Membrane Characteristics	Applications
3.	PVDF/HEMA	Electrospinning	Voltage = 30 kV and Speed = 150 rpm	Nanofibrous and hydrophilic surface with improved flux	MF (antifouling), water flux 63 L m^{-2} h^{-1}, and with a separation efficiency of 98% [73]
4.	PEEK/polyetherimide (PEI)	Extrusion (melt spinning)	40 wt.% PEEK with 60 wt.% PEI, at 360 °C	Hollow fiber membrane, hydrophilic, and recyclable	UF (industrial wastewater and antifouling), with a water flux of 216% [74]
5.	PVDF-bismuth oxychloride (BiOCl)	Phase inversion (blending)	(1:7.5) PVP/PVDF in NMP/EG at RT	Super hydrophilic, Cr (VI) adsorption, and with multifunctional features	UF (antifouling capability), with a water flux of 854 L m^{-2} h^{-1} bar^{-1} [75]
6.	Metal–organic framework (MOF)/ PA	IP	Dispersion of ZIF-L, monomer (EDA/TMC)	TFC (ultrathin)	Pervaporation dehydration [76]
7.	PA (PES support layer)	IP	Fluorinated amine/trimesoyl chloride with SDS as emulsifier	TFC (low surface free energy)	Nanofiltration (NF), wastewater treatment, and with antifouling [77,78]
8.	Polypropylene/PE	Stretching	0.5–3 bar pressure uniaxial/biaxial direction, and melting temperature	Good mechanical properties, thickness controlled, and large scale production	Uranium extraction from sea water [79]
9.	Graphene-PE terephthalate	Track etching	Irradiation/ion beams and acid or alkaline solution	Narrow pore size distribution, high porosity, and low cost	Bioseparation [80]
10.	nano silica-functionalized polydimethylsiloxane (PDMS) ink	3D printing	Homemade 3D printer, micronozzle diameter 150 µm, and curing at 120 °C	High precision, better resolution, low cost, and excellent control over thickness and porosity	Oil–water separation [81]

PA: polyamide; PVDF: polyvinylidene fluoride; UF: ultrafiltration; PVP: polyvinylpyrrolidone; TFC: thin-film composite; IP: interfacial polymerization; EG: ethylene glycol; MF: microfiltration; NMP: 1-methyl-2-pyrrolidinone; RT: room temperature; EDA: ethylenediamine; TMC: trimesoyl chloride; PES: polyethersulfone; SDS: sodium dodecyl sulfate; and PE: polyethylene.

2. Membranes Based on PA and Its Derivatives

In 1930, at DuPont Wallace, Carothers synthesized a PA, commonly known as nylon [82]. As engineering plastics, PAs are usually synthesized via step-growth polymerization and exhibit high durability and excellent thermal, barrier, and mechanical strengths owing to their rigid backbone. They are used in textiles, the automotive industry, carpets, kitchen utensils, and sportswear.

Figure 4 illustrates the typical structure of an aromatic PA, emphasizing its high-performance attributes and the presence of dense layers in a film or membrane.

Synthetic aromatic PAs are preferred over their aliphatic counterparts because the latter exhibits increased moisture susceptibility, resulting in mechanical weakness and dimensional deformities. Notably, aromatic PAs such as polyphthalamides (PPAs) are processable in the molten state and have accordingly attracted considerable commercial attention. Aromatic PAs have found significant utility exploited as membrane materials owing to their superior hydrolytic and permselective properties. Despite PAs being the gold standard in both academic and industrial efforts for membrane preparation, achieving precise tuning of the PA layer remains a challenging task (Figure 4). For example, the permeability and selectivity of PA-based membranes are directly related to their morphology. To produce the desired morphology wherein PA chains can form dense yet thin layers with a larger surface area (via nanopores), optimization of various parameters becomes crucial. These include the chemical structure of the monomer, fabrication processes, and the structural properties of the porous support. Notably, thin and dense layers are employed to improve permeation within these membranes. Previously, several strategies were introduced for fabricating dense and thin layers of PA chains, including layer-by-layer (LBL) assembly, the addition of interlayers or hydrophilic materials, and 3D printing [83,84]. Studies suggest that aromatic PAs can exhibit porosities ranging from approximately 15 ± 2% to 32 ± 4%. However, increasing the specific surface area via these nanopores enables high-mass crossflow and permeation. For precise nanopore creation, IP using emulsion and template methods is considered highly effective. In IP, a hydrophilic monomer is incorporated and polymerized at the interface, generating a high specific surface area in the resulting polymeric film. Furthermore, the presence of a hydrophilic additive or polymer affects solution viscosity and diffusion rate, directing the formation of a nanoporous morphology in the film [85]. Nevertheless, attempts are ongoing to fabricate defect-free PA membranes, featuring dense layers and ordered nanopores. The pursuit of a perfect membrane aims to establish an optimized water transport pathway, wherein the limits of permeability and selectivity would surpass those of previously reported membranes.

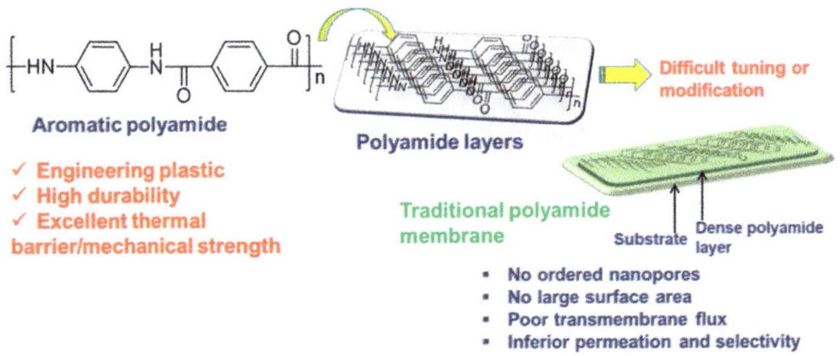

Figure 4. Chemical structure of a representative aromatic PA, categorized as a high-performance or engineering plastic due to its high durability and excellent thermal and mechanical properties. The layers of PA chains in a film are dense and inert to further modifications, necessitating advanced and optimized methodologies for the same.

Recently, a remarkable approach for fabricating asymmetric PA films with well-ordered nanopores and dense layers was reported. This method involves the successive utilization of a PA-based dendrimer and IP of two different monomers to create an efficient double-layered membrane for NF [86]. Figure 5 illustrates a schematic representation of this innovative approach. In the study, PA dendrimers of different generations were employed to create the lower porous layer of a membrane. Since PA dendrimers are hyperbranched and possess a nearly spherical structure, they display peculiar features such as well-defined functional periphery and uniform intramolecular voids or cavities. For this purpose, PA dendrimers with 32-amine terminal groups at the periphery were synthesized and subsequently treated in a salt solution (pH = 1) on the surface of a PSF support. This treatment facilitated their coupling with each other via a diazotization-coupling reaction involving a sodium nitrite solution, culminating in a covalent assembly (Figure 5). Consequently, the surface of the support becomes hydrophilic with minimal defects. In the IP step, these surface-supported covalent assemblies of the PA dendrimer are placed in a solution of piperazine (PIP) for 10 min, followed by immersion in a TMC/n-hexane solution (0.15 *w*/*v*%) for another 30 or 60 s to create a traditional or asymmetric PA dense film. Interestingly, the polar and nanosized cavities of the dendrimers assist the diffusion of PIP molecules, consistently allowing the IP process to smoothly create a uniform film of PA with ordered nanovoids or nanopores [86].

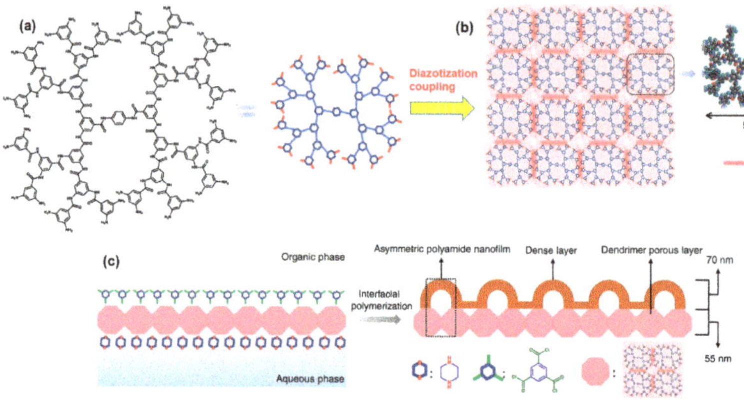

Figure 5. Strategy for creating a double-layered membrane comprising the following: (**a**) hyper-branched–amine-terminated dendrimer which, upon further coupling via diazotization, forms (**b**) a porous layer on a support, followed by (**c**) IP of two different monomers, namely PIP and TMC, conducted on the porous layer. This process aims to produce a dense and thin asymmetric successive layer of PA with an optimized transport pathway and ordered nanovoids. Reprinted with permission from Nature Comm. [86], Copyright 2020 Nature Portfolio. The article is licensed under a Creative Commons Attribution 4.0 International License, which permits use, sharing, adaptation, distribution, and reproduction in any medium or format, as long as you give appropriate credit to the original author(s) and the source, provide a link to the Creative Commons license, and indicate if changes were made.

Morphological investigations of these double-layered membranes using transmission electron microscopy (TEM) clearly delineated two different layers corresponding to the covalent assembly of dendrimers and asymmetric PAs produced after IP. Figure 6 displays the TEM micrographs showcasing the single-layer PA nanofilms, revealing PA aggregations throughout. However, following the IP process, the structural features of the membrane transformed significantly, characterized by the appearance of uniform nanovoids. At high resolution, an array of interconnected, ordered nanovoids (width of 70–120 nm) was also observed. Further analysis with atomic force microscopy (AFM) and X-ray techniques confirmed the formation of an asymmetric PA membrane, comprising a porous layer of spherical dendrimers and a dense PA layer. Such features are targeted to reduce the transmembrane resistance of water and increase the permeation flux. With regard to the separation performance of these membranes, studies have demonstrated a 3.7- to 4.3-fold higher water flux for various salt solutions (2000 ppm) compared to traditional PA membranes. To further examine the superiority of these double-layered membranes, their permeability and selectivity were tested and compared with the data estimated for different types of PA membranes. These included commercial PA membranes, crumpled PA membranes (IP with crumpled structures), PA membranes prepared using interlayers, traditional IP membranes, and mixed-matrix PA membranes (Figure 6).

The experiments revealed that the permeability (7.9×10^{-6} cm^2 s^{-1}) and selectivity (23,877.2) of the membrane for magnesium sulfate or $MgSO_4$ exceeded those of the afore-mentioned membranes. A similarly improved performance (permeability 7.5×10^{-6} cm^2 s^{-1} and selectivity 27,561.3) was also observed for the sodium sulfate or Na_2SO_4 solution. As depicted in Figure 6e, the high performance of the membrane can be attributed to the presence of a porous dendrimer layer and a thinner PA layer on the top (Figure 6e). This configuration provides a uniform transformation pathway, facilitating water flux while improving selectivity and salt permeation. In addition, the larger surface area of the nanovoids allows water to permeate efficiently, with increased salt rejection rates (Figure 6e) [86].

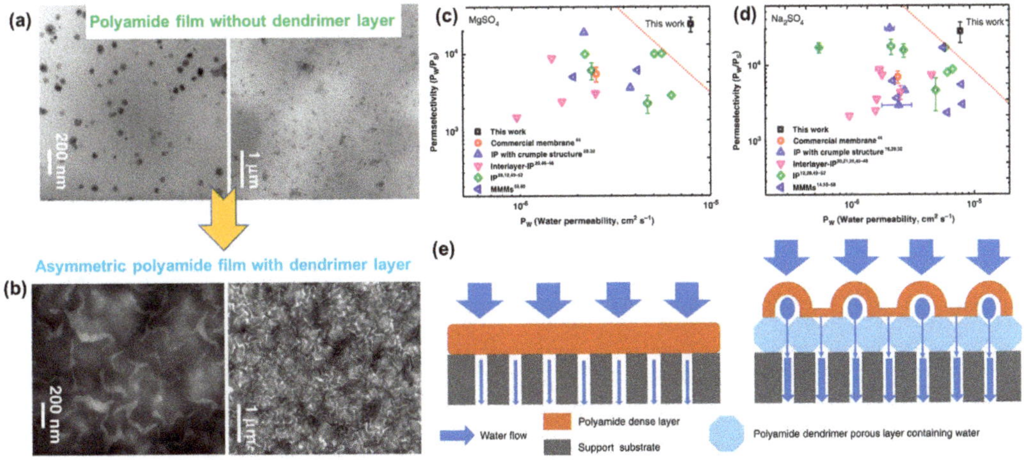

Figure 6. Morphological features of an asymmetric PA membrane: (**a**) TEM micrographs without the dendrimer; (**b**) with the covalent assemblies of dendrimer; (**c**,**d**) permeation and selectivity of an asymmetric membrane produced via IP on dendrimer and other available membranes reported in the literature, including commercial ones; and (**e**) schematic representation illustrating the transportation of water (salt water) across the membrane without and with a dendrimer porous layer, as fabricated via diazotization prior to IP. Reprinted with permission from Nature Comm. [86], Copyright 2020 Nature Portfolio. The article is licensed under a Creative Commons Attribution 4.0 International License, which permits use, sharing, adaptation, distribution, and reproduction in any medium or format, as long as you give appropriate credit to the original author(s) and the source, provide a link to the Creative Commons license, and indicate if changes were made.

In a similar approach, researchers utilized graphitic carbon nitride (g-C_3N_4) to construct a double-layered PA membrane through IP, aiming to achieve ultrapermeability and excellent selectivity for water purification [87]. Here, g-C_3N_4 functions similarly to the dendrimer mentioned earlier, influencing the IP process by reducing the monomer diffusion at the interface. This reduction results in the formation of a nanoscale–hollow-cone structure within the fabricated membrane.

Figure 7 illustrates a schematic representation of this process. To support PES, a suspension of g-C_3N_4 and PIP in an aqueous solution is deposited via dip-coating. After drying, the support coated with g-C_3N_4 and PIP is immersed in a solution of TMC in n-hexane, with IP performed by heating the entire assembly to fabricate a porous and ultrapermeable membrane. The surface morphological studies using SEM reveal an ordered structure in the PA membrane post IP (Figure 7c,d). The tilted nanosheets of g-C_3N_4 and their interlayer spaces noticeably influence the diffusion of the reacting molecules, resulting in a uniform structure with hollow channels within the membrane (Figure 7d). The ultrapermeability and separation selectivities of the membranes have also been investigated, revealing a superior water permeance of 105 L m^{-2} h^{-1} bar^{-1} and a high selectivity of 130 for Cl^- over SO_4^{2-} (Figure 7e). These outcomes offer promising opportunities for rapid and precise separation of pollutants from water [87]. Further advancements in PA-based membranes include the use of a fluorinated amine monomer ($CF_3(CF_2)_6CONH(CH_2CH_2NH)_2CH_2CH_2NH_2$) to fabricate membranes via IP with TMC for NF [88]. The perfluoroalkyl groups in the active layer of the PA are responsible for the decreased surface energy. Consequently, the membrane exhibits superior self-cleaning and antifouling properties. These membranes have shown potential suitability for wastewater treatment [89,90].

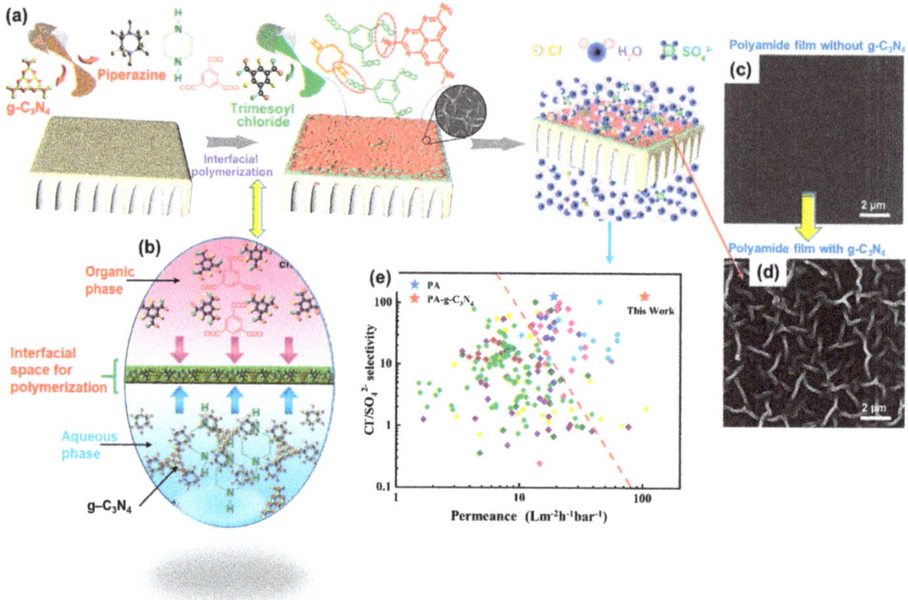

Figure 7. Fabrication process of a membrane utilizing a layer of porous g–C_3N_4: (**a**) Deposited on a support in order to conduct IP of two monomers, namely PIP and TMC, leading to polycondensation within the interfacial space. (**b**) Representation of TMC molecules dissolved in the organic phase, with the PIP (remaining in the aqueous phase) adsorbed on the g–C_3N_4 surface, resulting in a slow diffusion of PIP molecules through the pores. (**c**) Scanning electron microscopy (SEM) image of a tradition PA film without g–C_3N_4 and no porous morphology. (**d**) SEM image of a porous and double-layered PA membrane after IP, displaying a superstructure with ordered nanovoids. (**e**) permeance and selectivity of the membrane. Reprinted with permission from Nature Comm. [87], Copyright 2023 Nature Portfolio.

In pursuit of a superior membrane based on PAs, a recent report introduced a scalable and versatile fabrication method that combines phase inversion and IP methods for large-scale membrane production [91]. Figure 8 depicts a schematic representation describing this process for fabricating a membrane with ultrathin selective layers. In the first step, a monomer is dissolved in an aqueous phase together with polymer powder (PES) to prepare a homogenous solution that is cast onto a glass substrate. After drying, the glass substrate coated with the monomer and polymer powders is subjected to phase inversion by soaking in a coagulation bath containing water (Figure 8a). A solidified membrane is formed as a result of the solvent/nonsolvent exchange process. A solution of TMC in n-hexane is poured onto the membrane surface to initiate the IP process, resulting in the formation of a TFC membrane (Figure 8a). For comparison, in a separate study, the PIP monomer was replaced with cyclodextrin, which is also soluble in the aqueous phase. The chemical structures of the PIP and cyclodextrin membranes are shown in Figure 8b. The filtration performance tests of these membranes showcase superior water permeability and salt rejection (Figure 8c). The membranes exhibiting excellent performance surpass conventional membranes fabricated exclusively via a single method [91].

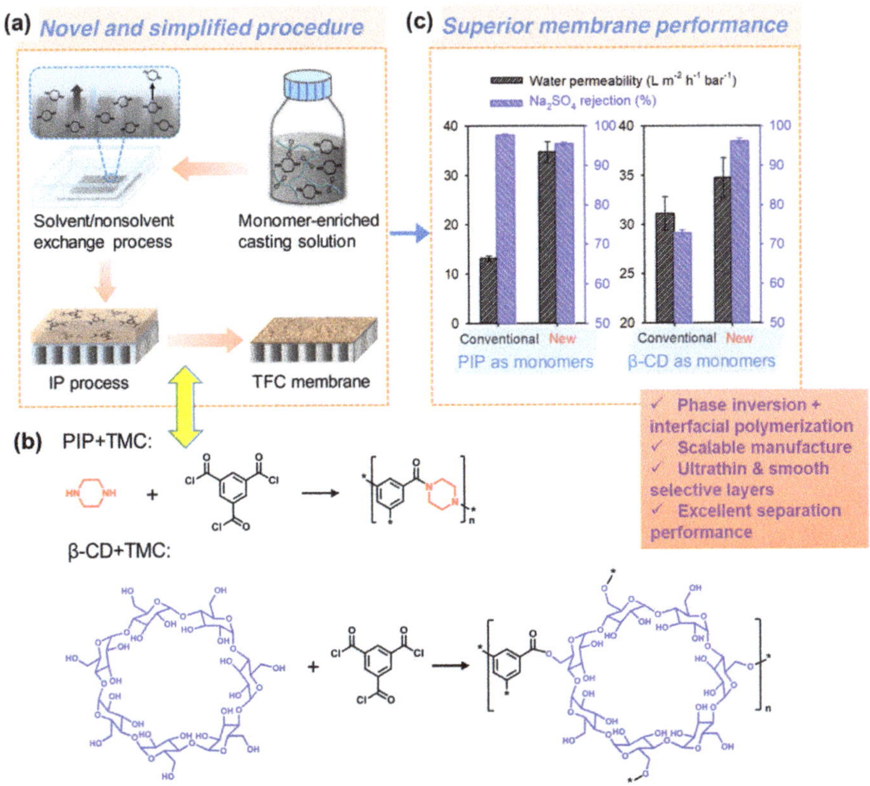

Figure 8. (**a**) Schematic representation illustrating the scalable manufacture of TFC membranes by combining state–of–the–art phase inversion and IP in which (**b**) PIP and β–CD are used as aqueous monomers for conducting IP with a TMC monomer dissolved in the organic phase. (**c**) Performance evaluation of the membrane concerning water permeability and salt rejection. Reprinted with permission from Environ. Sci. Technol. [91], Copyright 2020 American Chemical Society.

PA has been the material of choice for fabricating membranes, particularly when employing aromatic amine monomers like m-phenylenediamine (MPD) for initiating IP on diverse surfaces such as PSF and nylon, by integrating them with a 2D material known as MXene. MXene is known for its large surface area, abundant nanochannels, hydrophilicity, and robust mechanical properties. This led to the development of MXene ($Ti_3C_2T_x$)–PA membranes fabricated via IP. These membranes exhibit enhanced water flux, antifouling, and chlorine resistance, showcasing benefits for water desalination. These TFC membranes have found application in forward osmosis, water purification, and organic solvent recovery [92,93].

3. Membrane Based on Polyvinylidene Fluoride (PVDF) and Its Derivatives

Following the extensive utilization of PAs, PVDF (–$(CH_2CF_2)_n$–) has emerged as another favored polymer for fabricating membranes due to its impressive mechanical strength, chemical resistance, and thermal stability [94]. Its solubility in various common solvents and easy processability allow the manufacture of flat sheets, hollow fibers, or tubular membranes suitable for MF or UF.

As an advancement, a current modified-phase-inversion fabrication of a PVDF-based membrane is shown in Figure 9. Following this concept, a blend of PVDF, KCl, and $Bi(NO_3)_3$ in EG and NMP is coated on a glass substrate (Figure 9a) [75,95,96]. Before and after the addition of PVDF, the blend appears white and opaque (Figure 9a1–a4). The formation of a complex between Bi^{3+} and EG, and its further interactions with the PVDF backbone may have contributed to this color change. Furthermore, the coated substrate is immersed in DI water to initiate the phase inversion process. This process induces phase separation, resulting in the precipitation of BiOCl and the formation of a superhydrophilic membrane (Figure 9b). Performance tests conducted on these membranes suggest their successful use in water filtration, particularly with substances like HA and bovine serum albumin (BSA), further confirming their roles as separators and antifouling agents (Figure 9c,d, respectively).

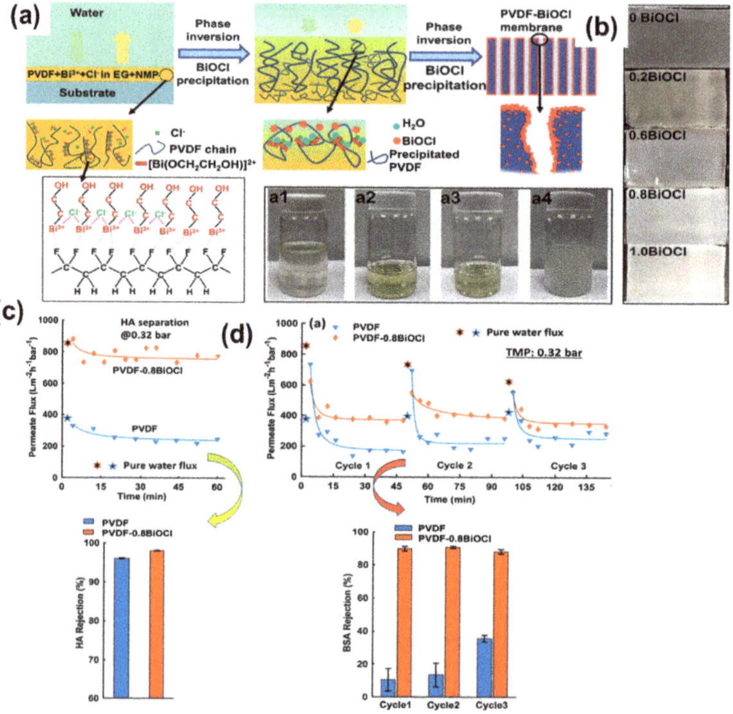

Figure 9. (**a**) Schematic representation illustrating the modified-blending-phase inversion and its mechanism for fabricating PVDF–BiOCl membranes: (a1) Bi^{3+} + Cl^- + EG + NMP, (a2) EG + NMP + PVP + PVDF, (a3) Bi^{3+} + EG + NMP + PVP + PVDF, and (a4) Bi^{3+}+Cl^- + EG + NMP + PVP + PVDF represent various solutions utilized in the studies. (**b**) Images of the membranes with varying BiOCl content formed from the reaction of ethylene glycol, potassium chloride (KCl), and bismuth(III) nitrate ($Bi(NO_3)_3$). (**c**) Changes in permeate flux and rejection ratios of the PVDF and BiOCl-0.8BiOCl membranes using 0.1 g/L humic acid (HA) solutions. (**d**) Permeation flux and rejection for BSA illustrating antifouling properties. Reprinted with permission from Separation and Purification Technology [75], Copyright 2021 Elsevier.

Further progress towards the development of PVDF membranes has explored more innovative and unique approaches, including the utilization of thermosalient (TS) crystal monomers such as 1,2,4,5-tetrabromobenzene (TBB). This compound is recognized as an energy-efficient–dynamic crystalline material due to its phase transition occurring above room temperature (Figure 10a) [97,98]. These dynamic functional crystals can activate the membrane via mechanical or heat responses, thus increasing the mass transfer. To fabricate these smart, responsive membranes, a layer of PVDF was integrated with a polyvinyl alcohol (PVA) hydrogel (Figure 10a). The fabrication process involves incorporating dynamic crystals of TBB into an aqueous solution containing PVA and glutaraldehyde (GA), resulting in a suspension. Treatment of this suspension with hydrochloric acid or HCl leads to its casting on the porous surface of the PVDF film. Subsequent polymerization between TBB and GA was performed to obtain a smart PVDF–PVA membrane with dispersed TBB crystals (Figure 10a,b). Experiments conducted to evaluate the transmembrane flux of these membranes showcase improved and superior performance compared with the undoped PVDF–PVA membrane (Figure 10c). Even after the fifth operational cycle, the average flux of TBB-doped membranes remains consistently higher, surpassing 160% (Figure 10c). Notably, the phase transition of TBB crystals likely impacts the transport properties upon heating and cooling. Applications of these membranes include distillation and antifouling. Additionally, with regard to hybrid PVDF membranes, another example involves the fabrication of polyaniline (PANI) layers within a PVDF matrix to optimize the hydrophilicity, porosity, antifouling properties, solvent content, and water flux. Figure 10d illustrates a schematic for manufacturing a PVDF–PANI hybrid membrane and its performance characteristics. These membranes have proven effective in the removal of harmful dyes from textile wastewater [99].

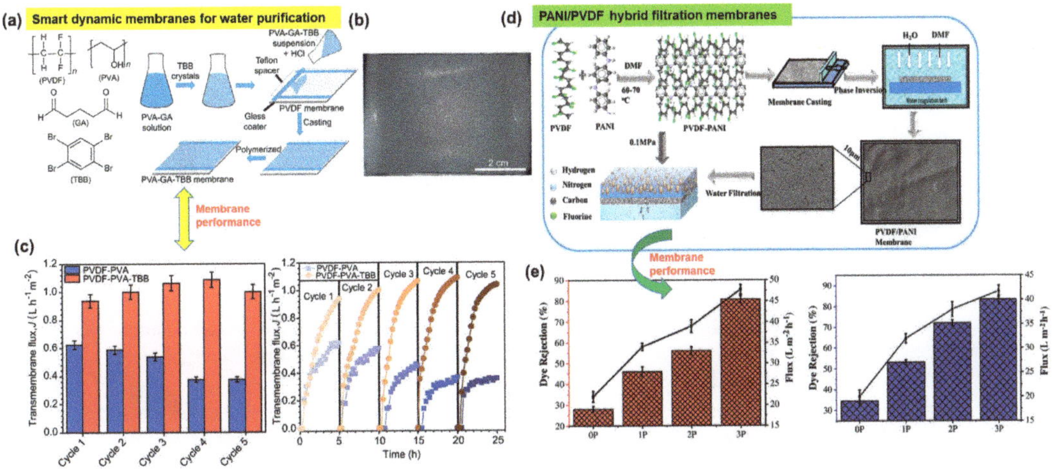

Figure 10. Hybrid and smart membranes based on PVDF. (**a**) PVDF film integrated with a different polymer and dynamic crystalline materials that are capable of sudden expansion or motion under thermal stimulation. (**b**) Microscopic image of the membrane. (**c**) Membrane water flux and durability. (**d**) Hybrid membranes of PANI and PVDF exhibiting (**e**) high dye rejection performance. Reprinted with permission from Nature Commun. [97], Copyright 2023 Nature Portfolio, and Ad. Eng. Mate. [99], Copyright 2022 Wiley-VCH GmbH, respectively. The Nature Communication article is licensed under a Creative Commons Attribution 4.0 International License, which permits use, sharing, adaptation, distribution, and reproduction in any medium or format, as long as you give appropriate credit to the original author(s) and the source, provide a link to the Creative Commons license, and indicate if changes were made.

In the study, a phase-inversion method is employed to fabricate PVDF membranes with various PANI concentrations (Figure 10d). The addition of PANI (PANI (1–4 wt.%) increases the hydrophilicity of the resulting membrane with reduced pore size, as evidenced by the contact angle and SEM measurements. The operational experimental results demonstrate an enhancement in water flux from 28 to 47 L m^{-2} h, signifying increased hydrophilicity upon addition of PANI to PVDF. Furthermore, the performance tests exhibit high rejection of dye molecules (allura red and methyl orange) (Figure 10e).

Graphene oxide (GO), a widely recognized 2D nanomaterial, has been extensively utilized in various research studies, including its integration with PVDF-based membranes. This integration is due to GO's advantageous attributes, such as large surface area, chemical inertness, and robust mechanical properties [100,101]. The functional moieties in GO enable surface modification in membranes, essential for simultaneous removing and recycling of industrial organic dyes. To achieve these functional features in a membrane, a simple and straightforward electrospinning approach was employed. This technique facilitated the fabrication of versatile porous structured PVDF–graphene oxide (PVDF–GO) nanofibrous membranes (NFMs) (Figure 11a). The scheme in Figure 11a illustrates the electrospinning of the solution of PVDF and GO that yielded the nanofibers, which were then collected on a rotating collector [102].

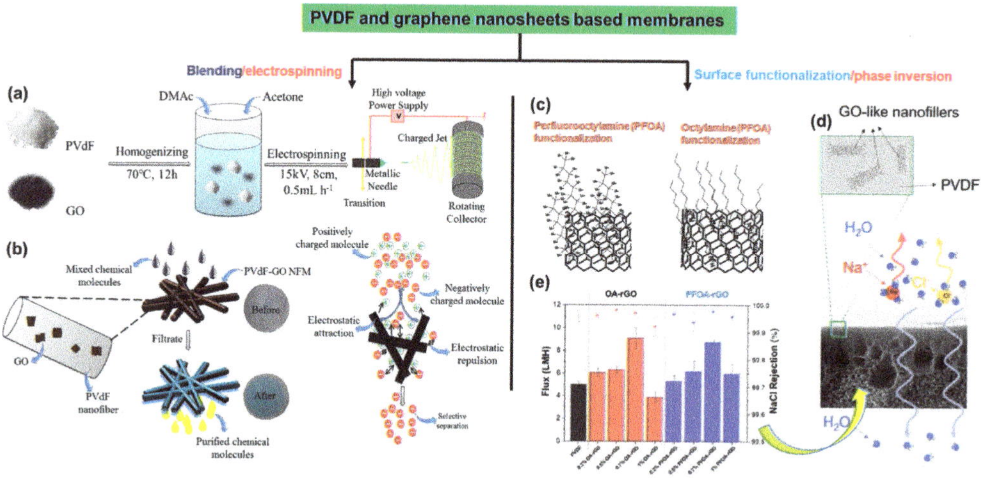

Figure 11. Illustration of two different strategies wherein (**a**) the nanosheets of GO in combination with PVDF were electrospun to generate (**b**) the membranes with knots and with preferential affinity towards positively charged dye molecules. (**c**) In another method, functionalization of graphene nanosheets was implemented to control the degree of hydrophobicity, and subsequently a membrane with (**d**) a PVDF solution is cast via a phase inversion step. (**e**) The performance of the hydrophobic PVDF–GO membrane. Reprinted with permission from Environ. Sci. Technol [102], Copyright 2018 American Chemical Society, and Journal of Environmental Chemical Engineering [103], Copyright 2023 Elsevier, respectively. The article is open access, distributed under the terms of the Creative Commons CC-BY license, which permits unrestricted use, distribution, and reproduction in any medium, provided the original work is properly cited.

In these membranes with porous and uniform geometries, the 3D pores are interconnected through triangular junctions, playing a significant role in the selective separation of various organic dyes (Figure 11b). During filtration, the membranes showcase a selective affinity towards positively charged molecules such that negatively charged organic dye molecules could pass through it and allow selective separation, as is needed for industrial water treatment (Figure 11b). These NF membranes exhibit 99% selectivity for positively charged dyes and 100% rejection of negatively charged dyes from mixed solutions [102].

In a similar study focusing on hydrophobicity and its influence on water flux and salt rejection, GO nanosheets were functionalized covalently and subsequently integrated with PVDF via a phase inversion method to fabricate a hydrophobic membrane with higher efficiency (Figure 11c,d) [103]. Two grafting agents, octylamine (OA) and perfluoroctylamine (PFOA), with distinct degrees of hydrophobicity, were anchored onto the GO surface (Figure 11c). As a result, the increased contact angle and porosity in the membrane directly impact its flux and rejection, reaching 8.8 L m^{-2} h^{-1} and 99.9%, respectively (Figure 11e). These membranes demonstrate promising applications in water [103–106].

4. Membrane Based on Poly (PEEK) and Its Derivatives

PEEK, a high-performance polymer, exhibits high performance, excellent mechanical properties, chemical resistance, thermal stability, and self-lubrication properties (Figure 12) [107–109]. Owing to its processability in various solvents, it has also been used in multiple sectors, spanning aerospace, chemical industry, biomedicine, and membrane technology [108–111]. By introducing carbon fibers, carbon nanotubes, and various nanoparticles into its matrix, the properties of PEEK can be further tailored, with the resulting composites finding applications in membrane production (Figure 12).

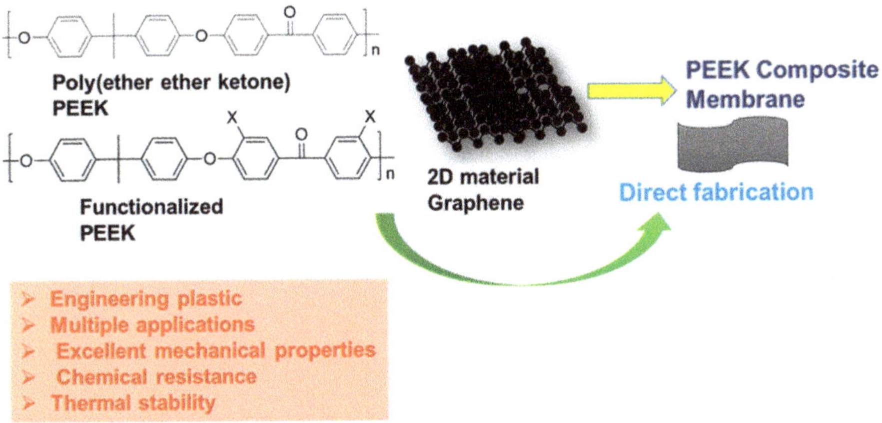

Figure 12. Typical PEEK polymer with its functionalized and composite derivatives that are used for membrane fabrication.

In a recent innovative approach, a PEEK hollow fiber membrane (PHFM) was initially fabricated from the blend of PEEK and PEI via melt spinning, as shown in Figure 13a [112]. Next, the ketone groups located on the surface of the hollow fibrous membrane were reduced to alcohol functionalities to anchor an active unit or initiator to create hydrophilic brushes via a controlled polymerization reaction, namely, atom transfer radical polymerization (ATRP) (Figure 13c,d). The analysis revealed that these hollow fibrous membranes were decorated with a hydrophilic layer of 2-hydroxylethyl acrylate (HEA) chains that directly affected the permeability, separation, and antifouling performance of the membranes [112].

Performance tests for these hollow PEEK membranes indicate a high water flux and BSA rejection ratio (Figure 13e,f). The hydrophilic channels resulting from the HEA chains increase the permeability of the membrane. Moreover, the membrane exhibited enhanced antifouling properties attributed to intermolecular repulsion between the fouling agent and the membrane. These modified PEEK base membranes, fabricated via melt spinning followed by covalent functionalization of the surface, are considered promising candidates for wastewater separation and purification.

In 2023, another novel approach presented a membrane with highly stabilized polymeric interfaces and 2D GO nanosheets, as shown in Figure 14 [113]. Several layers of polyaniline were created via in situ polymerization within the assemblies of PEEK and GO to regulate the pore size of the membrane (Figure 14a). Oxidative polymerization with ammonium persulfate or $(NH_4)_2S_2O_8$ led to the anchoring of polyaniline layers. As a result, a modulated microporous structure was developed within the membrane. Membranes with stabilized interfaces and modulated levels of hydrophobicity exhibited flexibility and high separation efficiency (Figure 14b). A rejection of 85% for RB was estimated (Figure 14e). These PANI@GO/PEEK membranes were durable for up to 100 h with a well-maintained water flux (Figure 14f). The performance of these membranes suggested their potential role in large-scale water purification.

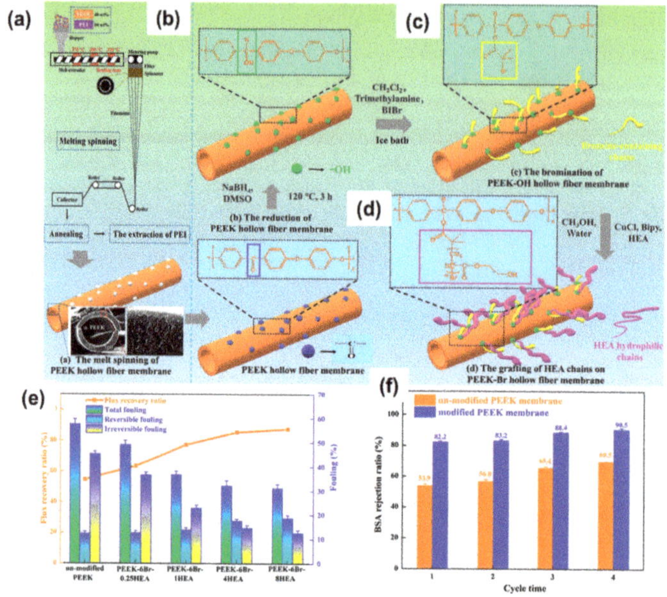

Figure 13. (**a**) Scheme of melt spinning involving a blend of PEEK and PEI to obtain the hollow fibrous membrane (bottom image). This membrane is then subjected to (**b**) reduction to convert the ketone groups into alcohol under sodium borohydride or $NaBH_4$, followed by (**c**) esterification on the surface to afford the ATRP active hollow fibrous membrane, which upon further treatment with a hydrophilic monomer, HEA, and a catalyst produces (**d**) brushes (polymeric chains of HEA) on the membrane surface with improved performance, for example, (**e**) improved water flux and (**f**) rejection ratio for BSA. Reprinted with permission from Chemical Engineering Journal [112], Copyright 2023 Elsevier.

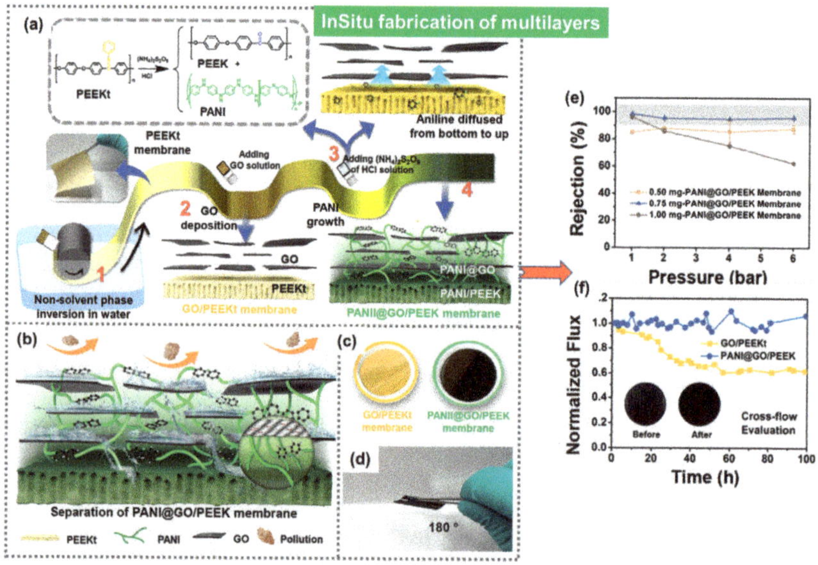

Figure 14. Schematic representation of the fabricating process for the PEEK-based–composite-separating membrane with multilayers made up of (**a**) different chemical components, that is, PANI,

PEEKt, and graphene nanosheets integrated in four successive steps; wherein, (1) PEEKt was fabricated via phase inversion, (2) a layer of GO nanosheets was integrated to PEEKt, and (3) layers of PANI were created in situ accompanied with the reformation of PEEKt into PEEK, leading to the (4) PANI@GO/PEEK membrane with multiple interfaces. (**b**) Schematic diagram illustrating the separation process through multiple interfaces. (**c**,**d**) Depiction of the free standing film of the precursor, GO/PEEKt, and the final product, PANI@GO/PEEK, after hydrolysis and oxidative coupling polymerization of PEEKt. (**e**,**f**) The rejection of Rose Bengal sodium (RB) and crossflow of the PANI@GO/PEEK membranes, respectively. Reprinted with permission from the Adv. Sci. [113], Copyright 2023 Wiley-VCH GmbH. The article is open access, distributed under the terms of the Creative Commons CC BY license, which permits unrestricted use, distribution, and reproduction in any medium, provided the original work is properly cited.

5. Membranes Based on Porous Organic Polymers and Covalent or Metal Organic Frameworks (COFs/MOFs)

A new category of hybrid porous materials that combine advantages of both organic polymers and crystalline MOFs have emerged and been utilized for fabricating membranes with greater efficiency for water treatment. These hybrids have garnered attention due to their well-defined supramolecular architectures derived from fundamental building blocks, such as metal clusters and organic ligands. These materials are being explored for the manufacture of membranes designed for water treatment. In these hybrids of polymers and MOFs, a suitable organic ligand is coordinated with a specific metallic center, producing coordinated supramolecular assemblies with long-range order (Figure 15a). These molecular assemblies can be extended into two or three dimensions depending on their linkages [114].

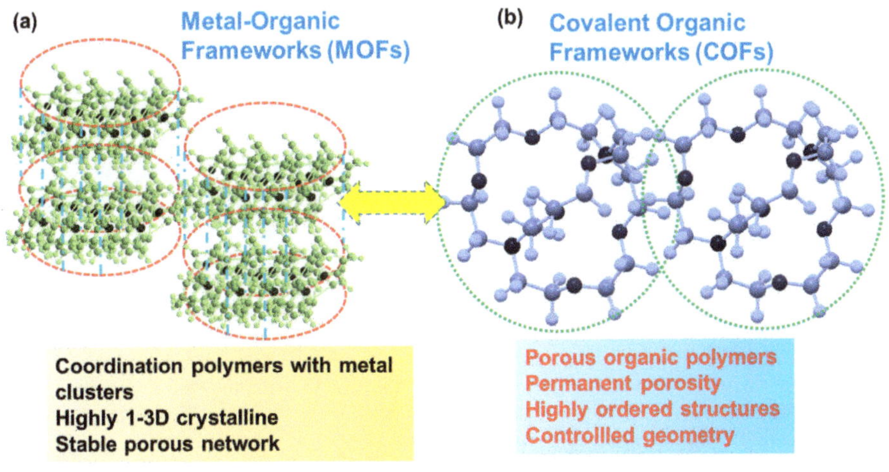

Figure 15. Coordinated and porous organic polymers categorized as (**a**) MOFs and (**b**) COFs, respectively, exhibiting common features such as stable porosity, crystallinity, high order, and controlled geometry.

The key features of MOFs or coordinated macromolecules include ordered open channels, large porosities, predictable pore sizes, and adjustable chemical environments, which offer great opportunities for separation applications. There is a subcategory of polymers that, while lacking metal centers, consist of covalently linked 2D scaffolds with the desired geometries and multiple reactive sites. Correspondingly, it has also attracted considerable attention [115]. These covalently linked 2D scaffolds are known as COFs or porous organic polymers, offering a molecular platform for tailor-made functional materials for membrane development (Figure 15b). Usually, these COFs are crystalline with stable and permanent porosity and are organized in an extended polygon network. These significant features of COFs have allowed researchers to exploit them for membrane manufacturing. Focusing on MOFs, a research group has devised a scalable fabrication method for producing rolls of membranes by using ultra-high-molecular-weight polyethylene, in which MOF particles were interwoven by combining a TIPS and hot pressing (TIPS–HoP) strategy [116]. The combined TIPS–HoP strategy, as shown in Figure 16a, involves several steps: first, blending MOF crystals with high-density polyethylene (HDPE, M_W > 40,000), melt index = 2.2 g per 10 min) and ultra-high-molecular-weight polyethylene (UHMWPE, M_W > 1,500,000) in paraffin at a higher temperature of 200 °C, allowing the HDPE and UHMWPE to melt; second, placing this molten blend on a moving belt while temperature is brought to ~90 °C for a soft solidification; and third, pressing this soft bulk blend via a roll-to-roll hot pressing at 120 °C resulting in the formation of a membrane which was washed in methylene chloride or CH_2Cl_2 to remove the paraffin used in the previous step.

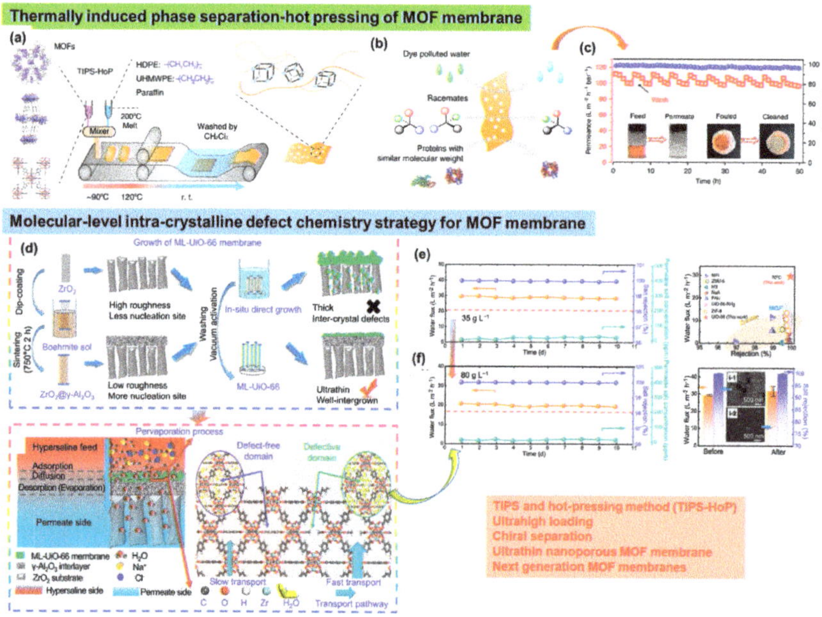

Figure 16. (**a**) TIPS and hot pressing were used to fabricate roll–roll membranes of MOF in which MOF crystals were loaded into the matrix of HDPE (M_W > 40,000, melt index = 2.2 g per 10 min) and UHMWPE (M_W > 1,500,000) in a molten state which, after cooling to 90 °C, was pressed to manufacture the membranes, with their (**b**) individual selectivity and permeation for different molecules such as dyes, racemates, and proteins in water; (**c**) high performance and recyclability of the membrane with a fouled and cleaned stage; (**d**) ultrathin nanoporous MOF membrane with intra-crystalline defects as created via missing linker within the MOF. (**e**,**f**) The ultrathin membrane with manipulated defects showed a higher water flux and salt rejection as compared to the membrane without any defects. Reprinted with permission from Nature Communi. [116], Copyright 2019 Nature Portfolio, and Nature Communi. [117], Copyright 2023 Nature Portfolio, respectively. These articles are licensed under a Creative Commons Attribution 4.0 International License, which permits use, sharing, adaptation, distribution, and reproduction in any medium or format, as long as you give appropriate credit to the original author(s) and the source, provide a link to the Creative Commons license, and indicate if changes were made.

The MOF–PE membranes fabricated via the TIPS–HoP procedure are highly flexible with micron-sized channels, wherein the MOF particles were linked via ultra-high-molecular-weight PE. This renders them as highly efficient separation membranes possessing both high selectivity and high flux (Figure 16b). In the study, several MOFs were used to manufacture roll-to-roll membranes to examine their individual selectivities and permeations for different molecules such as dyes, racemates, and proteins in water (Figure 16b). The permeance tests of these membranes have indicated only 10% decline after 5 h of operation with an outstanding antifouling effect (Figure 16c). The pollutants collected on the membrane surface were eliminated, and the performance of the membrane was further retained (Figure 16c). The high and consistent water fluxes of these membranes with significant rejection make them suitable for water treatment and antifouling sustainable agents [116].

In further advancements related with the MOFs-based membranes, a special strategy to manipulate and rationally design defects was followed to demonstrate how rationally designed defects in a MOF crystal can be used for high-performance membranes (Figure 16d) [117]. In the study, the team used an acetic acid or CH_3COOH modulator to create intracrystalline defects in the membrane by breaking the linker present in the MOF (UiO-66), which further assisted in producing an ultrathin layer of the modified MOF (Figure 16d, lower panel). As per the performance tests, these ultrathin membranes with well-designed defects showcased an improved water flux, as high as ~29.8 L m^{-2} h^{-1}, thereby surpassing the efficiency of other existing membranes for water treatment (Figure 16e,f). Furthermore, the membrane maintained salt rejection (approximately 99.8%) for a longer period of time and can be operated for 12 y at a typical content of salt (Figure 16f). Nevertheless, membranes with rationally designed defects were added to boost the membrane profiles for water purification [117].

With the emergence of new functional nanomaterials, further expansion of the above membranes has led to utilization of COFs. They are porous organic polymers which, owing to their robust stability and ordered network of nanochannels, offer superior selectivity and permeation [55,118]. Considering these unique properties, a next-generation membrane was fabricated via IP, wherein an oil–water–oil three-phase system was used as a method for preparing the COF nanosheets. These sheets were subsequently processed as a membrane via the filtration assembly method (Figure 17a) [119]. In the top organic phase, the diamine monomers with three kinds of acid groups, including the 2,5-diaminophenylphosphonic acid group (Pa–PO_3H_2), the 2,5-diaminobenzenesulfonic acid group (Pa–SO_3H), and the 2,5-diaminobenzoic acid group (Pa–CO_2H) were dissolved. Conversely, in the bottom organic phase, an aldehyde-based monomer (1,3,5-triformylphloroglucinol, Tp) was added, with the middle phase being pure water or an acidic or basic solution (Figure 17a,b).

As the monomers diffused towards the aqueous phase, they underwent polymerization in the interfacial space, resulting in the assembly of COF nanosheets. These nanosheets displayed a dimensional crystalline lattice, as observed with high-resolution TEM (Figure 17c). Subsequently, from the colloidal solution of these nanosheets, a membrane was fabricated via a filtration assembly using a porous support (PTFE with an average pore size of 200 nm) (Figure 17a). For a binary mixture of K^+/Li^+, these membranes have shown a high ion permeance with a selectivity of 4.2–4.7. The permeation performance of these membranes was high for single-cation transport, whereas it decreased significantly for binary mixtures (Figure 17d). Nevertheless, the study demonstrated a novel concept of confined cascade separation to successfully enhance the ion transport and separation processes [119].

In a rather simple and straightforward approach, COF membranes based on the 1D cellulose nanofibers (CNFs) and 2D COF nanosheets were fabricated by combining two simple steps: (1) mixed-dimensional assembly of a Schiff base-type COF $TpTG_{Cl}$ (1,3,5-triformylphloroglucinol, Tp and triaminoguanidinium chloride, TG_{Cl}) and cellulose nanofibers (CNFs), and (2) vacuum assisted self-assembly of the composites of COF $TpTG_{Cl}$@CNFs on to a PAN substrate (Figure 17e) [120]. The hydroxyl groups present on the surface of the CNFs interacted with the positively charged guanidinium units in the $TpTG_{Cl}$ (COF) framework, leading to dense interlocking in the nanosheets (Figure 17e). Interlocking further increases the mechanical strength and reduces the pore size of the membrane [121]. The separation performance of these membranes can be easily optimized by changing the CNF fraction during fabrication. These membranes have performed with a higher permeation flux of 8.53 kg m^{-2} h^{-1} and a separation factor ($\alpha_{water/n\text{-}butanol}$) of about 3876 for the water/n-butanol mixture (Figure 17f). According to the proposed transport mechanism for higher permeation flux, water molecules were preferentially adsorbed onto the membrane surface (Figure 17g). Subsequently, these molecules diffused faster through the well-organized channels of $TpTG_{Cl}$. A successive interlayer of CNF attracted these water molecules at a large concentration with enhanced selectivity via molecular sieving effects. Consequently, a large number of water molecules pass through the next interlayer of $TpTG_{cl}$ at a rapid rate. This entire process leads to a higher water flux, in which the molecules of the respective alcohols are rejected (Figure 17g). Superhydrophilic membranes with interlocked networks and reduced pore sizes are useful for solvent dehydration and the removal of dyes or salts from water [121].

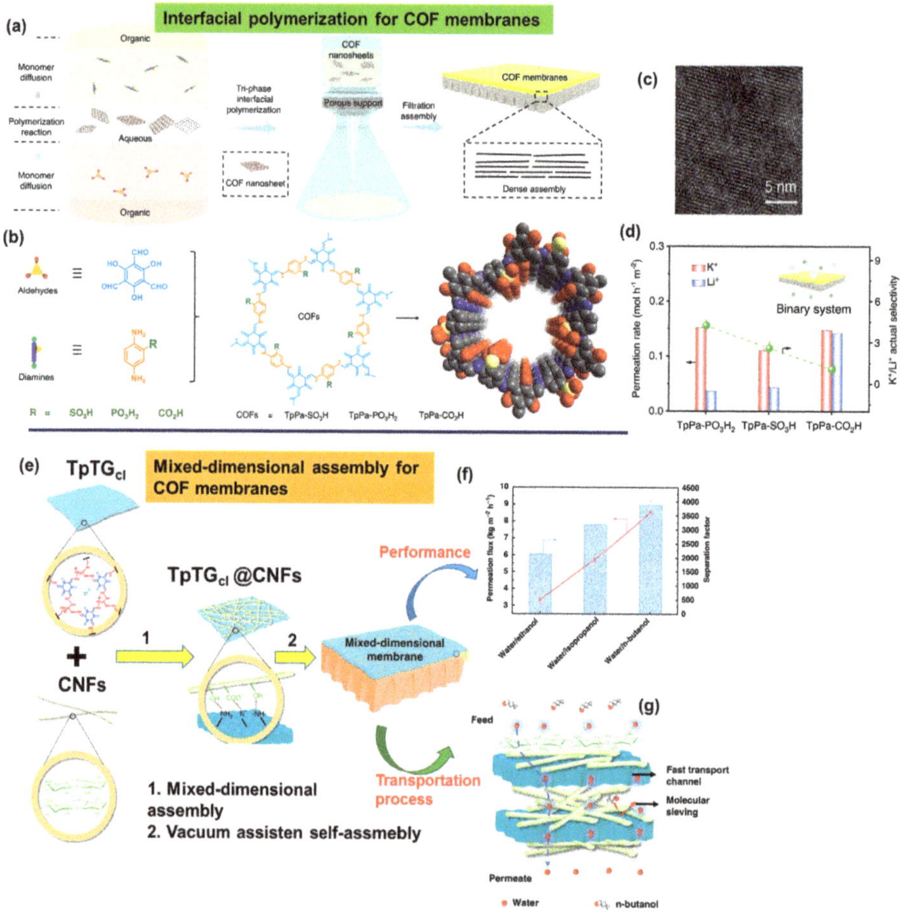

Figure 17. Fabrication of highly porous membranes utilizing (**a**) three–phase (organic–aqueous–organic) IP of (**b**) two different monomers: first, to produce the nanosheets of COFs subjected to a vacuum filtration assembly on a substrate, leading to the (**c**) formation of a dense membrane with a high degree of porous and crystalline networks. (**d**) Water flux ability and selectivity of a COF membrane in a binary mixture. (**e**) Membrane fabricated via the combination and synergism of mixed-

dimensional assembly (step 1) and vacuum assisted self-assembly (step 2). (**f**) Permeation properties of the membrane with (**g**) different transport channels in which the feed can run at various speeds or rates. Reprinted with permission from Nature Communi. [119], Copyright 2022 Nature Portfolio, and Nature Communi. [120], Copyright 2019 Nature Portfolio, respectively.

6. Conclusions

Drinking water scarcity has become a serious threat to human civilization worldwide. The influx of large-scale pollutants, such as organic dyes, detergents, toxic metals, and even byproducts of fertilizers (after their degradation), as well as next-generation nanoparticles or quantum dots, have adversely affected groundwater and rivers. Polluted water not only increases the risk for the extinction of numerous animal species, but also reduces drinking water availability. To treat polluted water or purify it to eliminate toxic and biohazardous elements, an operational filter device with an integrated membrane is required, which can work at a minimum energy cost (with high efficiency). As a membrane material, a number of well-known polymers such as PAN, PVDF, PES, PSF, PEEK, PA, PU, and poly(styrene) have been fabricated by following various methods and their improvised versions. These methods include IP, sputtering, solution casting, extrusion, melt pressing, phase inversion, and electrospinning. More advanced fabrication processes, such as mixed self-assembly followed by vacuum-induced self-assembly, have also been successfully attempted. With a focus on novel chemistry and controlled properties, such as pore size and interlocked networks, the latest materials, graphene, MOFs, and COFs, have been considered as alternatives to polymers owing to their supramolecular architectures, defects, and robustness. The manipulation of defects within these functional frameworks and the fabrication of membranes have shown promising outcomes. With regard to water flux and rejection of pollutants or salts, membranes based on the above frameworks have displayed superior performance. However, owing to their solid and supercrystalline nature, the processability of these materials remains challenging. Nevertheless, recent progress in polymeric materials and their derivatives for membrane fabrication, as highlighted in this review, has led to a scenario wherein their integration into a water filter device can be accomplished. The quest to conserve or produce drinking water either by purifying it from polluted resources or by desalination of seawater is still ongoing.

7. Challenges and Perspectives

Despite the significant advancements in membrane materials summarized above, several issues need to be addressed or minimized for their continued improvement and widespread applications. For example, the fouling and wetting of membranes can impose major problems with regard to their performance and durability. Over time, irreversible fouling can cause the membrane to disintegrate, necessitating a replacement. Membrane wetting is another concern that must be addressed as it can damage the membrane, reducing its lifespan. The sustainability of these membranes is a major focal point due to the negative or harmful effects on the environment. At the industrial scale, the production of membranes must be cost effective; thus, a rather cheaper raw material would serve this purpose. However, the design and development of a cheaper material for membrane fabrication that can work with its maximum water flux and can be deployed under any condition or weather remain major tasks that need to be accomplished. Further challenges may also involve reducing the greenhouse gases and industrial waste generated while producing these membrane materials; thus, more materials based on biomass could be alternatives. Precise control over pore size and functionality, which are directly associated with membrane selectivity, is another domain in which further improvements are required. Although there are a number of commercial membranes sold in the market for water treatment applications, they also suffer from similar problems, for instance, fouling and wetting. As a result, the production cost of more membranes increases significantly. This in turn puts an extra burden on the raw materials as far as their sources are concerned, which inflicts a severe toll on the environment. The energy consumption required to operate the filter device must be reduced. Accordingly, the focus needs to be shifted towards sunlight for supplying power to these filter devices at home or anywhere. Notably, a paradigm shift from polymers to 2D materials for membrane fabrication has already occurred, with some positive outcomes. Although, the materials such as graphene, MOFs, and COFs have fulfilled expectations to a large extent, the production of a commercial membrane based on these materials is still a pipedream. For example, manipulating defects in these materials and placing an appropriate chemical functionality within the pores or channels for selective permeation and rejection are other research-related barriers that need to be overcome. Moreover, these 2D nanomaterials are not easy to process and hence could pose additional challenges when fabricated at the industrial scale.

Funding: This work was supported by the National Research Foundation of Korea (2021R1A2C1003755, and 2022R1A6A1A03051158).

Institutional Review Board Statement: Not applicable.

Data Availability Statement: Data are contained within the article.

Conflicts of Interest: The authors declare that they have no competing interests.

Abbreviations

PASS, Poly(arylene sulfide sulfone); PES, Poly(ether sulfone); PVDF, Poly(vinylidene fluoride); PANI, polyaniline; GO, graphene oxide; ATRP, atom transfer radical polymerization; MOFs, metal–organic frameworks; COFs, covalent organic frameworks; CNFs, cellulose nanofibers; EDA, ethylenediamine; TMC, trimesoyl chloride; SDS, sodium dodecyl sulfate; PPAs, polyphthalamides; LBL, layer-by-layer; IP, interfacial polymerization; NF, nanofilteration; AFM, atomic force microscopy; SEM, scanning electron microscopy; and TEM, transmission electron microscopy.

References

1. He, C.; Liu, Z.; Wu, J.; Pan, X.; Fang, Z.; Li, J.; Bryan, B.A. Future global urban water scarcity and potential solutions. *Nat. Commun.* **2021**, *12*, 4667.
2. Flörke, M.; Schneider, C.; McDonald, R.I. Water competition between cities and agriculture driven by climate change and urban growth. *Nat. Sustain.* **2018**, *1*, 51–58.
3. Vörösmarty, C.J.; Green, P.; Salisbury, J.; Lammers, R.B. Global water resources: Vulnerability from climate change and population growth. *Science* **2000**, *289*, 284–288. [CrossRef] [PubMed]
4. Kataoka, T.; Nihei, Y.; Kudou, K.; Hinata, K. Assessment of the sources and inflow processes of microplastics in the river environments of Japan. *Environ. Pollut.* **2019**, *244*, 958–965. [CrossRef] [PubMed]
5. Zinabu, E.; Kelderman, P.; van der Kwast, J.; Irvine, K. Impacts and Policy Implications of Metals Effluent Discharge into Rivers within Industrial Zones: A Sub-Saharan Perspective from Ethiopia. *Environ. Manag.* **2018**, *61*, 700–715. [CrossRef] [PubMed]
6. Tkaczyk, A.; Mitrowska, K.; Posyniak, A. Synthetic organic dyes as contaminants of the aquatic environment and their implications for ecosystems: A review. *Sci. Total Environ.* **2020**, *717*, 137222. [CrossRef] [PubMed]
7. Zhitcovich, A. Chromium in Drinking Water: Sources, Metabolism, and Cancer Risks. *Chem. Res. Toxicol.* **2011**, *24*, 1617–1629.
8. Skidmore, M.E.; Sims, K.M.; Gibbs, H.K. Agricultural intensification and childhood cancer in Brazil. *PNAS* **2013**, *120*, e2306003120. [CrossRef]
9. Wang, M.; Abad, D.; Kickhoefer, V.A.; Rome, L.H.; Mahendra, S. Vault Nanoparticles Packaged with Enzymes as an Efficient Pollutant Biodegradation Technology. *ACS Nano* **2015**, *9*, 10931–10940. [CrossRef] [PubMed]
10. Dong, C.; Wang, Z.; Yang, C.; Li, X. Dual-functional single-atomic Mo/Fe clusters–decorated C3N5 via three electron-pathway in oxygen reduction reaction for tandemly removing contaminants from water. *PNAS* **2023**, *120*, e2305883120. [CrossRef]
11. Wu, J.; Cao, M.; Tong, D.; Finkelstein, Z.; Hoek, E.M.V. A critical review of point-of-use drinking water treatment in the United States. *NPJ Clean Water* **2021**, *4*, 40. [CrossRef]
12. Lewis, S.R.; Datta, S.; Gui, M.; Bhattacharya, D. Reactive nanostructured membranes for water purification. *PNAS* **2011**, *108*, 8577–8582. [PubMed]

13. Liu, Z.; Ma, Z.; Qian, B.; Chan, A.Y.H.; Wang, X.; Liu, Y.; Xin, J.H. A Facile and Scalable Method of Fabrication of Large-Area Ultrathin Graphene Oxide Nanofiltration Membrane. *ACS Nano* **2021**, *15*, 15294–15305. [CrossRef] [PubMed]
14. Sarkar, A.K.; Bediako, J.K.; Choi, J.W.; Yun, Y.S. Functionalized magnetic biopolymeric graphene oxide with outstanding performance in water purification. *NPG Asia Mater.* **2019**, *11*, 4. [CrossRef]
15. Abbo, H.S.; Gupta, K.C.; Khaligh, N.G.; Titinchi, S.J.J. Carbon Nanomaterials for Wastewater Treatment. *ChemBioEng* **2021**, *8*, 463–489. [CrossRef]
16. Adeleye, A.S.; Conway, J.R.; Garner, K.; Huang, Y.; Su, Y.; Keller, A.A. Engineered nanomaterials for water treatment and remediation: Costs, benefits, and applicability. *Chem. Eng. J.* **2016**, *286*, 640–662.
17. Andersson, J.; Castello, G.F.D.; Bilotto, P.; Hook, F.; Valtiner, M.; Dahlin, A. Control of Polymer Brush Morphology, Rheology, and Protein Repulsion by Hydrogen Bond Complexation. *Langmuir* **2021**, *37*, 4943–4952.
18. Hurst, P.J.; Graham, A.A.; Patterson, J.P. Gaining Structural Control by Modification of Polymerization Rate in Ring-Opening Polymerization-Induced Crystallization-Driven Self-Assembly. *ACS Polym. Au* **2022**, *2*, 501–509.
19. Huo, N.; Ye, S.; Ouderkirk, A.J.; Tenhaeff, W.E. Porous Polymer Films with Tunable Pore Size and Morphology by Vapor Deposition. *ACS Appl. Polym. Mater.* **2022**, *4*, 7300–7310. [CrossRef]
20. Nitti, A.; Carfora, R.; Assanelli, G.; Notari, M.; Pasini, D. Single-Chain Polymer Nanoparticles for Addressing Morphologies and Functions at the Nanoscale: A Review. *ACS Appl. Nano Mater.* **2022**, *5*, 13985–13997. [CrossRef]
21. Umermura, A.; Diring, S.; Furukawa, S.; Uehara, H.; Tsuruoka, T.; Kitagawa, S. Morphology Design of Porous Coordination Polymer Crystals by Coordination Modulation. *J. Am. Chem. Soc.* **2011**, *133*, 15506–15513. [CrossRef] [PubMed]
22. Malile, B.; Chen, J.I.L. Morphology-Based Plasmonic Nanoparticle Sensors: Controlling Etching Kinetics with Target-Responsive Permeability Gate. *J. Am. Chem. Soc.* **2013**, *135*, 16042–16045. [CrossRef] [PubMed]
23. Li, C.P.; Zhou, H.; Chen, J.; Wang, J.J.; Du, M.; Zhou, W. A Highly Efficient Coordination Polymer for Selective Trapping and Sensing of Perrhenate/Pertechnetate. *ACS Appl. Mater. Interfaces* **2020**, *12*, 15246–15254. [CrossRef] [PubMed]
24. Zhu, Y.; Gui, L.; Wang, R.; Wang, Y.; Fang, W.; Elimelech, M.; Lin, S.; Jin, J. Regulation of molecular transport in polymer membranes with voltage-controlled pore size at the angstrom scale. *Nat. Comm.* **2023**, *14*, 2373.
25. Abe, M.; Kametani, Y.; Uemura, T. Fabrication of Double-Stranded Vinyl Polymers Mediated by Coordination Nanochannels. *J. Am. Chem. Soc.* **2023**, *145*, 2448–2454. [CrossRef] [PubMed]
26. Mardle, P.; Chen, B.; Holdcroft, S. Opportunities of Ionomer Development for Anion-Exchange Membrane Water Electrolysis. *ACS Energy Lett.* **2023**, *8*, 3330–3342.
27. Getachew, B.A.; Kim, S.R.; Kim, J.H. Self-Healing Hydrogel Pore-Filled Water Filtration Membranes. *Environ. Sci. Technol.* **2017**, *51*, 905–913.
28. Ma, H.; Burger, C.; Hsiao, B.S.; Chu, B. Highly Permeable Polymer Membranes Containing Directed Channels for Water Purification. *ACS Macro Lett.* **2012**, *1*, 723–726. [CrossRef]
29. Sun, Q.; Aguila, B.; Song, Y.; Ma, S. Tailored Porous Organic Polymers for Task-Specific Water Purification. *Acc. Chem. Res.* **2020**, *53*, 812–821. [CrossRef]
30. Luo, H.; Chang, K.; Bahati, K.; Geise, G.M. Engineering Selective Desalination Membranes via Molecular Control of Polymer Functional Groups. *Environ. Sci. Technol. Lett.* **2019**, *6*, 462–466.
31. Benkhaled, B.T.; Montheli, T.; Lapinte, V.; Monge, S. Hydrosoluble phosphonic acid functionalized poly(2-ethyl-2-oxazoline) chelating polymers for the sorption of metallic cations. *J. Polym. Sci.* **2020**, *58*, 2875–2886. [CrossRef]
32. Song, Y.; Phipps, J.; Zhu, C.; Ma, S. Porous Materials for Water Purification. *Angew. Chem. Int. Ed.* **2023**, *62*, e202216724. [CrossRef]
33. Xu, X.; Yang, Y.; Liu, T.; Chu, B. Cost-effective polymer-based membranes for drinking water purification. *Giant* **2022**, *10*, 100099.
34. Ahn, H.; Kim, Y.M.; Yang, H.S.; Yeo, S.Y.; Yu, W.R.; Lee, B.S. Moisturized Polyacrylonitrile Copolymer for Stronger Precursor Fibers. *ACS Appl. Polym. Mater.* **2021**, *3*, 6285–6293.
35. Yu, S.; Guerrero, J.M.; Tai, Y.; Yang, S.; Choi, Y.Y.; Nam, J.; Myung, N.V. Maximizing Polyacrylonitrile Nanofiber Piezoelectric Properties through the Optimization of Electrospinning and Post-thermal Treatment Processes. *ACS Appl. Polym. Mater.* **2022**, *4*, 635–644. [CrossRef]
36. Facure, M.H.M.; Mercante, L.A.; Correa, D.S. Polyacrylonitrile/Reduced Graphene Oxide Free-Standing Nanofibrous Membranes for Detecting Endocrine Disruptors. *ACS Appl. Nano Mater.* **2022**, *5*, 6376–6384. [CrossRef]
37. Falireas, P.G.; Ladmiral, V.; Debuigne, A.; Detrembleur, C.; Poli, R.; Ameduri, B. Straightforward Synthesis of Well-Defined Poly(vinylidene fluoride) and Its Block Copolymers by Cobalt-Mediated Radical Polymerization. *Macromolecules* **2019**, *52*, 1266–1276. [CrossRef]
38. Meringolo, C.; Mastropietro, T.F.; Poerio, T.; Fontananova, E.; De Filpo, G.; Curcio, E.; Di Profio, G. Tailoring PVDF Membranes Surface Topography and Hydrophobicity by a Sustainable Two-Steps Phase Separation Process. *ACS Sustain. Chem. Eng.* **2018**, *6*, 10069–10077. [CrossRef]
39. Sadare, O.O.; Daramola, M.O. Blended Polysulfone/Polyethersulfone (PSF/PES) Membrane with Enhanced Antifouling Property for Separation of Succinate from Organic Acids from Fermentation Broth. *ACS Sustain. Chem. Eng.* **2021**, *9*, 13068–13083.
40. Bernstein, R.; Anton, E.; Ulbricht, M. UV-Photo Graft Functionalization of Polyethersulfone Membrane with Strong Polyelectrolyte Hydrogel and Its Application for Nanofiltration. *ACS Appl. Mater. Interfaces* **2012**, *4*, 3438–3446.

41. Ben-Haida, A.; Colquhoun, H.M.; Hodge, P.; Williams, D.J. Synthesis of a Catechol-Based Poly(ether ether ketone) ("o-PEEK") by Classical Step-Growth Polymerization and by Entropically Driven Ring-Opening Polymerization of Macrocyclic Oligomers. *Macromolecules* **2006**, *39*, 6467–6472. [CrossRef]
42. Anderson, L.J.; Yuan, X.; Fahs, G.B.; Moore, R.B. Blocky Ionomers via Sulfonation of Poly(ether ether ketone) in the Semicrystalline Gel State. *Macromolecules* **2018**, *51*, 6226–6237. [CrossRef]
43. Mulhearn, W.D.; Stafford, C.M. Highly Permeable Reverse Osmosis Membranes via Molecular Layer-by-Layer Deposition of Trimesoyl Chloride and 3,5-Diaminobenzoic Acid. *ACS Appl. Polym. Mater.* **2021**, *3*, 116–121. [CrossRef] [PubMed]
44. Zhou, W.; Zhang, X.; Gong, X.; Ding, M.; Yu, J.; Zhang, S.; Ding, B. Environmentally Friendly Polyamide Nanofiber Membranes with Interconnective Amphiphobic Channels for Seawater Desalination. *ACS Appl. Mater. Interfaces* **2022**, *14*, 35287–35296. [CrossRef] [PubMed]
45. Phillips, S.J.; Ginder, N.C.; Lear, B.J. Rapid Photothermal Synthesis of Polyurethane from Blocked Isocyanates. *Macromolecules* **2022**, *55*, 7232–7239. [CrossRef]
46. Alberti, S.; Dodero, A.; Sartori, E.; Vicini, S.; Ferrati, M.; Castelleno, M. Composite Water-Borne Polyurethane Nanofibrous Electrospun Membranes with Photocatalytic Properties. *ACS Appl. Polym. Mater.* **2021**, *3*, 6157–6166.
47. Sato, K.; Kajita, T.; Noro, A. Synthesis of a Cross-Linked Polymer Electrolyte Membrane with an Ultra-High Density of Sulfonic Acid Groups. *ACS Appl. Polym. Mater.* **2023**, *5*, 3480–3488.
48. Zhang, Y.; Almodovar-Arbelo, N.E.; Weidman, J.L.; S-Corti, D.; Boudoris, B.W.; Phillip, W.A. Fit-for-purpose block polymer membranes molecularly engineered for water treatment. *NPJ Clean Water* **2018**, *1*, 2. [CrossRef]
49. Lu, T.D.; Wang, Q.; Gu, S.S.; Sun, S.P. Beyond Symmetry: Exploring Asymmetric Electrospun Nanofiber Membranes for Liquid Separation. *Adv. Funct. Mater.* **2023**. *early on-line view 2310218*. [CrossRef]
50. Bachler, S.; Ort, M.; Kramer, S.D.; Dittrich, P.S. Permeation Studies across Symmetric and Asymmetric Membranes in Microdroplet Arrays. *Anal. Chem.* **2021**, *93*, 5137–5144.
51. White, N.; Misovich, M.; Yaroschuk, A.; Bruening, M.L. Coating of Nafion Membranes with Polyelectrolyte Multilayers to Achieve High Monovalent/Divalent Cation Electrodialysis Selectivities. *ACS Appl. Mater. Interfaces* **2015**, *7*, 6620–6628. [CrossRef] [PubMed]
52. Wang, K.Y.; Ong, R.C.; Chung, T.S. Double-Skinned Forward Osmosis Membranes for Reducing Internal Concentration Polarization within the Porous Sublayer. *Ind. Eng. Chem. Res.* **2010**, *49*, 4824–4831. [CrossRef]
53. Le, N.L.; Nunes, S.P. Materials and membrane technologies for water and energy sustainability. *Sustain. Mater. Technol.* **2016**, *7*, 1–28. [CrossRef]
54. Zhu, Z.; Liu, Z.; Tan, G.; Qi, J.; Zhou, Y.; Li, J. Interlayered Interface of a Thin Film Composite Janus Membrane for Sieving Volatile Substances in Membrane Distillation. *Environ. Sci. Technol.* **2023**, *57*, 7612–7623. [CrossRef]
55. Koros, W.J.; Zhang, C. Materials for Next-Generation Molecularly Selective Synthetic Membranes. *Nat. Mater.* **2017**, *16*, 289–297.
56. Seo, M.; Moll, D.; Silvis, C.; Roy, A.; Querelle, S.; Hilmayer, M.A. Interfacial Polymerization of Reactive Block Polymers for the Preparation of Composite Ultrafiltration Membranes. *Ind. Eng. Chem. Res.* **2014**, *53*, 18575–18579. [CrossRef]
57. Nielen, W.M.; Willot, J.D.; de Vos, W.M. Aqueous Phase Separation of Responsive Copolymers for Sustainable and Mechanically Stable Membranes. *ACS Appl. Polym. Mater.* **2020**, *2*, 1702–1710. [CrossRef]
58. Clarizia, G.; Tasselli, F.; Simari, C.; Nicotera, I.; Bernardo, P. Solution Casting Blending: An Effective Way for Tailoring Gas Transport and Mechanical Properties of Poly(vinyl butyral) and Pebax2533. *J. Phys. Chem. C* **2019**, *123*, 11264–11272. [CrossRef]
59. Raje, A.; Koll, J.; Schneider, E.S.; Gerogopanos, P. A novel organic solvent-free method for manufacturing polyethersulfone hollow fiber membranes using melt extrusion. *J. Membr. Sci.* **2023**, *683*, 121837. [CrossRef]
60. Tekin, F.S.; Zeynep, P.; Emecen, C. Controlling Cellulose Membrane Performance via Solvent Choice during Precursor Membrane Formation. *ACS Appl. Polym. Mater.* **2023**, *5*, 2185–2194.
61. Lavielle, N.; Hebraud, A.; Schlatter, G.; Meyer, L.T.; Rossi, R.M.; Popa, A.M. Simultaneous Electrospinning and Electrospraying: A Straightforward Approach for Fabricating Hierarchically Structured Composite Membranes. *ACS Appl. Mater. Interfaces* **2013**, *5*, 10090–10097. [PubMed]
62. Avossa, J.; Herwig, G.; Toncelli, C.; Itel, F.; Rossi, R.M. Electrospinning based on benign solvents: Current definitions, implications and strategies. *Green Chem.* **2022**, *24*, 2347–2375.
63. Wittmar, S.M.A.; Koch, D.; Prymak, O.; Ulbricht, M. Factors Affecting the Nonsolvent-Induced Phase Separation of Cellulose from Ionic Liquid-Based Solutions. *ACS Omega* **2020**, *5*, 27314–27322. [PubMed]
64. Zhang, T.; Hao, S.; Xiao, J.; Jia, Z.Q. Preparation of Poly(4-methyl-1-pentene) Membranes by Low-Temperature Thermally Induced Phase Separation. *ACS Appl. Polym. Mater.* **2023**, *5*, 1998–2005. [CrossRef]
65. Cha, B.J.; Yang, J.M. Preparation of poly(vinylidene fluoride) hollow fiber membranes for microfiltration using modified TIPS process. *J. Membr. Sci.* **2007**, *291*, 191–198. [CrossRef]
66. Xue, J.; Wu, T.; Dai, Y.; Xia, Y. Electrospinning and Electrospun Nanofibers: Methods, Materials, and Applications. *Chem. Rev.* **2019**, *119*, 5298–5415. [CrossRef] [PubMed]
67. Zhao, D.; Chen, S.; Wang, P.; Zhao, Q.; Lu, X. A Dendrimer-Based Forward Osmosis Draw Solute for Seawater Desalination. *Ind. Eng. Chem. Res.* **2014**, *53*, 16170–16175. [CrossRef]
68. Zhao, Q.; Chen, N.; Zhao, D.; Lu, X. Thermoresponsive Magnetic Nanoparticles for Seawater Desalination. *ACS Appl. Mater. Interfaces* **2013**, *5*, 11453–11461. [CrossRef]

69. Ge, Q.; Su, J.; Amy, G.L.; Chung, T.S. Exploration of polyelectrolytes as draw solutes in forward osmosis processes. *Water Res.* **2012**, *46*, 1318–1326. [CrossRef]
70. Liu, Z.Y.; Wei, Z.M.; Wang, X.J.; Zhang, G.; Long, S.R.; Yang, J. Preparation and Characterization of Multi-layer Poly(arylene sulfide sulfone) Nanofibers Membranes for Liquid Filtration. *Chin. J. Polym. Sci.* **2019**, *37*, 1248–1256. [CrossRef]
71. Van der Bruggen, B. Chemical modification of polyethersulfone nanofiltration membranes: A review. *J. Appl. Polym. Sci.* **2009**, *114*, 630–642. [CrossRef]
72. Ahmad, A.L.; Abdulkarim, A.A.; Ooi, B.S.; Ismail, S. Recent development in additives modifications of polyethersulfone membrane for flux enhancement. *Chem. Eng. J.* **2013**, *223*, 246–267.
73. Anwar, S.; Nasreen, A.S.; Sundarrajan, S.; Sayed Nizar, S.A.; Balamurugan, R.; Ramakrishna, S. In situ polymerization of PVDF-HEMA polymers: Electrospun membranes with improved flux and antifouling properties for water filtration. *Polym. J.* **2014**, *46*, 167–174.
74. Cheng, X.; Li, M.; Wang, J.; Yang, J.; Wei, K.; Zhang, Y.; Chai, W.; You, J. Crystallization-template-induced PEEK membranes for particulate matter capture at high temperature and separation of emulsion containing corrosive component. *J. Environ. Chem. Eng.* **2022**, *10*, 107469. [CrossRef]
75. Deng, W.; Li, Y. Novel superhydrophilic antifouling PVDF-BiOCl nanocomposite membranes fabricated via a modified blending-phase inversion method. *Sep. Purif. Technol.* **2021**, *254*, 117656. [CrossRef]
76. Zhang, X.; Zhan, Z.M.; Cheng, F.Y.; Xu, Z.L.; Jin, P.R.; Liu, Z.P.; Ma, X.H.; Xu, X.R.; Van der Bruggen, B. Thin-Film Composite Membrane Prepared by Interfacial Polymerization on the Integrated ZIF-L Nanosheets Interface for Pervaporation Dehydration. *ACS Appl. Mater. Interfaces* **2021**, *13*, 39819–39830. [CrossRef] [PubMed]
77. Liu, L.; Chen, X.; Feng, S.; Wan, Y.; Luo, J. Enhancing the Antifouling Ability of a Polyamide Nanofiltration Membrane by Narrowing the Pore Size Distribution via One-Step Multiple Interfacial Polymerization. *ACS Appl. Mater. Interfaces* **2022**, *14*, 36132–36142. [PubMed]
78. He, B.; Peng, H.; Chen, Y.; Zhao, Q. High performance polyamide nanofiltration membranes enabled by surface modification of imidazolium ionic liquid. *J. Membr. Sci.* **2020**, *608*, 118202.
79. Tan, X.; Rodrigue, D. A Review on Porous Polymeric Membrane Preparation. Part II: Production Techniques with Polyethylene, Polydimethylsiloxane, Polypropylene, Polyimide, and Polytetrafluoroethylene. *Polymers* **2019**, *11*, 1310. [CrossRef]
80. Madauß, L.; Schumacher, J.; Ghosh, M.; Ochedowski, O.; Meyer, J.; Lebius, H.; Ban-d'Etat, B.; Toimil-Molares, M.E.; Trautmann, C.; Lammertink, R.G.H.; et al. Fabrication of nanoporous graphene/polymer composite membranes. *Nanoscale* **2017**, *9*, 10487–10493.
81. Lv, J.; Gong, Z.; He, Z.; Yang, J.; Chen, Y.; Tang, C.; Liu, Y.; Fan, M.; Lau, W. 3D printing of a mechanically durable superhydrophobic porous membrane for oil–water separation. *J. Mater. Chem. A* **2017**, *5*, 12435–12444.
82. Magat, E.E.; Faris, B.F.; Reith, J.E.; Salisbury, L.F. Acid-catalyzed Reactions of Nitriles. I. The Reaction of Nitriles with Formaldehyde. *J. Am. Chem. Soc.* **1951**, *73*, 1028–1031.
83. Rajesh, S.; Zhao, Y.; Fong, H.; Menkhaus, T. Nanofiber multilayer membranes with tailored nanochannels prepared by molecular layer-by-layer assembly for high throughput separation. *J. Mater. Chem. A* **2017**, *5*, 4616–4628. [CrossRef]
84. Ostwal, M.; Wazer, E.; Pemberton, M.; McCutcheon, J.R. Scaling electrospray based additive manufacturing of polyamide membranes. *J. Membr. Sci. Lett.* **2022**, *2*, 100035. [CrossRef]
85. MacRitchie, F. Mechanism of interfacial polymerization. *Trans. Faraday Soc.* **1969**, *65*, 2503–2507. [CrossRef]
86. Yuan, B.; Zhao, S.; Hu, P.; Cui, J.; Niu, Q.J. Asymmetric polyamide nanofilms with highly ordered nanovoids for water purification. *Nat. Commun.* **2020**, *11*, 6102. [PubMed]
87. Zhao, C.; Zhang, Y.; Jia, Y.; Li, B.; Tang, W.; Shang, C.; Mo, R.; Li, P.; Liu, S.; Zhang, S. Polyamide membranes with nanoscale ordered structures for fast permeation and highly selective ion-ion separation. *Nat. Commun.* **2023**, *14*, 1112. [CrossRef] [PubMed]
88. Li, Y.; Su, Y.; Zhao, X.; Zhang, R.; Liu, Y.; Fan, X.; Zhu, J.; Ma, Y.; Liu, Y.; Jiang, Z. Preparation of Antifouling Nanofiltration Membrane via Interfacial Polymerization of Fluorinated Polyamine and Trimesoyl Chloride. *Ind. Eng. Chem. Res.* **2015**, *54*, 8302–8310. [CrossRef]
89. Li, Y.; Su, Y.; Zhao, X.; He, X.; Zhang, R.; Zhao, J.; Fan, X.; Jiang, Z. Antifouling, High-Flux Nanofiltration Membranes Enabled by Dual Functional Polydopamine. *ACS Appl. Mater. Interfaces* **2014**, *6*, 5548–5557. [CrossRef]
90. Li, Y.; Su, Y.; Zhao, X.; Zhang, R.; Zhao, J.; Fan, X.; Jiang, Z. Surface Fluorination of Polyamide Nanofiltration Membrane for Enhanced Antifouling Property. *J. Membr. Sci.* **2014**, *455*, 15–23. [CrossRef]
91. Liu, Y.; Zhu, J.; Zheng, J.; Gao, X.; Wang, J.; Wang, X.; Xie, Y.F.; Huang, X.; Van der Bruggen, B. A Facile and Scalable Fabrication Procedure for Thin-Film Composite Membranes: Integration of Phase Inversion and Interfacial Polymerization. *Environ. Sci. Technol.* **2020**, *54*, 1946–1954. [CrossRef] [PubMed]
92. Wu, X.; Ding, M.; Xu, H.; Yang, W.; Zhang, K.; Tian, H.; Wang, H.; Xie, Z. Scalable Ti_3C_2Tx MXene Interlayered Forward Osmosis Membranes for Enhanced Water Purification and Organic Solvent Recovery. *ACS Nano* **2020**, *14*, 9125–9135. [CrossRef] [PubMed]
93. Wang, X.; Li, Q.; Zhang, J.; Huang, H.; Wu, S.; Yang, Y. Novel thin-film reverse osmosis membrane with MXene Ti_3C_2Tx embedded in polyamide to enhance the water flux, anti-fouling and chlorine resistance for water desalination. *J. Membr. Sci.* **2020**, *603*, 118036. [CrossRef]
94. Zapsas, G.; Patil, Y.; Gnanou, Y.; Ameduri, B.; Hadjichristidis, N. Poly(vinylidene fluoride)-based complex macromolecular architectures: From synthesis to properties and applications. *Prog. Polym. Sci.* **2020**, *104*, 101231. [CrossRef]

95. Deng, W.; Zhao, H.; Pan, F.; Feng, X.; Jung, B.; Abdel-Wahab, A.; Batchelor, B.; Li, Y. Response to comment on "Visible-Light-Driven Photocatalytic Degradation of Organic Water Pollutants Promoted by Sulfite Addition". *Environ. Sci. Technol.* **2018**, *52*, 1677–1678. [CrossRef] [PubMed]
96. Li, J.; Yu, Y.; Zhang, L. Bismuth oxyhalide nanomaterials: Layered structures meet photocatalysis. *Nanoscale* **2014**, *6*, 8473–8488. [CrossRef] [PubMed]
97. Pantuso, E.; Ahmed, E.; Fontananova, E.; Brunetti, A.; Tahir, I.; Karothu, D.P.; Alnaji, N.A.; Dushaq, G.; Rasras, M.; Naumov, P.; et al. Smart dynamic hybrid membranes with self-cleaning capability. *Nat. Commun.* **2023**, *14*, 5751. [CrossRef]
98. Ahmed, E.; Karothu, D.P.; Warren, M.; Naumov, P. Shape-memory effects in molecular crystals. *Nat. Commun.* **2019**, *10*, 3723. [CrossRef]
99. Nawaz, H.; Umar, M.; Nawaz, I.; Ullah, A.; Tauseef Khawar, M.; Nikiel, M.; Razzaq, H.; Siddiq, M.; Liu, X. Hybrid PVDF/PANI Membrane for Removal of Dyes from Textile Waste water. *Adv. Eng. Mater.* **2022**, *24*, 2100719. [CrossRef]
100. Crock, C.A.; Rogensues, A.R.; Shan, W.; Tarabara, V.V. Polymer nanocomposites with graphene-based hierarchical fillers as materials for multifunctional water treatment membranes. *Water Res.* **2013**, *47*, 3984–3996. [CrossRef]
101. Mi, B. Graphene oxide membranes for ionic and molecular sieving. *Science* **2014**, *343*, 740–742. [PubMed]
102. Ghaffar, A.; Zhang, L.; Zhu, X.; Chen, B. Porous PVdF/GO Nanofibrous Membranes for Selective Separation and Recycling of Charged Organic Dyes from Water. *Environ. Sci. Technol.* **2018**, *52*, 4265–4274. [CrossRef]
103. Luque-Alled, J.M.; Leaper, S.; Abdel-Karim, A.; Skuse, C.; Gogojo, P. PVDF membranes containing alkyl and perfluoroalkyl-functionalized graphene nanosheets for improved membrane distillation. *J. Environ. Chem. Eng.* **2023**, *11*, 109898.
104. Fouladivanda, M.; Karimi-Sabet, J.; Abbasi, F.; Moosavian, M.A. Step-by-step improvement of mixed-matrix nanofiber membrane with functionalized graphene oxide for desalination via air-gap membrane distillation. *Sep. Purif. Technol.* **2021**, *256*, 117809.
105. Silva, T.L.S.; Morales-Torres, S.; Figueiredo, J.L.; Silva, A.M.T. Multi-walled carbon nanotube/PVDF blended membranes with sponge- and finger-like pores for direct contact membrane distillation. *Desalination* **2015**, *357*, 233–245. [CrossRef]
106. Leaper, S.; Caceres, E.O.A.; Luque-Alled, J.M.; Cartmell, S.H.; Gorgojo, P. POSS-functionalized graphene oxide/PVDF electrospun membranes for complete arsenic removal using membrane distillation. *ACS Appl. Polym. Mater.* **2021**, *3*, 1854–1865. [CrossRef]
107. Wei, L.; Wang, G.; Cai, M. Synthesis and properties of novel poly(ether ketone ketone)s containing 1,4-naphthylene units by electrophilic solution copolycondensation. *Polym. Int.* **2021**, *70*, 1048–1056.
108. Senra, M.R.; Vieira Marques, M.F.; Monteiro, S.N. Poly (Ether-Ether-Ketone) for Biomedical Applications: From Enhancing Bioactivity to Reinforced-Bioactive Composites—An Overview. *Polymers* **2023**, *15*, 373.
109. Markova, D.; Kumar, A.; Kalpper, M.; Mullen, K. Phosphonic acid-containing homo-, AB and BAB block copolymers via ATRP designed for fuel cell applications. *Polymer* **2009**, *50*, 3411–3421. [CrossRef]
110. Kumar, A. Self Assemblies of Poly (ether ether ketone) Block Copolymers for Biomedical Applications. *ChemistrySelect* **2021**, *6*, 9060–9068. [CrossRef]
111. Aristizábal, S.L.; Chisca, S.; Pulido, B.A.; Nunes, S.P. Preparation of PEEK Membranes with Excellent Stability Using Common Organic Solvents. *Ind. Eng. Chem. Res.* **2020**, *59*, 5218–5226. [CrossRef]
112. Zhang, S.; Chen, C.; Su, Z.; Qin, X.; Jiang, M.; Liu, P. High efficiency and flexible construction of hydrophilic polymer brushes on polyether ether ketone hollow-fiber membrane surface for improving permeability separation, and anti-fouling performances. *Chem. Eng. J.* **2023**, *454*, 140176. [CrossRef]
113. Lin, Z.; Zhing, J.; Sun, R.; Wei, Y.; Sun, Z.; Li, W.; Chen, L.; Sun, Y.; Zhang, H.; Pang, J.; et al. InSitu Integrated Fabrication for Multi-Interface Stabilized and Highly Durable Polyaniline@Graphene Oxide/Polyether Ether Ketone Special Separation Membranes. *Adv. Sci.* **2023**, *10*, 2302654.
114. Li, J.R.; Sculley, J.; Zhou, H.C. Metal-organic frameworks for separations. *Chem. Rev.* **2012**, *112*, 869–932.
115. Geng, K.; He, T.; Liu, R.; Dalapati, S.; Tan, K.T.; Li, Z.; Tao, S.; Gong, Y.; Jiang, Q.; Jiang, D. Covalent Organic Frameworks: Design, Synthesis, and Functions. *Chem. Rev.* **2020**, *120*, 8814–8933. [PubMed]
116. Wang, H.; Zhao, S.; Liu, Y.; Yao, R.; Wang, X.; Cao, Y.; Ma, D.; Zou, M.; Cao, A.; Feng, X.; et al. Membrane adsorbers with ultrahigh metal-organic framework loading for high flux separations. *Nat. Commun.* **2019**, *10*, 4204. [CrossRef]
117. Wang, X.; Lyu, Q.; Tong, T.; Sun, K.; Lin, L.C.; Tang, C.Y.; Yang, F.; Guiver, M.D.; Quan, X.; Dong, Y. Robust ultrathin nanoporous MOF membrane with intra-crystalline defects for fast water transport. *Nat. Commun.* **2023**, *13*, 266.
118. Shefer, I.; Peer-Haim, O.; Leifman, O.; Epsztein, R. Enthalpic and entropic selectivity of water and small ions in polyamide membranes. Environ. *Sci. Technol.* **2021**, *55*, 14863–14875. [CrossRef]
119. Wang, H.; Zhai, Y.; Li, Y.; Cao, Y.; Shi, B.; Li, R.; Zhu, Z.; Jiang, H.; Guo, Z.; Wang, M.; et al. Covalent organic framework membranes for efficient separation of monovalent cations. *Nat. Commun.* **2022**, *13*, 7123.
120. Yang, H.; Yang, L.; Wang, H.; Xu, Z.; Zhao, Y.; Luo, Y.; Nasir, N.; Song, Y.; Wu, H.; Pan, F.; et al. Covalent organic framework membranes through a mixed-dimensional assembly for molecular separations. *Nat. Commun.* **2019**, *10*, 2101. [CrossRef]
121. Wang, L.; Boutielier, M.S.H.; Kidambi, P.R.; Jang, D.; Hadjiconstantinou, N.G.; Karnik, R. Fundamental transport mechanisms, fabrication and potential applications of nanoporous atomically thin membranes. *Nat. Nanotechnol.* **2017**, *12*, 509–522. [CrossRef] [PubMed]

MDPI AG
Grosspeteranlage 5
4052 Basel
Switzerland
Tel.: +41 61 683 77 34

Polymers Editorial Office
E-mail: polymers@mdpi.com
www.mdpi.com/journal/polymers

www.ingramcontent.com/pod-product-compliance
Lightning Source LLC
Chambersburg PA
CBHW061510210926
51850CB00008B/1868
9783725859153